Global Perspectives on Law and the Environment

Volume 1

The book series "Global Perspectives on Law and the Environment" publishes cutting-edge research that explores the environment from a variety of interdisciplinary perspectives, including law, biodiversity, climate change, energy regulation and policy, public health, sustainability, and more. Rooted in the law, this series recognizes the crucial need for a multidisciplinary approach to addressing environmental challenges and emphasizes the importance of exploring the complex intersections between these diverse fields.

The series invites contributions from both academics and practitioners, encouraging proposals that bridge the gap between theory and practice. It seeks global submissions, with a particular interest in works that highlight emerging regional trends and innovations.

Serving as an invaluable resource, this series is designed for legal professionals, policymakers, judges, legislators, scholars, and students alike, offering insights into key areas such as environmental law, human rights, climate science, biodiversity conservation, sustainability, public health, and energy regulation and policy.

Dennis Oghenerobor Agelebe
Editor

Climate Ethics in Africa

Structuring a Standard for Trade and Development in Africa

Springer

Editor
Dennis Oghenerobor Agelebe
Environmental Law Center
Faculty of Law
University of Cologne
Cologne, Nordrhein-Westfalen, Germany

ISSN 3091-3608 ISSN 3091-3616 (electronic)
Global Perspectives on Law and the Environment
ISBN 978-3-032-08756-0 ISBN 978-3-032-08757-7 (eBook)
https://doi.org/10.1007/978-3-032-08757-7

This Springer imprint is published by the registered company Springer Nature Switzerland AG
The registered company address is: Gewerbestrasse 11, 6330 Cham, Switzerland

Foreword

The intersection of climate ethics, trade, and development is one of the most pressing concerns of many Global South countries. As Africa stands at the crossroads of rapid industrialization and the urgent need for environmental sustainability, the discourse on climate ethics is not just necessary—it is imperative. This book, Climate Ethics: Structuring a Standard for Trade and Development in Africa, is a timely and significant contribution to this crucial conversation.

Dr. Agelebe, a distinguished lawyer and professor specializing in environment, energy, international trade, and African affairs, and his co-authors bring a wealth of expertise to this work. Their deep understanding of climate governance's regulatory, economic, and ethical dimensions uniquely positions them to analyze comprehensively how Africa can develop robust frameworks that balance economic growth with environmental responsibility.

This book methodically explores the ethical foundations of climate policies and their implications for trade and development. It delves into the responsibilities of nations, corporations, and policymakers in ensuring that Africa's economic trajectory aligns with global sustainability goals. By dissecting legal frameworks, international agreements, and case studies, the author offers a compelling argument for why ethical considerations must be embedded in trade policies and development strategies across the continent.

The depth of this volume is evident in its exploration of key themes, including the role of climate ethics in Africa's development framework, the impact of climate change on economic and social structures, and the importance of transboundary cooperation. Chapters address significant topics such as the African Continental Free Trade Area (AfCFTA) as a mechanism for green tourism and climate action, the role of climate change adaptation strategies in economic development, and the intersection of intellectual property with access to green technology. Furthermore, the book critically assesses regional legal structures, including the relevance of the African Court on Human and Peoples' Rights in addressing climate change.

What makes this book particularly valuable is its practical relevance to a broad audience, including lawyers, scholars, policymakers, social scientists, and public administrators. It serves as an academic resource and a guiding document for those engaged in shaping Africa's trade and environmental policies. The insights presented here challenge readers to think critically about the role of climate ethics in fostering a just and equitable future.

As a colleague of the editor, I admire his commitment to impactful scholarship and drive for sustainable development. He pursues his dreams for a developed Africa with an uncommon dedication to climate justice and a strong awareness of the urgent need for ethical governance in Africa's development and climate change affairs. This book is a testament to that dedication.

I am confident this book will stimulate more robust dialogue and policy innovation in Africa's interest. It is an essential read for anyone interested in Africa's future and the global pursuit of sustainable development.

Prof. Dr. Omphemetse S. Sibanda
Executive Dean, Faculty
of Management and Law
University of Limpopo
Polokwane, South Africa

Preface

Climate change continues to put the world on edge despite the level of awareness and the intensity of deliberations to address and counteract the unprecedented environmental and health hazards the global community is vulnerable to. The relationship between climate change and development is generally acknowledged and discussed in the deliberations, leading to various solution pathways, mitigation and adaptation measures, low emission targets, and climate financing obligations. But most times, such discussions are rimmed with the context of how development through industrialization has and keeps contributing to global warming, or how climate change measures can contribute to an emerging and environment-friendly development.

In recent years, the dichotomy between the level of development of some countries and that of others has fueled the call for compensation from under-developed nations by developed countries. African countries have argued that considering their colonial history, the intersection of climate change, trade, and development complicates their roles and commitments to pursuing the global targets to address the challenge. Africa is more exposed to its vulnerabilities by the prevailing socio-economic inequalities, high development deficit, and reliance on climate-sensitive energy models for industrialization. As much as state and non-state actors in the fight against climate change generally believe in adaptation and mitigation measures, the premise for the measures and implementation tend to resonate more with the regulatory approach. Practice and ethics related to climate change measures and their interplay with other areas of a nation's social and economic life have not yet evolved to become customary. This book, "Climate Ethics in Africa: Structuring a Standard for Africa's Development," explores the intersections of climate change, trade, and development and examines an ethical approach to framing practices and laws that will satisfy the peculiarity of the climate change situation in Africa.

The contributors to this book are drawn from the science, social science, and legal fields. They present a multidisciplinary approach to understanding how climate ethics can advance Africa's development. Their scholarly contributions also present comparative contexts on the impacts of climate change on Africa's drive for socio-economic emancipation. As environmental and economic disruptions continue, we are witnessing an increase in violent conflicts between herders and farmers in certain countries. Furthermore, the pace of development is hampered as climate change interferes with planned projects, diverting scarce resources toward financing mitigation measures. To address these challenges effectively, it is essential to implement transformative policies that promote ethical practices in tackling climate change, particularly about African trade, intellectual property rights, migration, clean technology, and regional collaborations.

The motivation for bringing my colleagues together to author this book was born out of my desire to see African scholars brainstorm on how Africa's approach to climate change and development can be integrated with traditional ethical principles. Some such tenets, like Ubuntu, which is rooted in the African philosophy of interconnectedness and communal well-being, are globally acclaimed as influencing approaches to justice, governance, and social integration. As a co-author and editor of this book, my experience as a lawyer, researcher, and teacher has played a significant role in my initiative to first host a conference on Climate Ethics in Africa before inviting scholars to pen their thoughts on how Africa can demonstrate a stronger commitment to bridging the gap between policy and practice by ensuring ethical considerations underpin Africa's response framework and mechanisms to climate change. Therefore, this book reflects the collective scholarly inquiry of intellectual personalities, extending a call for ethical climate actions to policymakers, academics, and stakeholders across the continent and beyond.

I appreciate the contributors' insightful chapters covering different areas connected to climate change. Through this book, their passion for addressing the multidimensional and multidirectional issues related to climate ethics has enriched the topical discussions on Africa's development goals in the face of climate change challenges. I look forward to reading feedback that will open minds to deeper perspectives for future scholarly work.

Cologne, Germany Dennis Oghenerobor Agelebe
September 2025

Acknowledgements

The publication of "Climate Ethics in Africa: Structuring a Standard for Trade and Development in Africa" is the culmination of significant support from numerous individuals and institutions. I am profoundly indebted to Professor Kirk Junker, Director of the Environmental Law Center, whose mentorship is unparalleled. His unwavering support and the provision of a conducive intellectual environment were crucial in structuring a project focused on Afro-centric legal and ethical standards. I offer special thanks and deep appreciation to Professor Omphemetse Sibanda, Executive Dean, Faculty of Management and Law, University of Limpopo, South Africa, for his insightful "Foreword," which frames the vital discourse presented in this volume.

This research into African climate ethics was significantly bolstered by the Global South Studies Center, under the distinguished leadership of Professor Clemens Greiner. Their sponsorship of a lecture series on Climate Ethics in Africa and generous research grant support underscore their vital commitment to advancing critical Global South and African scholarship.

To the dedicated contributing authors, I extend my deepest gratitude. Your scholarly rigor and cooperative spirit have shaped this complex project, ensuring that the volume offers a comprehensive and multi-faceted standard for trade and development across the African continent.

I must also recognize my colleagues at the Environmental Law Center: Dr. Philippe Matthew Roy, Emma Shensher, Saskia Militz, Saskia Münster, Moritz Röhrs, Alejandra Kessler, and Marie Pflüge. Your prompt support and consistent availability have been invaluable. A special thanks is reserved for Marie Pflüge and Alejandra Kessler for their meticulous and crucial proofreading and editing of this manuscript.

My heartfelt appreciation goes to my priceless pearl, Sarah. Your support, encompassing every technical, emotional, and spiritual challenge of this academic endeavor, has been the bedrock of my strength. Your indomitable zeal and partnership continue to inspire me.

I am thankful to Springer Nature, Switzerland, the publishers of this work, for embracing this volume. Their professional excellence and efficient production process have been integral to its fruition. I offer special thanks to Dr. Anja Trautmann for her guidance and professionalism throughout the process.

Finally, I am deeply grateful to my spiritual father, Reverend Olusola Areogun, and my spiritual mother, Reverend Mrs. Oyenika Areogun, whose words of faith have provided spiritual sustenance for this entire journey.

The collective contributions acknowledged here have left an indelible mark on this work and on the future of climate-conscious trade and development policy in Africa.

Dennis Oghenerobor Agelebe

Contents

Climate Ethics in Africa's Development Framework: Examining the Status Quo ... 1
Dennis Oghenerobor Agelebe

Trade and Development in Africa: Does Climate Ethics Matter? ... 25
Patricia Iyore Ajayi and Olusanya Elisa Olubusoye

Socio-Economic and Environmental Impacts of Climate Change in the Horn of Africa: High Need for Climate Action and Transboundary Cooperations ... 37
Gemedo Dalle

Climate Change-Induced Conflicts: Examining the Interplay Between Natural Resource Depletion and Human Displacement Due to Insecurity in Africa ... 47
Kolawole Afuwape and Sarah Ayanwale

Application of Free Trade Pact for Climate Action: The Agreement Establishing the African Continental Free Trade Area ... 83
Semia Sama

AfCFTA: A Catalyst for Green Tourism in Africa ... 103
Collins Odote and Elizabeth Gitaru Mitaru

Climate Change and Nigeria's Narrative: Impacts, Policies, and Challenges ... 121
Olayinka Oluwamuyiwa Ojo, Oluwaseyi Adurapemi Olasoji, and Dennis Oghenerobor Agelebe

Work Practices in Nigeria: Environmental Health and Development Implications ... 155
Akeem Ayofe Akinwale

The Efficacy of Climate Change Adaptation Strategies for Enhancing Trade and Economic Development in Kenya 171
Aloyce Peter Ndege

Climate Change Policies in Southeast Asia and African Countries: Approaches to Trade and Development 197
Ibnu Sitompul and Dennis Oghenerobor Agelebe

Implementing Climate Standards in the African Continental Free Trade Area (AfCFTA): Progressive and Precautionary Lessons from the European Union 215
Deborah O. Ayodele-Olajire, Olaitan O. Olajire, and Adeniyi S. Gbadegesin

Intellectual Property Rights and Access to Green Technology in Economic Community of West African States: Issues, Challenges and the Way Forward 237
Solomon Ezike and Dennis Oghenerobor Agelebe

The Relevance of the African Court on Human and Peoples' Rights in Addressing Climate Change in Africa: A Critical Assessment 259
Bhavna Mahadew

Abbreviations

AATHP	ASEAN Agreement on Transboundary Haze Pollution
AFOLU	Agriculture, Forestry, and Other Land Use
AFSEC	Africa Electrotechnical Standardisation Commission
ANSI	American National Standards Institute
ARSO	African Organisation for Standardisation
ASCC	ASEAN Socio-Cultural Community
ASEAN	Association of Southeast Asian Nations
AU	African Union
AUGRAP	African Union Green Recovery Action Plan
BIAT	Boosting Intra-African Trade
BNRCC	Building Nigeria's Response to Climate Change
CBAM	Carbon Border Adjustment Mechanism
CBD	Convention on Biological Diversity
CDM	Clean Development Mechanism
CESR	Centre for Economic and Social Rights
CFTA	Continental Free Trade Area
CIE	International Commission on Illumination
CLARE	Climate Adaptation and Resilience Programme
COMESA	Common Market for Eastern and Southern Africa
COP	Conference of the Parties
COPOLCO	Committee on Consumer Policy
CSST	Climate Services Support Tool
CU	Customs Union
DCC	Department of Climate Change
DEVCO	Developing Country Matters
DRR	Disaster Risk Reduction
ECOWAS	Economic Community of West African States
EGD	European Green Deal
EMCA	Environmental Management and Coordination Act
EPI	Environmental Performance Index
ETP	Energy Transition Plan

EU	European Union
FCT	Federal Capital Territory
FDI	Foreign Direct Investment
FEPA	Federal Environmental Protection Agency
FLEGT	Forest Law Enforcement, Governance and Trade
FTA	Free Trade Agreement
GAP	Global Action Plan
GATS	General Agreement on Trade in Services
GATT	General Agreement on Tariffs and Trade
GCOS	Global Climate Observing System
GDP	Gross Domestic Product
GGFRP	Global Gas Flaring Reduction Partnership
GHG	Greenhouse Gas
GHRM	Green Human Resource Management
ICCC	Intergovernmental Commission on Climate Change
ICJ	International Court of Justice
IEC	International Electrotechnical Commission
IEEE	Institute of Electrical and Electronics Engineers
IFC	International Finance Corporation
IFF	Intergovernmental Forum on Forests
ILRI	International Livestock Research Institute
INDC	Intended Nationally Determined Contributions
IP	Intellectual Property
IPCC	Intergovernmental Panel on Climate Change
IPF	Intergovernmental Panel on Forests
ISO	International Organization for Standardization
ITU	International Telecommunication Union
JGGSDA	Joint Green Growth Strategy for Developing Africa
KCSAS	Kenya Climate Smart Agriculture Strategy
KEPSA	Kenya Private Sector Alliance
MDG	Millennium Development Goals
MEND	Movement for the Emancipation of the Niger Delta
MEPS	Minimum Energy Performance Standards
NAP	National Action Plan
NAPA	National Adaptation Programmes of Action
NASPA-CCN	National Adaptation Strategy and Plan of Action for Climate Change in Nigeria
NCCAP	National Climate Change Action Plan
NCCC	National Climate Change Council
NCCP—RS	National Climate Change Policy—Response Strategy
NCCRS	National Climate Change Response Strategy
NDMA	National Disaster Management Authority
NDP	National Development Plan
NEEDS	National Economic Empowerment and Development Strategy
NEFCP	National Environment Forest Conservation Policy

NEP	National Environmental Policy
NESEFCR	National Environmental Standards Enforcement and Climate Regulations
NESREA	National Environmental Standards and Regulations Enforcement Agency
NMDPRA	Nigerian Midstream and Downstream Petroleum Regulatory Authority
NNPC	Nigerian National Petroleum Corporation
NPBM	National Product Branding Mechanism
NPDD	National Programme to Combat Desertification
NSC	National Standards Commission
ODA	Official Development Assistance
ODRMC	Office of Disaster Risk Management Coordination
OIE	World Organisation for Animal Health
OUP	Oxford University Press
PAQI	Pan African Quality Infrastructure
PCR	Post-Conflict Reconstruction
PIA	Petroleum Industry Act
PRISE	Pathways to Resilience in Semi-Arid Economies
REC	Regional Economic Communities
REMP	Renewable Energy Master Plan
SABE	Strategic Advisory Body on the Environment
SADC	Southern African Development Community
SCBD	Secretariat of the Convention on Biological Diversity
SDG	Sustainable Development Goals
SE4ALL-AA	Sustainable Energy for All-Africa Agenda
SERAC	Social and Economic Rights Action Center
SO2	Sulfur Dioxide
SPS	Sanitary and Phytosanitary Measures
TAC	Treaty of Amity and Cooperation
TEC	Technical Expert Committee
TLAS	Timber Legality Assurance System
TNC	Third National Communication
TRIPS	Trade-Related Aspects of Intellectual Property Rights
UDA	United Democratic Alliance
UNCCD	United Nations Convention to Combat Desertification
UNCTAD	United Nations Conference on Trade and Development
UNDRR	United Nations Office for Disaster Risk Reduction
UNEP	United Nations Environment Programme
UNFCCC	United Nations Framework Convention on Climate Change
USMCA	United States-Mexico-Canada Agreement
VPA	Voluntary Partnership Agreements

WHO	World Health Organization
WMO	World Meteorological Organization
WTO	World Trade Organization

Climate Ethics in Africa's Development Framework: Examining the Status Quo

Dennis Oghenerobor Agelebe

Abstract Africa faces a unique challenge in balancing economic development with the ethical imperatives of climate change mitigation and adaptation. This paper explores the intersection of climate ethics and African development, highlighting the continent's vulnerability to climate change despite its minimal contribution to global greenhouse gas emissions. It examines the ethical principles of justice, equity, and responsibility, emphasising the need for developed nations to support Africa's climate resilience efforts. The paper also delves into the historical context of colonial exploitation, the role of Foreign Direct Investment, and the importance of integrating indigenous knowledge and community participation in climate action. By adopting an interdisciplinary approach, this study aims to comprehensively understand the ethical dimensions of climate change in Africa and propose sustainable and equitable development pathways.

1 Introduction

Africa, the second-largest continent, struggles to overcome the continual and effectively impassable development gap that has plagued Africa and its people since before the colonial period. With such hope that brought about the independence of African countries from the colonial hold, the world expected Africa to be politically and economically independent and volunteer for the welfare of the global society. However, the rate of development has been slow compared to the steadily rising needs of Africa's population and environment. A report by the World Bank predicts that approximately 60% of the continent's overall population is below the age of 25, and the population of Africa will double by 2050.[1] The estimated male unemployment rate is 6.6%, and the female unemployment rate is 8%. The president of the African

[1] World Bank (2020).

D. O. Agelebe (✉)
Dennis Oghenerobor Agelebe, Environmental Law Centre, Faculty of Law, University of Cologne, Cologne, Germany
e-mail: dageleb1@uni-koeln.de

D. Oghenerobor Agelebe (ed.), *Climate Ethics in Africa*, Global Perspectives on Law and the Environment 1, https://doi.org/10.1007/978-3-032-08757-7_1

Development Bank (AfDB), Akinwunmi Adesina, noted that Africa needs up to $100 billion per year to finance infrastructure development.[2] By 2025, the continent will require about $170 billion yearly to close the huge infrastructure gap.[3] However, there is concern about what reversing these developmental challenges in Africa might cost our environment, climate, and planet Earth. Even now, Africa is already bearing the consequences of other continents' development regarding climate change's impact, though it is one of the least emitters of greenhouse gases that contribute to climate change.[4]

Africa, with its numerous natural resources and various ecosystems, is driving economic development as it addresses the multiple ethical issues of climate change. The adverse effects of climate change on the continent, with an amplified toll paid by the affected communities and weak infrastructures, exacerbate the existing socio-economic challenges. An ethical issue requires the right action on the part of human beings because it is wrong for the few to continue polluting the environment while subjecting the rest of the world to the unfair consequences of our greedy ways. Climate change may have originated in other parts of the world. Still, there is an abundance of evidence that indicates that the continent of Africa is already feeling the bite of climate change despite being a far less emitter of greenhouse gases when compared with the emissions from more industrialized continents. Remarkably, those who have contributed the least to climate change are now facing its most severe consequences, especially with rising temperatures across the African continent. Globally, thermal conditions are proliferating, resulting in more frequent droughts and the development of desert-like environments. Agriculture, on which most African societies rely as their primary source of livelihood, is under pressure because arable land is gradually deteriorating in quality. People's ways of life become compromised through the loss of water sources and hence reduced ability to feed the nation. Several African countries have been severely affected by climate change, both in the social and economic domains and the physical. Due to the inability to adapt, vulnerable communities are experiencing a higher level of food insecurity, water scarcity, and increased health risks. Moreover, these conditions are further entrenching inequalities in societies that are already socially divisive due to environmental injustice across the lines of gender and socioeconomic classification, particularly against women and children. Additionally, economies that rely on agriculture are struggling to keep their economy afloat as the impact of climate change progresses rapidly across sectors, thereby highlighting the urgent need for immediate action.

Being a challenge with a global dimension and devastating implications for the current and future population, climate change is not simply an environmental concern but a problem that affects the social and cultural life of the people. Therefore, understanding the causes and the measures necessary for addressing the challenge is related to studying the contribution of human practices and perceptions to the cause of the change. In the same vein, the approach to making humans live with the change

[2] United Nations (2011).

[3] Wilson Centre (2023).

[4] IPCC (2023).

while reducing the impacts of changes can be derived from observing what practices humans allow, restrict or prohibit but have environmental implications. In the context of African development, therefore, climate ethics are critical in addressing issues peculiar to the African continent and the people of Africa. This chapter, thus, examines the connection between how the development of Africa is being prosecuted in line with regional and national climate change frameworks to understand the extent of the application of African ethics in addressing the challenge of climate change. Against this backdrop, the chapter highlights the ethical imperatives, challenges, and potential policy and legal pathways in driving the sustainable and equitable development agenda of the African Union. Thus, an examination of the socio-scientific factors underlining the growth ambitions of Africa is undertaken by acknowledging the historical context that shaped the evolution of modernizing the development of Africa and the trajectory of its development in the post-colonial era. Such aspect of the study is essential in light of the contribution of influence of external actors and factors-state and non-state, that continue to perpetuate the legacy of colonialism under cover of the dynamics of the global order known as the so-called globalization. These factors are interrelated because contemporary global challenges are always the subject of concerted discussion bordering on the vulnerability of areas and not much of the responsibility of actors in finding an equitable solution to challenges such as climate change. The question arises: How can Africa balance the development process and ethically address change in such a climatically sensitive world?

An interdisciplinary approach is adopted in assessing the extent to which climate ethic principles are considered or integrated into the continent's regional development frameworks. This approach draws from African ethics, global commitments, and national measures. Furthermore, it identifies potential pathways for fostering sustainable development that respects the environment and upholds justice, equity, and inclusivity principles.

Therefore, a section explores key African ethical theories for managing climate change and prevalent environmental challenges in Africa. These include what climate change means to different communities, how global historical emissions and contributions are related to current vulnerability, and how ethical issues arise in applying relevant measures for dealing with climate change. Critical analysis of these aspects will help advance the discussion of the moral issues of climate change and sustainable development in Africa.

Collectively, this chapter underscores the need to give ethical considerations the central place in Africa's developmental story. As the continent tries to deal with the various environmental complexities, ethics provides the framework through which development does not come hand-in-hand with environmentally unfriendly practices and social injustices by casting some light on possible ethical imperatives that can help Africa to strive for a better future, free from exploitation and closer to sustainability and climate justice.

2 Climate Change and Climate Justice: Understanding the Key Concepts

At the heart of the discussion on the world's frontline challenge of climate change is "who" and "what" in solving the onerous knotty issue of responsibility. While scientists have, with the help of science, determined what is responsible for climate change, the State and non-state actors that are known to be actively involved in activities generating the volume of emission heating the global sphere have refused to accept most of the results directed at them as being responsible for global warming. Several attempts have been made to agree on a responsibility-sharing formula that connotes acceptance of contribution to the cause of worldwide warming and agreement to pay a fair share of the cost of reversing the negative trend of climate change. The Kyoto Protocol[5] and the Paris Agreement[6] Are international instruments that demonstrate some of the global actions taken to determine the factor of "responsibility" on the platform of international law as provided by the United Nations Framework Convention on Climate Change. As the world continues to grapple with the increasing effect of climate change, as seen in the record extreme weather patterns, rise in sea level, droughts, and floods, the potential of witnessing multiple disruptions in the regular activities that define the very existence of our livelihood, even at different standards. Activities such as trade, cultural practices, learning, family life, and recreational activities have witnessed disruptions that have necessitated the initiation and adoption of adaptive approaches to surviving the challenging trend.

There is the question of how Justice as a rule of law concept interplays with the causes and effects of climate change. Whether we understand Justice in the context of climate change as an avenue to an equitable distribution of burden and compensatory approaches to responding to the challenge of climate change or seen as an institutional vehicle for the delivery of mitigative and adaptive measures in addressing climate change remains a subjective argument, that could sway the Global North and Global South dichotomy in the reasoning of climate change. Regardless of how we understand climate justice, the peculiarity of the vulnerability of those most affected by climate change means these people are vulnerable to the systemic pattern of neglecting those living in the least developed countries. It is not just about the strength of their economy but how they are disproportionately affected because of a combination of geographical and economic factors determining their level of fragility.[7] Precisely, the consideration is directed more at who needs Justice and why they deserve it. Furthermore, the approach is based on their humanitarian ties rather than their social relationship.[8]

However, scholars have argued that climate change events have forged relational ties that allow the negotiation of justice in a way that the concept of justice was not

[5] The Kyoto Protocol (1997).

[6] The Paris Agreement (2015).

[7] Levy and Patz (2023).

[8] Sangiovanni (2007, pp. 3–39).

envisaged. How they came to that conclusion may be from the conditional premises that sovereign jurisdictions lay bare when discussing from the point of advantage, disadvantage and interest at the negotiating table to determine the allocation of burden and benefits in addressing the effects of climate change. This argument is also strengthened when climate justice is viewed as a "global species" of justice. Patrick Hayden, arguing in line with the concept of John Rawls on Justice, observes that climate change stimulates a rallying forum that creates relationships that defy differing political leanings.[9]

3 Foundational African Ethical Principles

Cultural, religious and philosophical African ethics provide the basis for developing legal principles. For example, principles that find their premise in communal norms, such as respect for elders, age-grade cooperation, common economic practice, harmony with nature, and the Ubuntu philosophy, provide a moral compass that resonates within communities.[10] These principles, often intrinsic to African societies, contribute to formulating legal norms that reflect the people's values and beliefs. The central concept of Ubuntu is a shared sense of morality that governs the behaviours of individuals and communities in consistently living, interacting, and transacting with the awareness of a value system that, while unwritten, is legally enshrined in the traditional corpus governing the typical African society.[11] Nonetheless, Africa is known for bearing common values and upholding moral principles, demonstrating community diversity and tolerance. It is also noteworthy that these societies keep evolving. Among these values, which are recognized by populations related by kinship, language, or religion, some influence other cultural practices beyond the primary communities that give normative credence to such ethical values. An illustrative example is the Ubuntu concept, which traditionally aligns with the cultural traits of sub-Saharan Africa. In contrast, Northern African countries, marked by religious and ethnic demography, exhibit solidarity, communal cohesion, and a moral framework deeply rooted in religiosity.

However, one of the prominent ethnic groups in Ghana is the Akan people. Their ethics and culture have impacted other regional societies and beyond. According to research scholars, Ghana Akan ethics substantially represent common traits of moral standards indicative of the multiplicity of belief systems across sub-Sahara Africa.[12] These ethics are the foundation for ethical behaviour, interpersonal relationships, and group peace. They are derived from their traditional religious beliefs, cultural practices, and philosophical perspectives. An essential tenet of Akan ethics is "Sankofa," which means "go back and get it." This principle, which reflects the Akan belief in

[9] Hayden (2010, pp. 351–72).

[10] Louw (2006).

[11] Magesa (1997, pp. 1–29).

[12] Agada (2023).

the interconnection of time and the wisdom garnered from history, emphasizes the significance of learning from the past to build a better future.

Two broad values of Akan ethics are communitarianism and environmental stewardship values. The principle of communitarianism prioritizes the well-being of the community over individual interests. In Akan society, individuals are expected to contribute to the collective welfare through acts of generosity, cooperation, and mutual support. This ethos is encapsulated in the Akan proverb: "Nsuo baako ntɛm," which means "One head does not take a decision." It underscores the importance of consensus-building and collective decision-making in the community. The principle of environmental stewardship requires a strict adherence to respect for nature. According to the belief of the Akan people, the earth is a living entity and deity. Therefore, it should be revered and cared for by all its inhabitants as part of ways of showing high reverence to the earth and preserving it for generations, which resonates with the principle of intergenerational equity in international environmental law; the people practice sustainable agriculture, conservation of natural resources, and rituals honouring the earth.

4 Exploring the Interplay Between African Ethics and the African Legal Systems

An examination of the connection between African ethics and the development of contextual legal systems compels a critical discussion that interrogates how cultural practices intertwine with societal roles while having legal consequences for non-compliance. The ethics of African people is a phenomenon that is difficult to define due to the socially contextualized nature of their ethical perceptions. These African ethical ideologies are grounded in the continent's society, language, and history and thus embrace personality and human life within the reciprocal context of a typical African society.[13] It has played a role in evolving legal customary norms in the continent's cultures and institutions. To study the general properties of ethical rules in Africa, one must approach them using beliefs and norms' values. Therefore, this supports the sociological theory that morality is a basic structure of acceptable conduct throughout African societies' varying tiers of interaction or social power dynamics.[14] On the communal level, it predominantly relies on standards inherited from ancestry records while constantly reconstructing the fundamental aspects which underpin the rudimentary development of the conventional system of laws.[15]

The laws of Africa are as diverse as its geography and people. This diversity can be traced back to the remnants of colonialism, the acknowledgement and implementation of Indigenous customary laws, and the adoption of modern legal systems. These legal systems can be broadly categorized into three main types: customary, civil,

[13] Bujo (2003).

[14] Bujo (2001).

[15] Wiredu (1995).

and common law.[16] Customary law originates from African societies' traditional culture, norms, and practices.[17] It comprises a set of traditional practices together with standards regarding the organization of a corporation and its different facets, including, but not limited to, marriage customs, inheritance practices, ground property acquisition, and, most importantly, the method of solving disputes. Also, it is informal and passed on from one generation to the other: it develops with society and cultures over time.[18] Because of its flexibility and adaptability, the formal process of applying the law is less emphasized. Furthermore, customary courts, composed of traditional leaders and elders, play a crucial role in applying customary law and resolving disputes at the local level.[19]

African civil law systems originated from European legal systems, primarily France and Portugal. These systems have structural and written rules and regulations that express the rights, duties, and practices of various facets of law, including contract, property, and criminal law. Jurisdictions practising civil law have a hierarchical court system, with specialized tribunals assigned to handle different cases. Though the system is seen as contributing more to modernizing other national legal systems by promoting legal certainty, it has been criticized for appearing rigid, formalistic and not readily adaptable to local and cultural differences in society. African standard law systems can be attributed to British influence, particularly the English legal tradition.[20] These legal systems work on the principle of the statement of precedents, where the decision of one superior court becomes a precedent for another inferior court in making its decisions. In standard law systems, a unique role is given to judges, which makes legal practice based on principles well formulated in codes and legal precedents. As has been evidenced, the standard law system has been applied in many African countries. Still, it has opened deficiencies in the accommodation of Indigenous legal systems and the justice delivery systems to the most disadvantaged people.[21]

The application of normative values differs across cultures and systems, thereby demonstrating expressions and manifestations of humanism that vary according to the features of the norms. As can be observed in the solidarity with others' causes or the commitment to preserving nature, adherence to these nuanced values, particularly as a member of a specific culture, entails nurturing interpersonal connections. Such a degree of commitment is the key to creating a harmonious personality,[22] and it deservingly needs the honour of being actual and accurate. While moral philosophers naturally centre their philosophical endeavours on the ethical beliefs and circumstances prevalent in their societies, recognizing that specific social and

[16] Woodman (2015).

[17] Pimentel (2011, 14, 59).

[18] Ndulo (2011).

[19] Mboh (2021, 33–57).

[20] Heyl (2019).

[21] Astier (2012).

[22] Gyekye (1997).

cultural contextual interactions profoundly shape human experience, they do not suggest that the outcomes of their reflections are confined solely to their societies.[23]

Moreover, how the community laws support such cultural interactions is not considered to argue that the moral threshold is high enough to self-regulate the rationale behind solidarity for communal unity and nature protection.[24] In addition, they still think that because sentiments, purposes, responses, hopes, and aspirations are common to all individuals, individuals belong to one humanity in certain situations; they consider that conclusions reflected in their reflective activities necessarily involve the concerns of the vast human family.[25]

5 Legal Systems and the African Ethics

The relationship between African ethics and legal systems is complex and constantly evolving. It shows the ever-present tension between traditional law and African legal cultures on the one hand and received law adapted to local conditions on the other. As with many different aspects of life, it is important not to overemphasize the opposition between these two spheres of legal activity because convergence, adaptation, and hybridization characterize legal practices and institutions in Africa. Nevertheless, we can describe their relationship in some ways below:

5.1 Customary Law and Restorative Justice

In the culture of African ethics, its legal system has also adopted the Restorative Justice Approach. In many African societies, it is regarded improper and un-African to seek outside intervention to sort out a problem, let alone take a matter to court; instead, there are specific conflict-solving forums, dialogue, and reconciliation processes meant to restore order within the African community. Restorative justice aims at atoning the harm resulting from an infringement rather than penalizing a violation. They may also require the participation of community elders, religious leaders or any senior person in the society who will act as intermediaries between conflicting parties.

Restorative justice, part of customary law in African systems, is in sync with African ethical principles such as communalism, Ubuntu, and respect for older people.[26] In the event of a dispute, customary dispute settlement bodies use principles of mediation and reconciliation to grant justice to deserving parties within a community. This informal approach applies discussion, agreement, reconciliation

---

[23] Hogan (2011, pp. 224–229).

[24] Hama (2013, pp. 367–377).

[25] Said (2012) and Appiah (2005).

[26] Mabovula (2011, 38–47).

and restoration of loss and relationship rather than antagonistic legal setting and penalty.

5.2 *Legal Pluralism and Hybridity*

Legal pluralism is a system in which multiple legal systems operate within a jurisdiction, which is present in many African countries.[27] This pluralism reflects the integration and recognition of diverse cultures, religions, and ethnicities by different jurisdictions across Africa. In such circumstances, institutional personalities such as judges, lawyers, and traditional leaders often operate between formal legal systems and customary law; sometimes, they incorporate elements from both to accommodate various social groups. A consequence of this legal transplantation is the hybridization of the legal practices and the possible generation of creative ways of bringing the insights of African ethics into the legal context at the higher level of the norms.

5.3 *Challenges of Modernisation and Globalisation*

Contemporary African ethics and customary law have demonstrated resilience in adapting to the challenges posed by modernization, urbanization, and globalization. Colonization, international law, and globalization significantly influence African legal systems.[28] This influence has sometimes relegated Indigenous legal systems and reduced the effective operation of legal institutions inherent in the communities. Furthermore, the fluidity of social change and increasing economic development and technology rates have fueled tension and conflict within African society by challenging traditional culture and morality.

5.4 *Human Rights and Social Justice*

Discussions on human rights, social justice, and governance issues show how African ethics interact with legal frameworks. African countries signed regional and international human rights instruments and constitutional principles that embrace international human rights standards, human dignity and justice.[29] However, much discussion has been focused on reconciling human rights norms derived from Western systems with African culture and its ethical standards. People who are skeptical of the international human rights system believe it disregards Indigenous conceptions of

[27] Swenson (2018, 438–462).

[28] Heyns and Stefiszyn (2006).

[29] Viljoen (2012).

rights, duties, and justice, which can lead to tension between global and community approaches.

The interplay between African ethics and African legal systems is dynamic and complex. Given the constant interaction of different emerging areas with existing cultures and norms, the ethical standards of African societies continue to either adapt or challenge how these new areas, such as technological innovation and climate change, tend to affect them. Though reconciling indigenous ethical values with emerging norms will experience tense moments, there are also opportunities for dialogue, adaptation, and innovation that can enrich legal pluralism and promote social justice in Africa. As African societies continue to evolve, the interplay between ethics and legal systems will remain crucial to building inclusive, equitable, and accountable governance structures that reflect the aspirations and realities of diverse African communities.

6 Climate Ethics in Climate Action

Several ethical questions arise regarding the effectiveness of legal systems in distributing benefits and burdens. These questions are particularly relevant when addressing issues that emerge as governing bodies manage compensation for communities most affected by climate change, all while adhering to the principle of environmental justice. Such problems, for example, pertain to determining the moral and legal fulfilment of responsibility, fairness, and equity in compensating the affected communities. For example, evaluating how international law frameworks coordinate multilateral and unilateral responses to climate change within multilateral environmental agreements can provide insights into how national legal systems could function more effectively. Given the peculiar nature of international law, particularly its system of enactment and enforcement, should we expect so much from national systems? What other way is there to guarantee that state and non-state actors take on the duty of care for all human and ecological species currently impacted by climate change and will be impacted in the future if there isn't a systematic enforcement mechanism to implement international climate change law? John Broom believes that what we should do about climate change is primarily an ethical question.[30] However, this moral question has been examined more from the context of human-to-human and human-to-environment responsibilities and less from the understanding of the duties of state and non-state actors. The foregoing leaves a blank space for formulating climate actions with a socio-legal basis for implementation rather than science alone. Fundamentally, actors-state and non-state, and individuals are more exposed to law than the climate ethics of respecting the interest of the environment based on formulating policies and making laws that make ethical responsibility the baseline for a sustained drive towards achieving climate goals.

[30] Broome (2008, pp. 96–102).

Climate ethics, from a philosophical standpoint, assists us in rationalizing the moral principle that drives our awareness of the environmental imperatives for which legal and natural persons, as well as the state, should assume responsibility in responding to climate change. There are two main dimensions relating to responses to climate change challenges from the point of view of ethics. There is the one that pertains to the appropriateness of the action; it's known as deontological ethics.[31] According to ethical theory, adherence to moral rules determines whether an act is ethically good or bad.[32] The second dimension is consequentialist ethics, which considers the morality of an act to depend solely on its consequences. Byrne put it more straightforwardly: "It relates to the rightness of the effect or impacts of a particular course of action."[33]

In his profound exploration of philosophy, Immanuel Kant elucidated deontology as an ethical framework that scrutinizes duties, classifying actions as right or wrong. It is in the context of deontology that understanding the moral rationale for climate actions is valuable to develop a system that not only propounds the law to govern people's behaviour but also makes the people understand the appropriateness of the decisions and actions that are endorsed either by a community or by an individual, either public or private. The logical foundation of duty, grounded in moral perception, may offer feeble support for natural law systems striving to establish sufficient enforcement. If the consequences of actions are ignored when formulating environmental strategies and guidelines, the impacts of climate change and the human activities responsible for it extend to human rights issues. This suggests that a deontological approach may conflict with principles of justice and fairness unless human rights are safeguarded. The allocation of duties and responsibilities depends on identifying the most suitable method for protecting the overall well-being of the environment. As an illustration, when safeguarding fundamental human rights from the impacts of climate change—such as displacement due to environmental disasters and health issues arising from air pollution— it becomes crucial to ensure that implementing mitigation laws goes beyond merely meeting long-term objectives, like ensuring intergenerational equity. It involves allocating responsibilities by the principle of common but differentiated responsibilities, providing each party undertakes its share of the burden based on its unique circumstances and capabilities.[34]

Deontological ethics plays a crucial role in tackling the challenge of climate change through a quasi-regulatory approach. The regulatory approach involves integrating radical measures. Though its application may come at the expense of practical mitigation strategies, the approach establishes a hierarchical framework for enforcement methods that correspond to the urgency required by this global climate crisis. Also, the approach emphasizes strictly keeping to ethical duties and principles and providing a framework that justifies and supports the necessity for decisive and

[31] Alexander and Moore (2016).

[32] Ibid.

[33] Byrne (2011).

[34] Wallimann-Helmer (2019, pp. 751–768).

unconventional actions in the face of the pressing climate crisis. Given the process-focused approach of deontological ethics, it integrates theories that embody principles of climate justice from the context of equity in the distribution of burdens and benefits.

Furthermore, the concept of justice, particularly regarding rightness and fairness, converges with deontological ethics in climate ethics. We can see this intersection when evaluating the morality of the processes employed to address climate change's current or potential impacts, particularly their effects on marginalized, minority, and vulnerable populations. In their ethical analysis of making climate justice integral to the moral evaluation of strategies and policies concerning climate change, Deontologists emphasize the importance of reviewing the justness and fairness of actions.[35]

The second dimension, consequentialism, focuses on evaluating the appropriateness of the outcomes resulting from the chosen course of action. Consequentialists aim to understand the entities and factors responsible for the consequences of emission reduction measures and how the factors interact with the institutional structure that manages the entities' development. Moreover, they express concerns about scrutinizing the effects of the chosen emission reduction approach, delving into the broader consequences and implications of such strategies.[36]

Ethicists have been divided over what should qualify the moral theory that defines anthropogenic actions as tolerable and emission reduction methods that should be avoided. Some propose a no-average threshold for moral theories to test the relevance of the theory to controlling or curbing carbon emissions. Steven Gardiner suggests that moral theories should be subject to a do-or-die test. For instance, we should reject a moral theory if it fails to recognize that neglecting to address an urgent, human-made, and manageable global threat substantially critiques its adequacy. In other words, if a moral framework cannot effectively address and respond to a pressing and human-induced global challenge, it is deemed inadequate and should be discarded.[37] However, it could also be argued that the moral expectation placed on an ethical theory to provide a solution to a global challenge is not tenable. If a theory can identify with the challenge and the negative implication of not addressing it, its purpose has been satisfied. Perhaps consequentialists are not interested in determining the quantum of intervention needed to avert a lifestyle's foreseen or predicted outcome. The reasoning projected by some consequentialists is that, for example, if the quantum of intervention is to be considered, individual contributions to emissions reductions should be regarded as infinitesimal compared to corporate anthropogenic causes of climate change. Waltelr Sinnott-Armstrong supports the argument against the individual moral obligation to contribute to emission reduction. According to him, withdrawing himself from individual activities such as driving a car will make

[35] Bourban (2023, 551–573).

[36] Hourdequin (2010, 443–464).

[37] Gardiner (2011. 218).

no difference to the scale of emission reduction needed to reduce the occurrence of climate change anywhere in the world.[38]

The complexities embedded in consequentialism's theoretical foundations span various domains. Ultimately, there needs to be more consensus on the specific approach that consequentialists believe justifies a moral theory derived from the consequences of actions influenced by belief systems, cultural motives, and regulatory directives. Therefore, it is convenient to conclude that the diversity of perspectives within consequentialism reflects this ethical framework's varied interpretations and applications across different contexts and philosophical viewpoints.

7 Climate Ethics in the Context of African Development

The ethics of climate change, from the viewpoint of climate justice, is closely related to the principles of intergenerational justice and common but differentiated responsibilities because emissions of developed nations primarily drive climate change. At the same time, developing countries, though not guiltless, including those in Africa, disproportionately feel the impacts of global warming. This raises fundamental questions of fairness and responsibility, which makes ethical considerations require developed nations to take responsibility for their historical emissions and support developing countries in mitigating and adapting to climate change's effects. Our understanding of the above principles, as encapsulated in the concept of "climate justice," points to the moral obligation of wealthier industrial countries to assist African countries in their development and resilience-building efforts.

The impacts of climate change on African development are profound. The Executive Secretary of the U.N. Framework Convention on Climate Change (UNFCCC), Simon Stiell, in his address at a meeting of African ministers of environment, described the global climate crises as "an economic sinkhole, sucking the momentum out of economic growth."[39] According to him, Africa needs more investment in areas that will help the countries adapt to climate change, as the impact of climate change is costing African nations up to 5% of their economic output. Extreme weather events' rising frequency and intensity, including storms, droughts, and floods, significantly impact agriculture, the economy, and public health in various countries. Unfortunately, due to climate change-related conditions, African communities are losing their rich biodiversity and natural habitats. The situation's implications make the former Secretary General of the United Nations say, "Saving our planet, lifting people out of poverty, advancing economic growth—these are the same fight. We must connect the dots between climate change, water scarcity, energy shortages, global health, food security, and women's empowerment. Solutions to one problem must be solutions for

[38] Sinnott-Armstrong (2005, 285–307).

[39] Reuters (2024).

all."[40] Such actions are essential to protect the rights and well-being of both present and future generations.

However, for African countries, a call for climate justice also means a significant consideration for the continent's high development deficit in determining the contribution of African countries to the global low emission goals and the support that countries should receive to implement adaptation and mitigation measures. For African countries to effectively build resilience to climate change and integrate tailor-made response mechanisms, investment in technological resources must be on the front burner of sustainable development policies. Of course, this extends to the agitation of developing countries for increased climate financing by developed economies as part of the ongoing international cooperation in addressing the global climate crises. Many African countries need more financial and technological resources to address climate change effectively. However, negotiating the terms and inputs in the various international cooperation fora has been a tough line for African countries to stand on. The need for more climate finance and the complexities of negotiating international climate agreements remain significant hurdles in the path to climate justice. Moreover, the issue of adaptation versus mitigation is a complex ethical dilemma in navigating the ideological frame by which countries discuss, argue and arrive at what best suits their climate and development situation. Developing countries say they must prioritize adaptation to protect their vulnerable populations, while developed nations emphasize the importance of mitigating emissions. Finding a fair and ethical balance between these priorities is essential.

The transition to renewable energy sources and sustainable development practices is a key ethical imperative in addressing climate change in Africa. African nations have abundant renewable energy resources, such as solar and wind power, which can be harnessed for sustainable development. The argument on the ethical factors regarding the business models for investing in low or zero-emission energy development underscores the importance of initiating an international support partnership to help design a realistic pathway for transferring clean energy technologies and providing financial assistance to African countries, building a low-carbon economy. This contributes to mitigating climate change and supports the region's economic development and energy access.

8 The Challenges of Mainstreaming Climate Ethics in African Development Frameworks

No region in the world is or will be spared the effect of the climate change global crisis, but its impacts are particularly pronounced in African nations. The ethical dimensions of climate change demand urgent and robust responses, including integrating climate ethics into African development frameworks. However, mainstreaming climate ethics in African development is not without its challenges.

[40] United Nations (2011).

Integrating climate ethics with mainstream development objectives in Africa is a noble goal but a tricky one because it requires governments to navigate complex ethical dilemmas, including prioritization, resource allocation, historical responsibility, the adaptation-mitigation dilemma, support for vulnerable communities, technological transfer, capacity building, and addressing data and knowledge gaps. While these challenges are significant, they underscore the need for a global commitment to climate justice, prioritizing equity, fairness, and shared responsibility.

8.1 Prioritization and Resource Allocation

Africa faces significant challenges in balancing economic development with environmental sustainability, particularly regarding climate change. While efforts to integrate climate ethics into development frameworks are gaining momentum globally, African countries encounter unique hurdles in prioritizing and allocating resources to address climate-related issues. African countries face many challenges, including poverty, healthcare, and education. Balancing climate change mitigation and adaptation efforts with these immediate development needs presents an ethical dilemma. Governments must decide how to allocate limited resources, addressing the moral question of whether investing in climate action is justified when basic needs remain unmet.

Another significant challenge for mainstream climate ethics in African development is the challenge of resource allocation. Even when climate change is acknowledged as a priority, inadequate attention is paid to providing resources to address it. Several African countries lack the budgetary wherewithal to adequately allocate national revenue for addressing climate change due to a weak economy, external debt, and primary commodity exports, primarily natural resources. While the above factors are indeed part of the constraining factors, institutional factors such as corruption and poor fiscal governance systems in managing their national economies contribute to the constraints in resource allocation. The constraints in the allocation of resources for addressing climate change reduce the capacity of the government to invest in climate change response measures that integrate ethical doctrines in building a sustainable and climate-resilient economy. Additionally, institutional weaknesses, corruption, and lack of transparency exacerbate challenges in allocating resources effectively.

Though climate change affects indiscriminately, its impacts are heavier on the marginalized, the poor, women and Indigenous minorities where there is a lack of capacity to make the provision of resources for climate action a priority in a national budget and in the African Union development frameworks, thus deepening the prevailing social injustices. Furthermore, the lack of climate action undermines commitments to international agreements such as the Paris Agreement and other multilateral agreements that exist to draw commitments from state actors on achieving environmental targets for sustainable development.

Many African nations sign several such agreements but struggle to meet their commitments because of their limited financial resources and underdeveloped economies. The costs associated with climate mitigation and adaptation projects can be prohibitive and challenging when implementing climate-friendly policies and infrastructure.

8.2 *Historical Responsibility*

Africa's climate change woes also have links to its colonial history under the exploitation of Western countries. The European imperialists started exploiting Africa's natural resources in the early 16 century, exporting minerals and timber, among others, to the Western world. As adopted from the colonial enterprise, African countries' policies were extractive, favouring the colonial powers at the detriment of the citizens and the environment. The mining of resources was mainly done with little or no regard for the impacts on the environment and communities, especially the indigenous people. The colonial patrons were only interested in what was financially profitable without concern for their activities' social or environmental impact. Therefore, it also corroded traditional income sources and natural habitats, converting Africa into a theatre of environmental apocalypse together with a paradoxically timeless age of post-colonial socio-economic subjugation.

One of the primary ethical considerations is related to the responsibility of the significant contributors to global warming to pay more for the remedies for climate change, with particular attention to the role of Western countries who sponsored industrial and mining activities in African countries from the colonial era. Also, there is the question of how the standard of care in Western countries differs from what is allowed and negotiated by their proxy businesses in African countries in sectors most beneficial to their economy, particularly the extractive industry. When Western powers established the extractive sector in Africa, its regulations were designed to support the industrialization of their own countries, often at a high environmental cost to the people of Africa. That selfish and aggressive extractive drive also breached the ethical codes of African communities as resource-rich lands that were treated as sacred were either forcefully desecrated or the leaders induced to sell the lands despite the opposition of the people. Some people paid with their blood and lives to protect their lands because they have lived with the understanding that some lands should not be touched, even when no written codes expressly say so. In contemporary Africa, the harmful development philosophy of colonial masters still influences the approach of African governments. Governments have control over the land and ensure that laws allow them to determine what to do with the land for the "benefit" of the people. So, once the investment is in the "overriding public interest," the possible environmental consequence is not a factor. The result of the colonial incursion is what we are living with today. The Western countries are far more industrialized and continue to exploit our natural resources to sustain their developed economies, while Africa wallows in underdevelopment and multiple impacts of climate change.[41] According to Prof.

[41] UNFCCC (2020).

Petteri Taalas, Secretary General of the World Meteorological Organization, "Africa is responsible for less than 10 percent of global greenhouse gas emissions. However, it is the continent that is least able to cope with the negative impacts of climate change. Heatwaves, heavy rains, floods, tropical cyclones, and prolonged droughts are devastatingly impacting communities and economies, with increasing numbers of people at risk."[42]

It is well understood that Africa is the most vulnerable continent to climate change and the least prepared to address it. However, beyond the region's heightened vulnerability, which is partly influenced by its geographical positioning but primarily by its limited ability to adapt, stemming from deficiencies in national economies, education, healthcare, infrastructure, and governance systems, there is the historical culture of controlling the capacity of African countries. Having determined most African countries' geographical, social and political structures when granting independence, they ensured a certain measure of economic dependence was entrenched in the systems.[43]

Africa is currently not economically and socially positioned to defend itself against the disruptive effects of climate change because colonial powers neglected to invest in developing infrastructure, education, healthcare, and governance systems in their colonies, leaving African nations ill-equipped to cope with the challenges posed by climate change. Exploiting Africa's resources has weakened national economies, exacerbating poverty and inequality and increasing vulnerability to climate change impacts. According to the African Development Bank, "Africa will need investments of over $3 trillion in mitigation and adaptation by 2030 to implement its NDCs."[44] As Africa confronts the challenges of climate change, former colonial states are not making decisions on Climate Financing as a moral and ethical responsibility to provide financial support for climate change adaptation and mitigation efforts. Instead, they are debating their approach as a global response, which should be seen as an unethical disregard for the historical elements of their global responsibility in the context of financing climate response strategies in Africa

8.3 Foreign Direct Investment

Foreign Direct Investment (FDI) plays a significant role in the economic development of African countries as it brings in capital and technology that helps contribute to job creation, innovation, infrastructure development, human capital development, and diversification of the economy. Many African countries depend on FDIs to drive their development agenda because it opens their local products to the global market and increases their foreign earnings, which are needed to keep their economy stable for sustainable growth.[45] However, the impact of FDI on climate change in Africa is a

42 WMO (2022).

43 Grant (2015).

44 African Development Bank (2019).

45 UNCTAD (2014).

growing concern. Since the post-colonial era, African countries have been attracting FDI and looking for increasing inflows in different sectors of the economy. The focus has been on the economic gains—actual and prospective—of FDI in Africa, with minimal literature on the contribution of FDI to Africa's increasing vulnerability to climate change.

The scarcity of literature examining the environmental impact of Foreign Direct Investments (FDIs) in Africa is hardly surprising, given the pivotal role FDIs play in African nations' post-colonial development. Upon attaining self-governance, African countries faced limited alternatives for fostering indigenous economies capable of driving the requisite growth. This 'deliberate' legacy of 'dependency' by the colonial significantly contributes to this reliance on foreign investors by African countries who either perpetuated existing investments or facilitated the entry of new Western investors into African markets. Consequently, Africa's economic growth is now predominantly measured and influenced by the annual influx of FDIs.

The tensions between the FDI and climate ethics in Africa's development frameworks are evident in several forms. Foreign direct investment (FDI) inflows into sectors like extractive industries often lead to harmful environmental practices. Many operators of these investments show little concern for the damage their exploitative activities inflict on the environment.[46] Though the inflow of FDI could help stimulate economic growth in a reasonable period, the consequential result can be devastating for ecosystems and local communities, which offends the environmental principles of intergenerational equity and environmental justice. Africa sees FDI as a viable alternative to a slow and more unpredictable domestic investment based on the argument that beyond the seed investment, other externalities accompany FDI in expanding other economic potentials through knowledge transfer in technology, innovations and management. However, development in the context of Africa's needs for improved quality of living, financial independence, and viable and equitable social structure has been redefined by globally proclaimed goals like the sustainable development goals (SDG)[47] and the self-assumed commitments to achieve the stated targets under the Paris Agreement, 2015.[48]

Analysts have observed that Africa's economies have been attracting more FDI outside the sectoral circle of the extractive industry since 2016.[49] However, with the expansion of FDI in greenfield FDI, its translation into development that grows personal income and quality of life of the people of Africa is not evident when compared with the statistical analysis of expert commentators. The OECD reported that Africa's growth rate of 2.6% between 2000 and 2016 can be attributed to several factors: an increase in commodity prices, improved macroeconomic management, and diversification in national development strategies, especially in expanding national revenue sources. In addition, FDI spurred by new trade partnerships

---

[46] World Bank (2020).

[47] African Union (2024).

[48] African Union (2024).

[49] Ayodele Odusola (2018).

with China, India, and several other emerging economic partners has contributed immensely to the growth recorded during the same period.[50]

The proportion of what Africa expects to drive its development strategies still leans more on the inflow of foreign direct investment. However, for most African countries, reliance on FDI is insignificant in jumping-starting their economic development. Some countries have suffered from globally impacted incidents such as the past COVID-19 pandemic and the ongoing Ukraine-Russia war.[51] Therefore, they desperately need a fresh injection of FDI to either re-invent pre-crises national development plans or stimulate their economy to make it attractive for foreign investors looking for greenfield market economies to explore enterprising ventures at the conception or consolidation stages of the business. This will create a competitive market that fosters commercial practices, ultimately driving the state's economy toward achieving the goals outlined in the African Union Agenda 2063.

However, the influence of business entities that access and operate in African countries via FDI on the political system weighs much on the framing and implementing policies and regulations that should preserve ethical principles and doctrines inherent in the national system. The exclusivity some multinationals enjoy has made them too powerful to control the governing administration. This is causing the erosion of ethical standards that, over decades, are considered traditional and reflective of a society's commitment to environmentally sustainable practices. The relevant laws are either weakened in their enforcement or modified through legislative changes to facilitate more permissive market access and operational conditions. There is more literature on how FDIs, through the activities of multinational corporations (MNCs), are substantial contributors to the climate crises that Africa is suffering from. Also, there is very little contestation over the empirically backed studies that have attributed the negative impacts of FDIs on the quality of the environment in African countries. However, African governments have been more receptive to an international advisory that claims the sustainable recovery of Africa from its economic doldrums depends on an increased inflow of FDI. In the aftermath of the COVID-19 pandemic, the World Economic Forum in the year 2021 identified opportunities in key areas, including renewable energy, e-commerce and logistics, sustainable tourism, and circular economy, as sectors that should attract FDIs to drive the growth of the continent in the long and short term.[52] In contrast, the African Union in 2022 listed strategies that focused on building domestic investment-driven industrialization to accelerate the continent's economic growth. Looking at the policies and practices different African countries have adopted or observed to reach their development goals, you can see that there is still a struggle between making the investment environment suitable for domestic investment under a strict regulatory system and letting foreign direct investments (FDIs) into the economy under a less rigorous regulatory system that is more tolerant of actions that go against African ethics.

[50] African Union Commission/OECD (2018).

[51] World Bank (2020).

[52] World Economic Forum (2021).

8.4 Technological Capacity and Transfer

Unlike developing countries, western countries have the technological capacity to build systems that adequately respond to natural disasters linked to climate change. Due to years of investment in scientific research, innovation, and environmental protection, developed countries have established advanced early warning systems to predict weather events and enhance disaster preparedness. Climate ethics imply a responsibility to transfer these technologies and build the capacity of African nations to harness them. However, technological transfer and capacity-building initiatives are complex and often hindered by political, economic, and logistical challenges. The global body of knowledge evolves around national and international regimes that protect the proprietary interest of creators in their inventions. Protecting intellectual property rights is imperative to encourage innovation, protect consumers, and provide legal security for sharing knowledge. However, the commercialization of knowledge sharing can significantly limit the mainstreaming of climate ethics in the African continental frameworks, which should allow countries to cooperate and be guided on liberalizing access to critical technological research information.

Traditional and contemporary ethical practices must be receptive to the technological innovations Africa needs to accelerate its development. Similarly, technological innovations will help create environment-friendly businesses that maintain a high commitment to climate change regimes at the national and international levels. Integrating African environmental and ethical values into innovations is essential in building a receptive process of communal cooperation between host communities and investors. Still, the receptivity of framing environmental and ethical standards for investors at the regional level will depend on the national stakeholders' perception of the necessity of extending standards to innovative creations. It is much easier for people to embrace the argument for climate technologies and creative practices as the solutions being produced to solve climate change predicaments caused by unbridled industrialization. If the technology originates from the global north and is applied in Africa, It usually comes with a price. First, the price tag for using the technology, and second, the price that comes with what some experts call part of the 'Bright Green Lies.'[53]

The practice of protecting the intellectual property rights of innovators is justified by the legal theory of natural law, which states that the creator owns his creation to the extent of his intellectual input and labour.[54] However, when access to critical information that should drive research required to create solutions tailored to African ethical values is restricted due to high costs and proprietary barriers, it becomes challenging for African researchers to leverage shareable knowledge from their counterparts in developed countries.

The second price is distrust in the technology that is proposed to be applied. The distrust can be based on two premises: the technology's contribution to the climate

[53] Jensen et al. (2021).

[54] Hughes (1988, 287).

challenge it is designed to solve and what the technology will cost the environment in developing solutions for climate situations.

African nations often face data and knowledge gaps in assessing climate change impacts and formulating effective policies. These gaps hinder the development of evidence-based climate strategies and adaptation plans. Ethical considerations include the responsibility of developed nations to support capacity building and knowledge transfer to address these gaps, but this is easier said than done.

9 Conclusions

Engaging local communities and ensuring their participation in decision-making is essential to pursuing climate ethics in the context of African development. Indigenous knowledge and traditional practices can provide valuable insights into climate adaptation and resilience. Empowering local communities and respecting their rights align with the ethical principles of justice, equity, and inclusivity.

Climate ethics should be at the centre of the discussion on African development in the context of climate change. For the sake of climate justice, developed countries should play their part in paying for their emissions and supporting Africa in mitigating and adapting to climate change. Although there are challenges and complexities to be addressed in determining fair and ethical means of responding to the challenge of climate change, it is necessary to integrate ethical principles in determining roles, measures, business models, and regulatory approaches. If ethical considerations are maintained, we can build an environment where climate change cannot hinder development. This may also be done by solving ethical issues and mainstreaming climate ethics into each nation's developmental paradigm to pave the way to sustainability and justice.

References

African Developement Bank (2019) Climate change in Africa. https://www.afdb.org/en/cop25/climate-change-africa

African Union Commission/OECD (2018) Overview: harnessing Africa's development dynamics to realize Agenda 2063. In: Africa's development dynamics 2018: growth, jobs and inequalities. OECD Publishing, Paris/African Union Commission, Addis Ababa. https://www.oecd.org/en/publications/africa-s-development-dynamics-2018_9789264302501 en.html

African Union (2024) Linking agenda 2063 and the SDGs. https://au.int/en/agenda2063/sdgs. Accessed 25-05-24

African Union (2024) Meeting of the committee of African heads of state and government on climate change. https://au.int/en/newsevents/20220206/meeting-committee-african-heads-state-andgovernment-climate-change-cahoscc

Agada A (2023) Wiredu on the humanistic orientation of akan morality. In: Molefe M, Allsobrook C (eds) Human dignity in an African context. Palgrave Macmillan, Cham. https://doi.org/10.1007/978-3-031-37341-1_13

Alexander L, Moore M (2016) Deontological ethics. In: Zalta EN (ed) The Stanford encyclopaedia of philosophy, winter 2016 edition. Metaphysics Research Lab, Stanford University, London
Astier C (2012) The exponential growth of African business law and the spread of common law. Publicado em junho de
Ayodele Odusola (2018) Addressing the fdi paradox in Africa. Available online at https://www.un.org/africarenewal/web-features/addressing-foreign-direct-investment-paradox-africa
Bourban M (2023) Climate ethics. In: Wallenhorst N, Wulf C (eds) Handbook of the Anthropocene. Springer, Berlin, pp 551–573. https://doi.org/10.1007/978-3-031-25910-4_87
Broome, J (2008) The ethics of climate change. Sci Am 298(6):96–102. JSTOR. http://www.jstor.org/stable/26000646
Bujo B (2001) Foundations of an African ethic: beyond the universal claims of Western morality (Brian McNeil, trans.). Crossroad, New York
Bujo B (2003) Foundations of an Africa ethic, Paulines, Nairobi
Byrne J (2011) Climate ethics. See online at https://www.researchgate.net/publication/229090769_Climate_ethics/citation/download
Climate Change in Africa. https://www.afdb.org/en/cop25/climate-change-africa
Gardiner S (2011) A perfect moral storm: the ethical tragedy of climate change. Oxford University Press, New York, p 218
Grant R (2015) Africa: geographies of change. Oxford University Press, New York
Gyekye K (1997) Tradition and modernity: philosophical reflections on the African experience. Oxford University Press, New York
Hama A (2013) The righteous mind: why good people are divided by politics and religion. J Soc Polit Econ Stud 38(3):367–377
Hayden P (2010) The environment, global justice and world environmental citizenship. In: Brown GW, Held D (eds) The cosmopolitanism reader. Polity Press, Cambridge, pp 351–372
Heyl C (2019, July 29) The judiciary and the rule of law in Africa. Oxford Research Encyclopedia of Politics. Retrieved 23 Feb 2024, from https://doi.org/10.1093/acrefore/9780190228637.001.0001/acrefore-9780190228637-e-1352
Heyns CH, Stefiszyn K (eds) (2006) Human rights, peace and justice in Africa: a reader. Pulp, London
Hogan M (2011) The moral landscape: how science can determine human values by Sam Harris, pp 224–229
Hourdequin M (2010) Climate, collective action and individual ethical obligations. Environ Values 19(4):443–464. https://doi.org/10.3197/096327110X531552
https://unfccc.int/kyoto_protocol
https://unfccc.int/process-and-meetings/the-paris-agreement
https://www.worldbank.org/en/region/afr/overview
Hughes J (1988) The philosophy of intellectual property. Geo. LJ 77:287
IPCC, Fact sheet-Africa. https://www.ipcc.ch/report/ar6/wg2/downloads/outreach/IPCC_AR6_WGII_FactSheet_Africa.pdf
Jensen D, Keith L, Wilbert M (2021) Bright green lies: how the environmental movement lost its way and what we can do about it. Monkfish Book Publishing, London
Levy BS, Patz JA (2023) Climate change, human rights, and social justice. Ann Global Health 81:1
Linking Agenda 2063 and the SDGs. https://au.int/en/agenda2063/sdgs
Louw DJ (2006) The African concept of Ubuntu and restorative justice. In: Sullivan D, Tifft L (eds) Handbook of restorative justice: A global perspective. Routledge, New York
Mabovula NN (2011) The erosion of African communal values: a reappraisal of the African Ubuntu philosophy. Inkan J Hum Soc Sci 3(1):38–47
Magesa L (1997) African religion: the moral traditions of abundant life. Orbis Books, New York, pp 1–29
Mboh L (2021) An investigation into the role of traditional leaders in conflict resolution: the case of communities in the Mahikeng Local Municipality, North West Province, South Africa. Afr J Conf Resolut 21(2):33–57

Meeting of the Committee of African Heads of State and Government on Climate Change. https://au.int/en/newsevents/20220206/meeting-committee-african-heads-state-and-government-climate-change-cahoscc
Ndulo M (2011) African customary law, customs, and women's rights. Indiana J Glob Legal Stud 18(1):5
No. 3 (May–June 2015); Centre for international sustainable development law, "Human rights, the sustainable development goals and the law: international legal symposium", 09 November 23, available at: https://www.cisdl.org/human-rights-the-sustainable-development-goals-the-law-international-legal-symposium/
Odusola A. Addressing the FDI paradox in Africa. Available online at https://www.un.org/africarenewal/web-features/addressing-foreign-direct-investment-paradox-africa
Ongoing Regional Infrastructure Will Boost Free Trade in Africa. https://www.un.org/africarenewal/magazine/october-2022/ongoing-regional-infrastructure-will-boost-free-trade-africa#:~:text=There%20is%20an%20infrastructure%20challenge,cannot%20succeed%20without%20adequate%20infrastructure
Pimentel D (2011) Legal pluralism in post-colonial Africa: linking statutory and customary adjudication in Mozambique. Yale Hum Rights Develop Law J 14:59
Politics. Ethics. Elsevier, Oxford, pp 285–307
Reuters (2024, Sept 5) Climate change costs Africa up to 5% of GDP, UN climate head says. Retrieved from https://www.reuters.com/world/africa/climate-change-costs-africa-up-5-gdp-un-climate-head-says-2024-09-05/
Said EW (2012) Culture and imperialism. Appiah, K. A. (2005)
Sangiovanni A (2007) Global justice, reciprocity, and the state. Philos Public Aff 35(1):3–39. JSTOR. http://www.jstor.org/stable/4623779
Sinnott-Armstrong W (2005) It's not my fault: global warming and individual moral obligations. In: Sinnott-Armstrong W, Howarth R (eds) Perspectives on climate change: science, economics
Swenson G (2018) Legal pluralism in theory and practice. Int Stud Rev 20(3):438–462
The Kyoto Protocol to the United Nations Framework Convention on Climate Change (1997) Available online at https://treaties.un.org/pages/ViewDetails.aspx?src=TREATY&mtdsg_no=XXVII-7-a&chapter=27&clang=_en
The Paris Agreement (2015) Available online at https://unfccc.int/sites/default/files/english_paris_agreement.pdf
UNFCCC Article published 27th October 2020. https://unfccc.int/news/climate-change-is-an-increasing-threat-to-africa
United Nations Conference on Trade and Development. World Investment Report 2014: Investing in the SDGs Action Plan. UN (2014)
United Nations (2011, Jan 28) Secretary-general's remarks at the world economic forum, Davos, Switzerland. Retrieved from https://www.un.org
Viljoen F (2012) International human rights law in Africa. OUP, Oxford
Wallimann-Helmer I (2019) Justice in managing global climate change. In: Letcher TM (ed) Managing global warming: an interface of technology and human issues. Academic Press, London, pp 751–768. https://doi.org/10.1016/B978-0-12-814104-5.00026-0
What is consequentialism online at https://plato.stanford.edu/entries/consequentialism/
Wilson Center (2023) Modernizing infrastructure vital to fulfilling Africa's promise. https://www.wilsoncenter.org/article/modernizing-infrastructure-vital-fulfilling-africas-promise#:~:text=How%20much%20is%20the%20infrastructure,remaining%20one%2Dthird%20for%20maintenance
Wiredu K (1995) "Customs and morality", in his conceptual decolonization in African philosophy. Hope Publication, Ibadan
WMO. State of the climate in Africa report 2022 online at https://reliefweb.int/report/world/state-climate-africa-2022
Woodman GR (2015) African legal systems. https://doi.org/10.1016/B978-0-08-097086-8.10111-4

World Bank Report Assessing the Economic Impact of Covid-19 and Policy Responses in Sub-Saharan Africa online at https://thedocs.worldbank.org/en/doc/795471589952285352-0090022020/original/05222020AfricasPulse2020AprilCesarCalderon.pdf (pubdocs.worldbank.org)
World Bank (2020) World development report 2020: trading for development in the age of global value chains. World Bank Publications. Online at https://www.worldbank.org/en/publication/wdr2020
World Economic Forum (2021) Why foreign direct investment is key to Africa's sustainable recovery. https://www.weforum.org/agenda/2021/08/foreign-direct-investment-key-africa-sustainable-recovery/

Trade and Development in Africa: Does Climate Ethics Matter?

Patricia Iyore Ajayi and Olusanya Elisa Olubusoye

Abstract The African continent faces enormous environmental challenges, such as deforestation and degradation, which have damaged fragile ecosystems and caused water scarcity, floods, and mudslides in many communities. Likewise, air pollution and improper waste management pose severe problems in cities. Furthermore, rising temperatures and extreme weather conditions could damage the infrastructure needed for transportation, leading to slower and more costly intra-African trade. Given this, member countries must have legal and economic measures to circumvent these climate issues for the smooth running of trade. The African Continental Free Trade Area (AfCFTA) is expected to increase trade and competition among local firms, boost production, and facilitate the movement of labour and goods from one region to another. Increased industrialization and growing populations increase CO_2 emissions and other greenhouse gases from industry and automobiles. Studies have shown that African countries have done little to adhere to the Paris Agreement to reduce global CO_2 emissions. Therefore, for AfCFTA to succeed, climate ethics should be at the core of the agreement because it can steer action, facilitate arbitration, resolve conflicting interests, and establish priorities. Hence, this study seeks to examine ethical principles of climate change that reflect concerns about the fragility of the global climate in the conduct of trade and development. It is expected that the outcome of this study will aid the development of applicable climate ethics principles and standards for African Trade and Development.

P. I. Ajayi (✉)
Department of Economics, University of Ibadan, Ibadan, Nigeria
e-mail: patajayi71@gmail.com

O. E. Olubusoye
Department of Statistics, University of Ibadan, Ibadan, Nigeria

D. Oghenerobor Agelebe (ed.), *Climate Ethics in Africa*, Global Perspectives on Law and the Environment 1, https://doi.org/10.1007/978-3-032-08757-7_2

1 Introduction

African countries need to increase their trade with each other to foster interregional trade, cross-border investments in infrastructure, and industrialization. This will also result in resilience to global financial, food-related climatic, and pandemic shocks while contributing to a larger share of international trade.

Over the years, regional integration in Africa has been based more on collective self-reliance than a purely trade liberalization agenda. Thus, to promote a trade liberalization agenda, the African Continental Free Trade Area (AfCFTA) was established in 2015, and trade within its framework commenced in 2021. The AfCFTA aims to integrate Africa's markets while strengthening the economic relationships and building upon member countries' economic rights and obligations. It also aims at free movement of persons, capital, goods and services crucial for deepening economic integration and promoting agricultural development, food security, industrialization and structural economic transformation. Indeed, the AfCFTA is predicted to bring vast economic benefits to the continent as it is expected to boost consumer and business spending in Africa to over $6.7trillion by 2030.[1] It is also likely to boost Africa's exports by $560 billion by 2035, increase real income gain by 7% and lift 30 million Africans from extreme poverty.[2] However, for AfCFTA to be successful, there is a need to be mindful of environmental challenges that may arise due to the impacts of climate change during the implementation process. According to the Intergovernmental Panel on Climate Change (IPCC), Africa should expect adverse climate impacts on port facilities, buildings, roads, railways and bridges due to rising sea levels and extreme weather conditions.[3] Furthermore, deforestation and land degradation have damaged fragile ecosystems, resulting in biodiversity loss and water scarcity while causing impacts such as floods and mudslides on communities. Likewise, air pollution and improper waste management pose severe problems in cities. Thus, issues of climate change impacts and environmental degradation need to be seriously addressed not to undermine the predicted economic gains of AfCFTA.

In the past, Africa had its traditional ways of ensuring that the environment was well cared for, for example, by planting trees and shrubs, creating a diversion to help drainage, cutting down on waste by recycling wastes, etc.; these ways have been discarded due to civilization. To foster sustainable development, it is imperative to return to some of these ways with modifications to suit the modern world. We need to unlearn and relearn new ways to forge ahead as a continent to combat climate change and enhance trade and development. To achieve this, it is essential to understand each society's perception of nature, how it treats nature, and how it relates to nature to enhance the environment's health. Many environmental and health challenges today have been attributed to humans' unhealthy activities and misuse of the natural environment, leading to global warming.[4] Some activities that have contributed to the

[1] World Economic Forum (2023).

[2] Ibid.

[3] van der Ven and Signe (2021).

[4] Tosam (2019, 172–192).

degradation of the environment include dependence on wood as a source of fuel for cooking, over-cultivation of land, over-grazing and felling trees without planting new ones. Pollution from industries has also put untold pressure on the environment; all these have intensified global warming and climate change, exposing the environment to drought, storms, and floods, all of which disrupt economic activities and supply of safe drinking water, food and health services.[5]

1.1 Trade, Development and Climate Ethics

In simple terms, trade is the voluntary exchange of goods or services between economic actors such as consumers, companies and nations. Trade is carried out in different sectors of the economy. For instance, trade refers to purchasing and selling securities, commodities or derivatives in financial markets. Free trade means international exchanges of products and services without obstruction by tariffs or other trade barriers. Since transactions are consensual, trade is generally considered to benefit all parties. Trade can also be between 2 or more partners; when two or more countries engage in trade, it is called bilateral trade, while trade involving more than two nations is called multilateral trade. Trade also exists between regions because different regions may have a comparative advantage in producing some tradeable commodity. Over the years, trade openness has continued to increase among countries and areas in a fast, globalizing digital world.

Economic development is a process by which the economic well-being and quality of life of a nation and its citizens are improved according to targeted goals and objectives. Development is concerned with quality improvements, introducing new goods and services, risk, mitigation, and innovation and entrepreneurship dynamics. It is about positioning the economy on a higher growth trajectory. It is the product of long-term investments in generating new ideas, knowledge transfer and infrastructure; it depends on functioning social and economic institutions and cooperation between the public and private enterprises. Development requires collective action and large-scale, long-term investments that address the fundamental conditions necessary for the functioning of the economy. Development depends on education so workers can fully participate in the economic, social and cultural patterns of behaviour that encourage initiative and engagement. It requires a balance between education and support for industrial development. In essence, the development process leads to improved health and well-being of the population.

Climate ethics focuses on the ethical dimensions of climate change and concepts such as climate justice. The idea of climate ethics stemmed from ethics itself, and it raises moral questions about who is responsible for what actions, who should be held accountable and what should and should not be allowed. Because climate ethics is a large-scale issue, many approaches have been theorized that will have global benefits.

5 Tosam and Mbih (2015, 787–800).

Climate ethics addresses responsibilities that individuals owe the earth systems while considering mitigations already in place.

The link between trade, development and climate ethics shows that development resulting from trade expansion can directly impact the environment by increasing pollution or degrading natural resources. Besides, trade in services also brings about the transfer of communicable ailments and diseases through people migrating from one region to another. Since trade brings about specialization, it may lead to more pollution-intensive activities in some countries than others. Again, trade liberalization provides firms with incentives to adopt more stringent environmental standards. The consequences of climate change come from more frequent extreme weather events and rising sea levels; supply, transport and distribution chain infrastructure are likely to become more vulnerable to disruptions due to climate change. Since maritime shipping accounts for about 80% of global trade by volume, it could experience negative consequences, such as more frequent closures due to extreme weather conditions. Climate change can also decrease the productivity of all factors of production (that is, land, labour, and capital), which can result in substantial output losses and a decrease in global trade volume.

2 Background on Africa's Trade Policies and Development (AfCFTA)

African trade has existed since 10,000 BC, notwithstanding specific recent historical markers. Herodotus described the earliest evidence of African trade across the Sahara (c. 484-425BC). At that time, salt, ivory, animal skins, and enslaved people were traded northwards, while manufactured goods such as pottery, glass and metalwork were traded southwards. In the precolonial era, the Savannah south of the Sahel was identified as the breadbasket of West Africa's production of cereals and fish for long-distance markets. The domestication of African rice took place in the middle Niger, where the Niger River and its tributaries also produced fish. The river was used for shipping cereals and dried fish. Gold and kola nuts, two significant commodities for the far-off markets, were made in the forest to the south. Mali was renowned for its riches of gold; Ghana was also known as a land of gold, and gold was also found near Bure and Bambuk in the Senegal and Niger River basins, respectively.

As civilization emerged, new commercial trade routes were created. Regional and long-distance trade played an instrumental role in cultural exchange, introducing new technologies and regional integration. The African Union recognized the role that trade plays in emerging economies. This was promoted by the Agenda 2063, which acknowledged that trade is a potent driver of economic growth and development. Intra-African trade has remained relatively low over the past 10 years, with 2015 and 2016 recording the most significant levels with 19 and 20% of overall trade, respectively. The majority of intra-African commerce takes place between nations that belong to the same regional economic community (REC), mainly when those

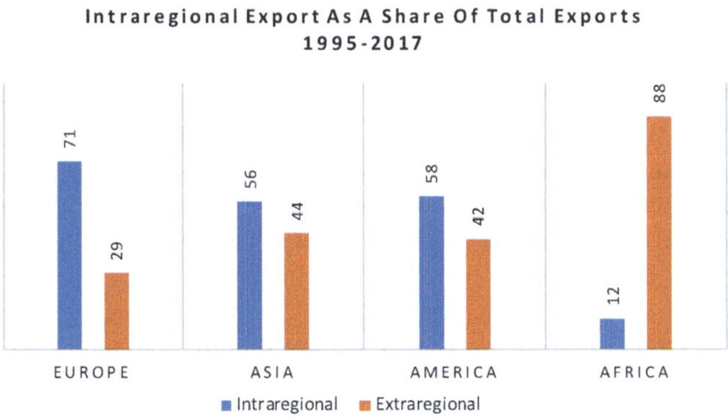

Fig. 1 Intraregional export as a share of total exports (UNCTAD 2021)

nations are participants in the REC's Free Trade Agreement (FTA) or Customs Union (CU). Intra-African exports surged by more than 410% between 1995 and 2017, going from $26.7 billion in 1995 to $137.1 billion in 2017. It consistently showed positive growth throughout these years, except 1998–2001, 2009, and 2015–2016.[6]

Commerce restrictions in Africa led to a fragmented regional economy, making the continent's contribution to global trade minimal. Africa's role in the worldwide trade market has also been chiefly that of a provider of raw materials (commodities) in exchange for manufactured goods. This made Africa sensitive to external shocks resulting from export dependence due to the difficulties brought on by constantly fluctuating commodity prices and the poor value addition of her natural resources. As a result of price shocks in global commodities, which continue to make up the majority of Africa's exports, the lack of growth or decline in growth between 1998 and 2001 and 2009 was primarily attributed to global recessions and the negative growth that followed in 2014 was also caused by price shocks in global commodities. Over the past forty years, the percentage of trade within Africa is still relatively small compared to trade overall, primarily with nations outside of Africa (see Fig. 1). In sum, African exports during the previous two decades have averaged a 12% intra-African share, compared to an intraregional share of 56% in Asia, 58% in the Americas, and 71% in Europe. Additionally, Africa still only makes up a small portion of global trade. The poor performance of intra-African commerce has several causes. One of them is that African exporters within the region frequently deal with relatively high tariffs, with an average protection rate of 8.5% compared to 2.5% when exporting to markets outside the area.[7] Furthermore, due to NTBs like burdensome and pointless document requirements, pointless delays, ineffective ports, and poor infrastructure, Africa has relatively thick borders (with restrictions that slow down trade), which raises the cost of dealing inside Africa.

[6] Coulibaly et al. (2022).

[7] United Nations Conference on Trade and Development (UNCTAD) (2021).

Strategic milestones in Africa's regional trade began in 1963, when the Organisation of African Unity (OAU), the precursor to the African Union, was launched. In 1970, the Regional Economic Communities were established, and in 1991, the Treaty of Abuja was signed to set up the African Economic Community. In 2002, OAU was reorganized and re-launched as the African Union (AU). 2004, the Pan African Parliament was established, and in 2012, the Action Plan on Boosting Intra-African Trade (BIAT) was endorsed. In 2014, the EAC-COMESA-SADC Tripartite Free Trade Area was finalized. In June 2015, in Johannesburg, the summit of the Africa Union, which led to the launch of the CFTA negotiations, was held. The key features of the CFTA were an agreement establishing the African Continental Free Trade Area, the Protocol on Trade in Goods, the Protocol on Trade in Services and the Protocol on Settlement of Dispute (see Table 1).

The African Union Assembly of Heads of State and Government's 18th Ordinary Session determined in January 2012 that the African Continental Free Trade Area (AfCFTA) would be established by approximately 2017 within its first year of operation. The Action Plan for Boosting Intra-African Trade, which outlined seven program clusters, including factor market integration, trade policy, trade facilitation, productive capacity, and infrastructure connected to trade, was also approved at the conference. AfCFTA sought to unite 54 African nations with a combined population of over 1.2 billion and a gross domestic product (GDP) of over $3 trillion. "Create a single continental market for products and services, with free mobility of business persons and investments," according to the objective. Even though it only makes up a small portion of all trade in Africa, intraregional trade is growing more significantly. The AfCFTA's ability to considerably accelerate economic development in the region is currently fostering the push for regional integration. Increasing intra-African commerce faces both significant potential and obstacles with implementing the AfCFTA. In the first 5 years of the AfCFTA agreement, governments agreed to eliminate tariffs on 90% of goods. Thereafter, tariffs on the remaining product groupings and nontariff trade obstacles would be eliminated.

There has always been a need for countries in Africa to compete and participate in the global economy to create value, reduce poverty and promote inclusion. The African Continental Free Trade Area (AfCFTA) provided that unique opportunity for the continent. The African Continental Free Trade Area (AfCFTA), the centrepiece project of Agenda 2063, is a continental geographic area where members of the AU can freely trade products and services. By offering a comprehensive and mutually beneficial trade agreement between the member nations that addresses trade in products and services, investment, intellectual property rights, and competition regulation, the AfCFTA seeks to increase intra-African commerce. Establishing the African Continental Free Trade Area (AfCFTA) and enacting the Agreement Establishing the AfCFTA (AfCFTA Agreement) on May 30, 2019, offers significant chances to advance intra-African trade and support sustainable development. The AfCFTA is essential to accomplishing the continental integration envisioned in the Agenda 2063: The Africa We Want of the African Union and the Organization for African Unity's Treaty Establishing the African Economic Community (Abuja Treaty) (Agenda 2063) is also anticipated to drive the necessary economic transition

Table 1 Key features of the AfCFTA agreement (UNCTAD 2021)

Agreement establishing the African continental free trade area	Protocol on trade in goods	Elimination of duties and quantitative restrictions on imports Imports shall be treated no less favourably than domestic products Elimination of non-tariff barriers Cooperation of customs authorities Trade facilitation and transit Trade remedies, protections for infant industries and general exceptions Cooperation over product standards and regulations Technical assistance, capacity-building and cooperation
	Protocol on trade in services	Transparency of service regulations Mutual recognition of standards, licensing and certification of services suppliers Progressive liberalization of services sectors Service suppliers shall be treated no less favourably than domestic suppliers in liberalized sectors Provision for general and security exceptions
	Protocol on dispute settlement	Modelled on the WTO dispute settlement system Dispute settlement body Appellate body
	Phase II negotiations	Intellectual property rights Investment Competition policies
	Phase III negotiations	E-commerce

to encourage the inclusive and sustained growth needed to African nations which will carry out the 2030 Agenda for achieving sustainable development (Agenda 2030) Goals for Sustainable Development (SDGs).[8]

[8] United Nations Conference on Trade and Development (UNCTAD) (2021).

3 Theoretical Linkage Between Climate Ethics and Trade

The World Trade Organisation (1995) and the Kyoto Protocol (1997) were multilateral institutions designed to address global trade and climate change. The WTO, which embodies the multilateral regime of rules governing international trade, was set up in Switzerland in 1995, while the Kyoto Protocol negotiations culminated in Kyoto, Japan, in 1997 pursuant to the earlier United Nations Framework Convention on Climate Change (UNFCCC) was established to bind numerical limits to industrialized countries emissions to reduce global warming. Although the two institutions have conflicts and complementarities, the interconnections between climate ethics and trade policy cannot be entirely ignored.[9] Before the General Agreement on Tariffs and Trade (GATT) was formed into the WTO, it was accepted that the links between trade and the environment needed to be examined. As such, trade and environmental links have been formally examined. The international community recognized the potential of climate ethics to promote sustainable development. The 1992 United Nations Conference on Environment and Development action plan, Agenda 21, emphasized the need to ensure that environment and trade policies are mutually supportive to achieve sustainable development.

The connection between trade and environmental issues has taken centre stage. Several NGOs (non-governmental organizations) have demanded that the World Trade Organisation (WTO) formally incorporate such a linkage regarding ecological standards. Bhagwati, in 1999, suggested a comprehensive framework for linking trade with the environment. He classified environmental damage as either local or global. He also split countries addressing environmental issues into egoistic (seeking their advantage) and altruistic (seeking others' advantage). Egoistic countries act in ways that are ultimately self-interested, while altruistic countries, which contrast with egoistic countries, are motivated to act in the interest of others and the globe.[10] Three primary relationships between trade and the environment have been recognized in recent literature: the scale effect, which implies that trade might harm environmental quality by expanding the scope of economic activity; the composition effect, which states that trade may affect environmental quality by shifting production to other countries when there are poorly enforced regulations, and it is unclear how this will affect the world generally and the technology effect, which states that trade may boost the acceptance and spread of cleaner technology.[11]

The many potential effects of trade on the environment, as shown in Table 2, can be divided into two groups: those that operate via GDP in a similar way to investment, technology, and other sources of economic growth and those that are unique to trade alone and thus hold even for a given level of GDP. There are beneficial and unfavourable effects within each category. The aspect of environmental

[9] Frankel (2010).

[10] Bhagwati (2000).

[11] Zagonari (2000).

Table 2 Is trade good or bad for the environment? (Frankel 2010)

Effects of trade on the environment	Via growth in income	Given a constant level of income
Harmful effects	Larger scale of economic activity	"Race to the bottom" in national regulation
Beneficial effects	Shifts to cleaner techniques and composition of economic activity	"Gains from trade": ratcheting up of standards, innovation, consumer power
Statistical evidence on the bottom line, looking across countries		
For sulfur dioxide (SO_2)	Statistical evidence on the bottom line, looking across countries	Other things equal, the beneficial effects of trade on SO_2 seem to dominate over the harmful effects
For carbon dioxide (CO_2)	There is no sign that total emissions will decline on their own (presumably because CO_2 is a global externality: little regulation is possible at the national level)	Trade, if anything, may increase CO_2 emissions, even for a given income level

quality at risk determines the overall outcome. Statistics studies provide a reasonably precise answer for sulphur dioxide (SO_2) concentrations, a common air pollutant linked to respiratory illness and other health effects and precursor to acid rain. The Environmental Kuznets Curve shows that continued economic growth typically produces cleaner air. The market doesn't completely fix the issue on its own; instead, as people become wealthier, they demand cleaner air, and in a nation with an attentive and capable government, the outcome is an efficient regulatory system. How about the impact of trade openness while maintaining the same income level? More trade opens up opportunities for innovation, a global push for higher standards, consumer empowerment, and other avenues that can help reduce air pollution. Compared to other growth drivers, trade openness seems to worsen emissions. It is not difficult to understand how trade and growth could lessen local air pollution while increasing greenhouse gas emissions. The latter is an externality on a global scale. Regulation at the national level is insufficient to handle the international "free rider problem" since each nation would have no motivation to reduce emissions because it would face the financial burden alone, even though benefits would go to others. To combat global climate change, we need not only the desire and the money but also the kind of international collaboration that results from a multilateral regulatory system.[12]

[12] Frankel (2010).

4 The Role of Climate Ethics in Trade

One of the main objectives of the AfCFTA Agreement is to create a single continental market for goods and services, including the free movement of persons and capital. This is expected to be achieved, among others, through better harmonization and coordination of trade liberalization and facilitation instruments across the many overlapping regional economic communities (RECs) and Africa in general.

Globally, Africa accounts for only a tiny share of greenhouse gas (GHG) emissions below 4%. However, according to the 2022 Intergovernmental Panel on Climate Change (IPCC) Report, the African continent will be most negatively affected by climate change. Even though trade is perceived as a major contributor to carbon emissions because greenhouse gases are released by producing and transporting traded goods and services, free trade agreements such as the AfCFTA must incorporate climate-friendly policies. A lot of the GHGs released due to trade and production activities account for an average of 20–30% of global GHG emissions. The international transport sector alone generates 12% of emissions.

Due to growth in global trade, it has become a lot easier to access a variety of goods and services; such goods are transported through low-cost transportation such as air freight. However, there has been a supply downturn due to COVID-19 and the Ukraine war, which has led to a reduction in food supplies. This has had a negative impact on Africa as the basic needs of food, housing, and access to energy have become even more difficult to meet. A vibrant food item market under AfCFTA will go a long way to addressing some of these challenges. Besides, it is also helpful to ensure that farmlands in Africa are not destroyed due to excess rainfall or drought, as the case may be.

It has been estimated that the manufacturing sector will be the primary beneficiary of the AfCFTA; this implies rising energy needs, among others. It is, therefore, essential for the continent to leverage new and less energy-intensive technologies. Thus, having a good energy market will be a better strategy for reducing emissions. The African continent is rich in renewable energy sources such as geothermal, wind, solar, hydroelectricity, and many others. Harnessing and investing in these renewable energy sources in different sectors will further reduce dependence on traditional fossil fuels, hopefully resulting in a cleaner trade environment. Africa certainly has the potential to lead the world in this sphere.

Given that Africa has the world's lowest per capita CO_2 emissions, the continent needs to achieve greater leeway in the COP27 negotiations. This can be done by incorporating ethical issues of climate change in implementing the AfCFTA alongside providing the necessary financing to help mitigate the negative impacts of climate change.

One way to address climate issues in the free trade area is to ensure that all people have the right to affordable and sustainable energy. Vulnerable and traditional groups need to be represented in policy and decision-making. Manufacturing companies should be required to pay a tax commensurate with the quantity of pollutants they emit.

5 Outlook for AfCFTA

Contrary to the WTO Agreement, the AfCFTA Agreement does not refer to sustainable development or the complementarity of trade and environmental policies. For AfCFTA to succeed, climate ethics should be at the core of the agreement because it can steer action, facilitate arbitration, resolve conflicting interests, and establish priorities. AfCFTA, Africa's regional trade institution, has to consider how existing trade measures affect the environment and development. Africa's environment and trade policies have to be mutually supportive. An open, multilateral trading system enables more effective resource allocation and utilization, which raises productivity and incomes while reducing environmental demands. A sound environment, on the other hand, provides the ecological and other resources needed to sustain growth and underpin ongoing trade expansion. An open trading system that is backed by the adoption of sensible environmental regulations will benefit the environment and promote sustainable development.

Globally, 143 nations have ratified the Paris Agreement, and 33 of those are in Africa, including Benin, Burkina Faso, Cameroon, Chad, Ethiopia, Gabon, Gambia, Kenya, Nigeria, Somalia, Tunisia, Uganda, and Zambia. That represents 60% of all African nations. These countries have made climate commitments by setting their Nationally Determined Contributions (NDCs).[13] The NDCs represent each nation's efforts to combat climate change. The majority of African nations made plans to prioritize climate-proofing development initiatives in their NDCs, particularly in key industries like agriculture and energy. Although Africa's urgent need is to accelerate socio-economic development, efforts in the area of clean technology and trade best practices should be adopted by all countries to achieve the regional demand for the development and global requirement for net-zero emissions. AfCTFA can play a role in ensuring that multinational firms are required to follow the environmental regulations of their home nations wherever they operate. AfCFTA and climate considerations are not mutually exclusive, and a collaborative and mutually reliant Africa has positive implications for trade, food and environmental security.

References

African Business (2021) A short history of African trade. https://african.business/2021/10/trade-investment/a-short-history-of-african-trade/

African Union, Trade and Industrial Development. https://au.int/en/trade-and-industrial-development

Bhagwati J (2000) On thinking clearly about the linkage between trade and the environment. Environ Develop Econ J 5:483–529

Coulibaly S, Kassa W, Zeutack AG (2022) Africa in the new trade environment: market access in troubled times. World bank group. https://openknowledge.worldbank.org/bitstream/handle/10986/36884/9781464817564.pdf?sequence=2&isAllowed=y

[13] Munang and Mgendi (2017).

Emmanuel Akyeampong (2015) History of African trade African export-import bank. https://elibrary.acbfpact.org/acbf/collect/acbf/index/assoc/HASH0100/039e055d/606b9c4a/9550.dir/History-of-African-Trade.pdf
Frankel J (2010) Climate and trade: links between the Kyoto protocol and WTO. Environ Sci Policy Sustain Dev. https://doi.org/10.3200/ENVT.47.7.8-21
Munang R, Mgendi R (2017) Paris agreement on climate change: How is Africa faring one year later? African Renewal. https://www.un.org/africarenewal/magazine/may-july-2017/paris-agreement-climate-change-one-year-later-how-africa-faring
Tosam (2019) African environmental ethics and sustainable development. Open J Philos 9:172–192
Tosam MJ, Mbih RA (2015) Climate change health and sustainability in Africa. Environ Develop Sustain Multidiscip Approach Theory Pract Sustain Develop 17:787–800. https://doi.org/10.1007/s10668-014-9575-0
Tralac (2021) Summary of intra-Africa trade 2019 trade law centre. https://www.tralac.org/documents/publications/trade-data-analysis/3982-summary-intra-africa-trade-2019/file.html
United Nations Conference on Trade and Development (UNCTAD) (2021) Implications of the African continental free trade area for trade and biodiversity: policy and regulatory recommendations. https://unctad.org/system/files/official-document/ditctedinf2021d3_en.pdf
van der Ven C, Signe L (2021) Greening the AfCFTA: it is not too late—Africa growth initiative of brookings. Policy Brief.
World Economic Forum (2023) AfCFTA: a new era for global business and investment in Africa available online. https://www3.weforum.org/docs/WEF_Friends_of_the_Africa_Continental_Free_Trade_Area_2023.pdf
Zagonari F (2000) A note on Bhagwati's 'On thinking clearly about the linkage between trade and environment.' Environ Develop Econ J 5:483–529

Socio-Economic and Environmental Impacts of Climate Change in the Horn of Africa: High Need for Climate Action and Transboundary Cooperations

Gemedo Dalle

Abstract Climate change exacerbates conflicts in vulnerable regions, particularly arid and semi-arid pastoral areas and across borders. Rising temperatures, erratic rainfall, and extreme weather accelerate rangeland degradation, deforestation, and water scarcity, leading to food and resource shortages that drive migration and violent conflicts. Increased global resource competition further heightens international tensions. In the Horn of Africa, drought has affected 36.4 million people, causing widespread crop failure and livestock deaths, with Ethiopia alone losing 6.85 million livestock since 2021. The economic toll is severe, especially for pastoralist communities reliant on animal products. Women and children are disproportionately affected, necessitating targeted interventions. Climate change also threatens Ethiopia's coffee industry, a key export sector, with projections showing a 60% decline in suitable areas for Arabica coffee cultivation by 2050. Additionally, climate change has shifted species' ecological ranges, enabling invasive species like *Prosopis juliflora* to spread, harming pastoral livelihoods and biodiversity. By 2050, highly suitable areas for its invasion are expected to expand by 63.6%. Urgent climate action is required to mitigate these risks. This review calls for early warning systems, environmental risk assessment mechanisms, livelihood diversification, improved resource management, and resilience-building to reduce conflicts and promote sustainability.

1 Introduction

Climate change has been affecting the lives and livelihoods of people in all countries with disproportionate negative impacts on those in Least Developed Countries such as Ethiopia. The countries in East Africa are highly vulnerable to climate variability and its impacts due to ecological and socio-economic factors, as well as low capacity for investment in adaptation and mitigation. The subregion has been affected by the worst drought in over four decades. According to the UN Office for the Coordination

G. Dalle (✉)
Center for Environmental Science, Addis Ababa University, Addis Ababa, Ethiopia
e-mail: gemedo.dalle@aau.edu.et

D. Oghenerobor Agelebe (ed.), *Climate Ethics in Africa*, Global Perspectives on Law and the Environment 1, https://doi.org/10.1007/978-3-032-08757-7_3

of Humanitarian Affairs (OCHA), at least 36.4 million people will be affected across the Horn of Africa due to the prolonged drought, with livelihoods lost due to crop failures and livestock deaths.[1] According to this assessment, communities in the Horn of Africa are facing the immediate threat of starvation following the fifth consecutive failed season in parts of Ethiopia, Kenya, and Somalia.

Rising temperatures and changes in precipitation have made crop failure a common phenomenon in many countries in East Africa, directly contributing to food insecurity and malnutrition. Climate-induced water, energy, and food stresses are critical challenges of the twenty-first century that require a comprehensive approach across all sectors and regions. Climate change threatens the sustainability of the livelihoods of people living in developing countries.

Climate change is another major factor contributing to the internal displacement of populations. Studies show that about 216 million people (of which 85.7 million are in sub-Saharan Africa) could be internally displaced due to climate change impacts by 2050.[2] Climate change has significantly contributed to environmental degradation, leading to human migration and violent conflicts.[3] The scarcity of resources driven by climate change has the potential to ignite conflicts. As climate change diminishes the availability of natural resources, it intensifies competition for essentials like food, water, and energy—resources of utmost concern. In response, African governments and international stakeholders are urged to take proactive measures to prevent and adapt to climate-induced displacement, particularly in fragile and conflict-affected regions. Practical actions are essential, including (i) increasing investments in adaptation, prevention, disaster preparedness, early warning, and early action; (ii) ensuring community-based participation in the formulation of responses; (iii) supporting an integrated, multi-stakeholder approach to the prevention and management fronts of climate displacement; (iv) supporting evidence-based approaches across the humanitarian-development-peace nexus; (v) addressing and strengthening gender-sensitive humanitarian protection and assistance for displaced communities; and (vi) addressing and accelerating financing, including the mobilization of climate finance mechanisms.[4]

Eastern African countries are disproportionately affected by climate change due to low adaptive capacity, frequent armed conflicts, and weak governance systems. These countries depend on rain-fed agriculture, and frequent droughts have been causing a significant decline in food production, the degradation of freshwater, and an increase in environmentally induced migration. Climate change will likely intensify droughts, storms, and floods, which may lead to environmental migrations and potential conflicts. Reuveny pointed out that regions that exhibit more ecological problems and depend more heavily on the environment for livelihood also exhibit more environmental migration episodes.[5]

---

[1] OCHA (UN Office for the Coordination of Humanitarian Affairs) (2022).

[2] Clement et al. (2021).

[3] Kolmannskog (2010, 46).

[4] CCCPA (International Center for Conflict Resolution, Peacekeeping and Peacebuilding) (2022).

[5] Reuveny (2007).

Existing data show that droughts are occurring more frequently in East African countries. According to Carleton, before 1999, a drought caused by a poor or failed rainy season happened once every five to six years, but that frequency has changed to two to three years since 1999.[6] The situation worsens in the Greater Horn of Africa with an ongoing fifth consecutive drought season in parts of Ethiopia, Kenya, and Somalia. These frequent droughts directly contribute to water and fodder shortages, increasing competition over natural resources. This competition affects social cohesion and triggers internal displacements. In some fragile countries, such competition may lead to conflicts that cost lives and livelihoods.

Ethiopia is among the most affected countries in East Africa as La Niña-induced frequent droughts have caused five consecutive failed rainy seasons since 2020.[7] The country faces extreme climate change risk as drought conditions have been the worst for 40 years. OCHA report of 2022 pointed out that more than 8 million people in Ethiopia are affected across southern and south-eastern parts of the country, including Somali (more than 3.5 million people), Oromia (more than 3.4 million), SNNP (more than 1.1 million), and Southwestern (more than 200,000 people) regions. This figure increased significantly in one year as 24 million people live in drought-affected areas, with 11 million people estimated to be food insecure.[8] The other bitter truth is that climate change affects women and children disproportionately. Reports show that 2.8 million children are suffering from acute malnutrition (ACAPS 2023).

This review summarises climate change's environmental, economic, and sociocultural impacts in drought-affected areas of the Horn of Africa, specifically focusing on Ethiopia. The paper identifies critical interventions that build resilience and promote development in those affected areas.

2 Impact of Climate Change on Agricultural Productivity: Crops

Climate change is expected to severely impact livelihoods, public health, food security, and water availability. Shifting environmental conditions significantly threaten agricultural productivity, leading to reduced food production, potential food crises, and an increased risk of conflict. Variations in temperature and rainfall patterns are particularly concerning, as they are likely to negatively affect cereal production in many developing countries, with the Horn of Africa being especially vulnerable. For example, the linkage between water and food security in the Nile basin is evident because water scarcity negatively impacts agriculture and food security. According to Kindiki, water scarcity is the biggest threat to food security worldwide, including in East African countries.[9] The impacts of climate change on African countries

[6] Carleton (2022).

[7] OCHA (2023).

[8] Ibid.

[9] Kindiki (2009).

are severe because most African countries depend on rain-fed agriculture and have limited capacity for climate change adaptation and mitigation.

In 2011 alone, climate change caused environmental and social crises in the Horn of Africa. East Africa was hit by a drought reported to be the worst in 60 years.[10] However, the 2020–2022 drought has now surpassed the horrific droughts in 2010–2011 and 2016–2017 in both duration and severity and will continue to deepen in the months ahead, with catastrophic consequences.[11] According to this assessment, crop harvest has failed for five consecutive seasons in the Horn of Africa, resulting in food shortages and the threat of starvation.

3 Impacts on Pastoral Livestock Production

Climate change and pastoralists' livelihoods are interlinked. Climate change severely affects pastoralists' livelihoods because they depend heavily on livestock, which, in turn, depend on forage and water. Variability in the weather pattern due to climate change has significant implications for pastoralist livelihoods and security. Pastoralists live in a hostile and arid environment prone to frequent drought. Low precipitation and high temperatures affect livestock availability of forage and water supplies.

Drought negatively impacts pastoralists because it causes livestock loss. It must be noted that pastoralists depend heavily on their livestock for their livelihood. According to Doti, drought periods positively correlate with increased ethnic conflicts over stiff competition for water and pasture, which sometimes extends across borders.[12] Pastoral insecurity determines grazing areas because when insecurity is high, livestock herds tend to concentrate in small, secure grazing zones while leaving large tracts of land unused. Furthermore, cattle raids among neighbouring communities increase during droughts.

Livestock are everything for the pastoralist communities, as these communities depend on their livestock for income, food, and nutrition. For the Borana pastoralists, livestock are a source of wealth and pride. According to a recent report, Borana pastoralists lost their livestock to prolonged drought. Reports show that about 6.85 million livestock deaths have occurred since late 2021[13] And about 25 million livestock are weakened and at risk of death.[14] The Oromia National Regional State Disaster and Risk Management Commission (ODRMC) reported that 1.3 million livestock died only in early 2022, and the estimated economic loss was 26 billion Ethiopian Birr, equivalent to 483,831,711 US Dollars.[15] The loss was much higher in 2023. Similarly, the Somali National Regional State reports showed that 683,465

---

[10] IRC (International Rescue Committee) (2011).

[11] OCHA (2022, n. 1).

[12] Doti (2010).

[13] OCHA (2023, n. 7).

[14] ACAPS (2023).

[15] ODRMC (Oromia Disaster and Risk Management Commission) (2022).

livestock died during the drought period from August 2021 to February 2022. The economic impact of livestock deaths at the national, regional, and household levels was significant.

4 Impacts on Women and Children

Climate change represents the ultimate resource scarcity perspective. It may be a source of conflict, as increasing temperatures, precipitation anomalies, and extreme weather are expected to aggravate situations in areas already experiencing high population pressure and resource scarcity.[16] These factors could force millions of people to migrate, leading to increased risks of inter-ethnic tensions and higher pressures on resources in destination areas. According to Reuveny, environmental problems such as extreme weather events (frequent storms, floods, and droughts), land degradation, and declining freshwater resources play a role in migration, putting women and children in complex and life-threatening situations.

Moreover, climate change has led to the loss of significant numbers of livestock, leaving thousands—if not millions—of households vulnerable to food insecurity and exposing children to malnutrition. In drought-affected areas, the limited availability of food and milk, combined with restricted access to health and nutrition services, has resulted in high rates of acute malnutrition among children under the age of five.[17]

According to assessment reports, since late 2020, drought has caused internal displacement of at least 590,000 people (IDPs) in Ethiopia: Oromia had registered over 275,000 IDPs, and Somali over 308,000 IDPs, but the actual numbers could be higher.[18]

Furthermore, the frequent and severe drought has forced many schools to be closed, forcing students, including children, to be out of school. According to the unpublished data from the Oromia Disaster and Risk Management Commission, 61 schools were closed, and nearly 10,000 students dropped out due to drought in only two zones (Borana and East Bale), among the many zones affected by drought in the region. The situation in the Somali Regional State was worse as 915 schools were affected: 316 schools were closed entirely, and 599 schools were partially closed. Parents are forced to take their children out of school. Girls' school dropouts are increasing because they have to assist their mothers with household work and in taking care of their siblings.

The water shortage caused by the drought has severely impacted local communities, with women and children bearing the brunt of its effects. Insufficient rainfall failed to replenish surface and underground water sources, leading to ponds and traditional deep wells drying in affected areas. This has resulted in a critical water scarcity

[16] Urdal (2008).

[17] OCHA (2022, n. 1).

[18] UNHCR (2022).

for both humans and livestock. Assessments reveal that women, including pregnant women, were forced to walk over 18 miles from their villages to the nearest water source, carrying 20-L jugs of water. Tragically, some of these women experienced miscarriages, while others gave birth along the arduous journey.

In many rural communities, particularly pastoral areas, women are primarily responsible for fetching water, often requiring them to walk even longer distances during drought. These frequent droughts force women, often on empty stomachs, to spend up to six hours fetching water for their families. Upon returning home, their responsibilities do not end—they must continue caring for their family members, shouldering multiple burdens despite their physical exhaustion.

5 Environmental Impacts of Climate Change

Biodiversity (the variability of life in all forms) underpins sustainable development. Poverty eradication, food security, fresh air and water provision, soil conservation, and human health depend on maintaining and using the world's biological diversity. Biodiversity is significant to people in developing countries whose livelihoods are directly linked to locally available resources. However, the loss of species and the degradation of ecosystems are accelerating at an alarming rate, threatening the very basis of sustainable development. Climate change is among the significant factors causing the loss of biological diversity. Climate change affects all aspects of biodiversity, posing a serious threat to many species and the well-being of humans.

Climate change affects individual organisms, populations, species distributions, and ecosystem composition and function both directly, for example, through temperature increases and precipitation changes. Mitigation measures are required to reduce global greenhouse gas emissions to ensure ecosystem stability. Biodiversity conservation and sustainable development are two interrelated branches focusing on social progress, economic growth, and environmental protection on one side and ecosystem conservation on the other.

Dryland includes arid and semi-arid areas, grasslands, and savannahs with rich biodiversity. During the twentieth century, a decrease in rainfall over large portions of the Sahel and an increase in rainfall in east central Africa was noted. An irregular weather trend is expected to continue, accompanied by rising sea levels. Some changes that will occur in ecosystems as a result of climate change include accelerated habitat loss, changes in the geographic range of many species, increased biodiversity extinction rates (many species that are already threatened are likely to become extinct), and changes in physiology such as changes in reproduction timings, duration of growth season for plants.[19]

Plants and animals are less able to adapt to the current phase of global warming. As a result, many species will be unable to adapt quickly enough to the new conditions or to move to regions more suited to their survival. In many countries, more people,

[19] ICLEI (Local Governments for Sustainability) (2008).

particularly those at lower income levels, are now forced to live in exposed and marginal areas (i.e. floodplains, exposed hillsides, and arid or semiarid lands), putting them at risk of the negative impacts of climate change. Even minor climate changes can disastrously impact the lives and livelihoods of these people. The same can be said of many species adapted to particular climatic conditions. A small change in these conditions could mean losing these species forever. While there is still more to understand about climate change, enough is known about the range of impacts, the magnitude of risks, and the potential for adaptation to act now.

According to the Secretariat of the Convention on Biodiversity (SCBD), impacts associated with increased temperature include a further rise in global mean sea level, more precipitation in temperate regions and Southeast Asia, associated with a higher probability of floods, less precipitation in Africa associated with a greater likelihood of droughts and an expanded range of some dangerous "vector-borne diseases," such as malaria.[20] These predictions have already taken place in the Horn of Africa.

Agriculture is fundamentally dependent on biodiversity and on ecosystem services: species of crops and livestock and their genetic diversity are the basis of agriculture; crop genetic diversity provides the materials for human societies to adapt to climate change and species of earthworms, fungi, soil micro-organisms, flora and fauna surrounding agricultural areas underpin ecosystem services that sustain agriculture, such as pollination and nutrient cycling. In other words, biodiversity is the foundation of ecosystem services to which human well-being is intimately linked. It is essential to improve sustainability and cope with climate change.

In this context, it is worth mentioning that coffee is one of Ethiopia's top export commodities. However, studies have shown that climate change negatively affects coffee production and productivity.[21] Coffee productivity has been declining and is predicted to decrease substantially. About 39–59% of the growing area could experience climatic changes significant enough to render it unsuitable for coffee farming.[22] In other words, the livelihood of about 15 million smallholder coffee producers and the economy of Ethiopia are at risk.

Furthermore, climate change increases the invasion of invasive alien species, directly threatening biodiversity and decreasing agricultural productivity. For example, under the current climatic scenario, 94.8% of Ethiopia is unsuitable for *Prosopis juliflora* establishment and invasion, while 0.4% (4.56 million ha) was highly suitable. However, in 2050, highly suitable areas for *Prosopis juliflora* are expected to increase by 55.6% and 63.6%, while moderately suitable areas are projected to increase by 33.3% and 42.9% under RCP4.5 and RCP8.5 climate scenarios, respectively.[23]

[20] SCBD (Secretariat of the Convention on Biodiversity) (2007).

[21] Chemura et al. (2021).

[22] Moat et al. (2017).

[23] Workeneh et al. (2020).

6 Summary and Recommendations

Climate change may aggravate conflict in Africa in the following ways: changing patterns in climate and agricultural yield could result in increased population movements and increased stresses on urban destination areas; the control of resources could become a political instrument; and Africa's natural resources may attract the attention of more prosperous, resource-poor countries. Also, climate changes could contribute to conflicts arising over competition for food, water, energy or land. For example, tension between farmers and herders is standard. Pastoral communities along borders are notoriously insecure, which could worsen in a more variable climate. Trans-boundary waterways may become flashpoints. Where rainfall decreases, runoff becomes more variable or inter-state tensions arise from competition for limited water resources.

Climate change causes food and water shortages, increases flood risks, and reduces key resources such as fodder for livestock, leading to conflicts among different groups and directly contributing to the internal displacement of people. Continued drought has caused many African dryland ecosystems to lose lives and livelihoods. As a result of the frequent drought, communities have lost their assets, and their adaptive capacity has decreased over time. There is a high and urgent need to build adaptive capacity for the local communities.

Biodiversity contributes to the mitigation and adaptation to climate change by reducing emissions from deforestation and other forms of habitat destruction, providing a 'safety net' of genetic resources for adaptation, providing protection(shields) against flooding and coastal erosion, providing goods and services such as clean water, energy, and food. Therefore, biodiversity conservation is part of the solution to climate change. It is a critical weapon in the fight against climate change, as proper management of biodiversity can reduce the impacts of climate change. Biodiversity is crucial to human well-being, sustainable development, and poverty reduction as it provides provisioning, regulating, and supporting cultural ecosystem services. Conservation and sustainable use of the diversity of cultivated plants and domestic animal breeds are also fundamental for achieving Sustainable Development Goals, specifically SDG 1 (No Poverty) and SDG2 (Zero Hunger).

Combating climate change is critical to achieve the sustainable development goals and to ensure peace and security in the world. The four dimensions of security in the context of Millennium Development Goals (MDG) and climate change are climate security, human security, economic security and energy security. The scarcity of natural resources triggers competition for the meagre resources available among individuals, communities, and even institutions, thus affecting human security on the African continent. Stability is essential for sustainable development and to ensure security. African countries in general and East African countries in particular need to improve governance to maintain stability and security by guaranteeing broad-based development, build a coalition for action through strengthened institutional architecture and regional cooperation to address conflict prevention and resolution, and improve disaster preparedness to reduce impacts.

Therefore, the following are recommended as a result of this review work:

1. Promoting Climate Smart Agriculture (CSA) for increased and sustainable agricultural productivity.
2. Building national and local adaptive capacity: To address the negative impacts of climate change adequately, there is a high need to strengthen or develop and implement a National, Regional (sub-national), and local adaptation strategy and action plan. Governments in the Horn of Africa and development partners should increase investments in adaptation, disaster preparedness, early warning, and timely interventions at local levels.
3. Strengthening gender-responsive interventions with a specific focus on women and children.
4. Promoting large-scale solar PV for productive use: installing solar-powered pumps to extract water from deep underground and irrigation.
5. Supporting rangeland rehabilitation and establishment and management of green fodder banks.
6. Strengthening transboundary collaboration for climate action and community-based development interventions.
7. Promoting and supporting the management of transboundary natural resources.

References

ACAPS (2023) ETHIOPIA. Impact of drought: Oromia and Somali regions. Thematic Report, February 2023. https://www.acaps.org/sites/acaps/files/products/files/20230207_acaps_thematic_report_horn_of_africa_analysis_ethiopia_impact_of_drought_oromia_and_somali_regions.pdf

Carleton E (2022) Climate change in Africa: What will it mean for agriculture and food security? International Livestock Research Institute (ILRI). Accessed on 12 March 2023 at: https://www.ilri.org/news/climate-change-africa-what-will-it-mean-agriculture-and-food-security#:~:text=Many%20crops%20foundational%20to%20African,decrease%20by%2010%20per%20cent

CCCPA (International Center for Conflict Resolution, Peacekeeping and Peacebuilding) (2022) The climate-displacement nexus in Africa: implications for sustainable development

Chemura A, Mudereri BT, Yalew AW et al (2021) Climate change and specialty coffee potential in Ethiopia. Sci Rep 11:8097. https://doi.org/10.1038/s41598-021-87647-4

Clement V, Rigaud KK, de Sherbinin A, Jones B, Adamo S, Schewe J, Sadiq N, Shabahat E (2021) Groundswell part 2: acting on internal climate migration. World Bank, Washington, DC. © World Bank. https://openknowledge.worldbank.org/handle/10986/36248. License: CC BY 3.0 IGO

Doti T (2010) Climate variability, pastoralists' vulnerability and options. The case of the Borana of Northern Kenya. In: Mwiturubani DA, van Wyk, J-A (eds) Climate change and natural resources conflicts in Africa. Monograph 170. Institute of security studies

ICLEI (Local Governments for Sustainability) (2008) Biodiversity and climate change

IRC (International Rescue Committee) (2011) Millions need lifesaving aid as the UN declares famine in Somalia. http://www.rescue.org/press-releases/millions-need-life-saving-aid-un-declares-famine-somalia-11268. Accessed on 25/07/2011

Kindiki K (2009) Water and food security in the Nile River Basin. Legislative, policy and institutional arrangements for cooperation. In: Tadesse D (ed) Climate change and transboundary water resource conflicts in Africa. Workshop report. 29–30 September 2009, Mombasa, Kenya

Kolmannskog V (2010) Climate change, human mobility, and protection: initial evidence from Africa. Refug Surv Quart 29(3):103–119 (17 pages). New Challenges of Migration in Africa. https://www.jstor.org/stable/45054464

Moat J, Williams J, Baena S, Wilkison T, Gole TW, Challa ZK, Demissew S, Davis AP (2017) Resilience potential of the Ethiopian coffee sector under climate change. Nat Plants 3:17081. https://doi.org/10.1038/nplants.2017.81

OCHA (UN Office for the Coordination of Humanitarian Affairs) (2022) Horn of Africa drought: regional humanitarian overview and call to action (Revised 28 November 2022). https://reliefweb.int/report/ethiopia/horn-africa-drought-regional-humanitarian-overview-call-action-revised-28-november-2022

OCHA (2023) Ethiopia: drought situation update #1. Accessed at https://reliefweb.int/report/ethiopia/ethiopia-drought-situation-update-1-10-march-2023

ODRMC (Oromia Disaster and Risk Management Commission) (2022). https://www.press.et/?p=71592

Reuveny R (2007) Ecomigration and violent conflict: case studies and public policy implications. Hum Ecol. https://doi.org/10.1007/s10745-007-9142-5

SCBD (Secretariat of the Convention on Biodiversity) (2007) Biodiversity—climate interactions: adaptation, mitigation and human livelihoods workshop. The International Framework for Biodiversity and Climate Change. Royal Society and GBSC, London, 12–13 June 2007

UNHCR (2022) Drought situation response update. Regional update #4. East and Horn of Africa, and the Great Lakes Region. November 2022. https://data.unhcr.org/en/documents/details/97731

Urdal H (2008) Demographic aspects of climate change, environmental degradation and armed conflict. United Nations Expert Group Meeting on Population Distribution, Urbanization, Internal Migration and Development. Population Division Department of Economic and Social Affairs. United Nations Secretariat New York

Workeneh S, Dalle G, Bobasa AF (2020) Impacts of climate change on current and future invasion of *Prosopis juliflora* in Ethiopia: environmental and socio-economic implications. Heliyon 6(8):e04596. https://doi.org/10.1016/j.heliyon.2020.e04596

Climate Change-Induced Conflicts: Examining the Interplay Between Natural Resource Depletion and Human Displacement Due to Insecurity in Africa

Kolawole Afuwape and Sarah Ayanwale

Abstract Energy security, climate change, and issues related to food and water are presenting significant challenges to the governance of international relations in a world that is already fraught with instability. Projections indicate that future climate change will likely reduce the availability of critical natural resources in various regions, potentially affecting the global landscape. The conflict related to climate change has largely been overlooked, even though it carries significant consequences. Through causal pathways involving natural disasters and declining or fluctuating resource availability, several analysts associate climate change with an increased probability of war. There is considerable consensus that climate change can play a role in exacerbating conflict, particularly under specific circumstances and through mechanisms. The existing literature indicates that climatic conditions can exacerbate conflict in agriculturally reliant areas, particularly when these conditions intersect with various socioeconomic and political factors, including low economic development and political marginalization. This study contends that the diminishing quality and availability of essential natural resources for sustaining livelihoods jeopardizes human security. This trend is expected to persist in the future due to climate change. Climate change is anticipated to diminish the capacity of governments to provide the vital opportunities and services necessary for individuals to sustain their livelihoods. This paper briefly assesses how natural disasters, resource scarcity, and global climate change may influence the probability of violent conflict in the future. This study subsequently explores strategies to improve the investigation of the link between climate change, conflict, and displacement while offering detailed policy recommendations to tackle this potential challenge.

K. Afuwape (✉)
Jindal Global Law School, O.P. Jindal Global University, Sonipat, India
e-mail: Kolawole.afuwape@jgu.edu.in

S. Ayanwale
Concordia University of Edmonton, Edmonton, Canada

D. Oghenerobor Agelebe (ed.), *Climate Ethics in Africa*, Global Perspectives on Law and the Environment 1, https://doi.org/10.1007/978-3-032-08757-7_4

1 Introduction

Understanding the impacts of severe weather and climate phenomena is essential for assessing climate change's repercussions on human communities and infrastructure.[1] Climate change has become a significant aspect of security studies in recent years. A rising consensus among security analysts suggests that the effects of climate change may intensify current socio-political divisions or serve as a catalyst for additional threats.[2] Most conflicts are over a resource perceived as scarce, at least when territory is counted as a resource.[3]

Natural hazards and violent conflicts result in disasters that impact numerous individuals globally. These distinct crises carry substantial political, economic, and social consequences, which can undermine developmental progress, exacerbate poverty and inequality, and heighten the likelihood of future crises.[4] A more profound understanding of this interconnection is crucial for enhancing efforts to reduce disaster risk and avert the emergence or recurrence of violent conflict. This requires that measures addressing disasters and disputes do not worsen current levels of risk and vulnerability.[5] This study investigates the relationship among disasters, resource limitations, and conflict. Its objective is to go beyond the 'do not harm' principle, which highlights the necessity of avoiding the aggravation of an already dire situation during program execution. The paper also aims to investigate how programming can create opportunities to tackle both disaster and conflict risks when they arise within the same environment.

This study delineates disaster-conflict interfaces as scenarios where a connection exists between disaster and conflict, or the reverse, beyond merely co-occurring in the exact location and impacting the same demographic.[6] The concept of 'crisis' encompasses both catastrophic events and conflicts.[7] Disasters are significant disturbances in a community or society's operations that result in substantial losses in human life, property, economic stability, or environmental integrity that surpass the affected community or society's capacity to manage with its available resources.[8] Disasters occur when a susceptible societal asset is subjected to a natural hazard, including earthquakes, tsunamis, windstorms, and floods. These events have historically been regarded as unavoidable occurrences beyond human control.[9] Today, experts acknowledge that human activities significantly influence the fundamental causes of hazards.[10]

[1] Forzieri et al. (2018, 97–107).

[2] Chin-Yee (2019, 2–31).

[3] Robins and Fergusson (2014, 1–9).

[4] Carter (2021).

[5] Peters (2021).

[6] Burger et al. (2019, 103).

[7] Tagarev and Ratchev (2020, 5147).

[8] Chaudhary and Piracha (2021, 1101–1131).

[9] Rivera and Kapucu (2015).

[10] Tröger and Braun (2024).

A natural hazard can only result in a disaster under two conditions: (a) there must be exposure of a community or societal asset to the hazard, and (b) the community or asset must possess a degree of vulnerability. For instance, an earthquake occurring in a region devoid of human habitation would likely have no repercussions for society, thus not qualifying as a disaster. Vulnerability refers to the extent to which a community is at risk from the effects of a hazard. This vulnerability is shaped by physical, social, economic, and environmental factors and processes.[11] A community's vulnerability to disaster risk can be reduced if it possesses coping capacity, including resources and skills.[12] Disasters may manifest as either a gradual onset, such as droughts, where the exposure to hazards extends over several months, or as a rapid onset, exemplified by earthquakes, where the duration of exposure is limited to mere minutes.

The theoretical connections between the depletion of natural resources, natural disasters, conflicts stemming from resource scarcity, and environmental quality are significant.[13] The finite nature of resources and the imperative to safeguard the environment represent two of our planet's most urgent challenges.[14] It is essential to comprehend the connection between the depletion of natural resources and environmental quality and the implications this relationship has for our society.[15] The depletion of natural resources has been a matter of concern for many years, and its importance is increasingly recognized within climate change research.[16]

The concept of 'natural resources' pertains to the practical benefits that social groups obtain from their surrounding environment and is commonly employed within human geography.[17] The term "natural resources" is inherently contentious, as the assignment of utility and value to the nonhuman environment serves as a fundamental mechanism through which prevailing social groups establish order and exert control over the world. A resource crunch can be defined as a critical movement that occurs due to the shortage, scarcity, or overuse of resources, leading to its complete depletion.[18] Resource depletion refers to the rate at which a resource is utilized, exceeding its natural replenishment rate. Natural resources are typically categorized into two groups: renewable and non-renewable.[19] The utilization of any of these resource types beyond their natural replenishment rate is classified as resource depletion. The exhaustion of natural resources, reliance on fossil fuels for energy production, and the aggregate emissions of greenhouse gases present substantial challenges that

---

[11] Surjan et al. (2016, 37–55)

[12] Gaillard et al. (2019, 863–876).

[13] Judge (2014, 199).

[14] Martin et al. (2016).

[15] Zhang et al. (2022).

[16] Lu and Wang (2023).

[17] Guo (2018, 23–29).

[18] Subramanian (2018, 8–19).

[19] Byaro et al. (2023).

necessitate prompt intervention. The growing global population and the progression of economies have resulted in an unparalleled surge in the demand for natural resources, especially fossil fuels.[20]

Analyzing the dynamics of conflict requires an examination of its underlying causes, the individuals involved, and the factors that may contribute to conflict resolution, often referred to as 'capacities for peace'.[21] Manifestations, readily identifiable occurrences indicative of social unrest, proximate causes or factors that exacerbate the fundamental causes of conflict and promote violent confrontations, and root causes, which refer to systemic issues that ignite conflict without necessarily leading to violence, represent various classifications of conflict causes.

The capacity to foster peace can play a significant role in maintaining lasting peace, diminishing violence, or averting its emergence.[22] Achieving enduring peace and sustainable development relies on the capacity of essential stakeholders and communities to establish a shared understanding of priorities at both national and local levels, engage in negotiations to resolve potential conflicts before they escalate into violence, and embrace diversity in the processes of planning and executing social development initiatives. Conflict is a complex and evolving phenomenon characterized by multiple causes rather than a straightforward trajectory. It encompasses various stages, ranging from latent tensions to overt acts of violence.[23]

2 The Challenge Posed by Finite Natural Resources and the Potential Consequences of Climate Change

The interconnectedness of resource scarcity, environmental catastrophes, and conflict underscores the complex challenges that climate change presents.[24] Resource scarcity, exacerbated by climate-induced natural disasters, can escalate into violence and conflict, particularly in regions with fragile governance.[25] Addressing these interconnected challenges requires integrated approaches that promote environmental sustainability, resilience, and cooperation.[26] One way to mitigate the effects of climate change and foster the growth of more resilient, stable, and peaceful societies is through the sustainable management of natural resources in conjunction with initiatives in disaster risk reduction and environmental peacebuilding.[27]

[20] Lu and Wang (2023).

[21] Avruch (2022).

[22] De Coning (2018).

[23] Maltarich et al. (2018, 6–37).

[24] Ide and Scheffran (2014, 263–279).

[25] Ide (2023, 50–78).

[26] Khan et al. (2024, 57–68).

[27] Kiyala (2022, 67–93).

2.1 Resource Crunch and Climate Change

Resource crunch denotes the significant reduction or acute shortage of vital natural resources, including water, cultivable land, minerals, and energy sources. This scarcity may arise from factors such as environmental degradation, unsustainable resource extraction, or the impacts of climate change. Increasing global temperatures have resulted in significant droughts, diminishing water resources, and expanding desert areas worldwide. Key natural resources, such as freshwater, are facing considerable strain because of the rising demand from an expanding global population and alterations in climate patterns that interfere with natural replenishment cycles. Climate change intensifies the depletion of natural resources by altering precipitation patterns, causing glaciers to melt, increasing sea levels, and leading to extreme weather phenomena, including extended periods of drought and flooding.[28] The consequences of these actions significantly diminish the accessibility of vital resources such as freshwater and arable land, which are crucial for agriculture and human survival.

Reduced availability of these vital resources disrupts agricultural productivity, leading to food insecurity.[29] Food and water scarcity can, in turn, fuel migration, economic instability, and social tensions, particularly in vulnerable regions where populations depend on agriculture for livelihoods.[30] In Sub-Saharan Africa and the Middle East, climate change has been associated with decreased agricultural productivity, increasing resource competition and worsening poverty levels.

The connection between climate change and resource depletion is complex and interdependent.[31] To adequately tackle these interconnected challenges, viewing climate change and resource depletion as components of a broader, unified crisis is essential. A thorough strategy to mitigate the effects of climate change, reduce resource consumption, and promote sustainable practices is critical.

Climate change denotes the long-term alterations in temperature and typical weather patterns that arise from human activities, primarily the burning of fossil fuels, deforestation, and a range of industrial processes.[32] This process releases greenhouse gases, such as carbon dioxide and methane, which trap heat in the Earth's atmosphere and raise global temperatures.[33] The effects of climate change include a rise in the frequency and severity of extreme weather phenomena, the thawing of polar ice caps, an increase in sea levels, and a notable reduction in biodiversity within diverse ecosystems.[34] Resource depletion, conversely, pertains to the excessive extraction and utilization of the Earth's limited natural resources, such as water,

[28] Yildiz et al. (2022, 17–25).

[29] Khan et al. (2022, 82).

[30] Adaawen et al. (2019, 15–31).

[31] Verma et al. (2024, 211–221).

[32] Abdallah et al. (2023, 2431–2446).

[33] Gahlawat and Lakra (2020, 14–23).

[34] Shivanna (2022, 160–171).

minerals, cultivable land, and fossil fuels.[35] Various actions contributing to climate change, such as burning fossil fuels and deforestation, also exhaust resources. As a result, resource depletion may exacerbate climate change's impacts.

The Industrial Revolution and the early stages of capitalism were when the idea of infinite economic expansion first emerged.[36] This notion was fueled by technological innovations, production processes, and the availability of seemingly abundant natural resources. As economies rapidly expanded, this growth was widely viewed as a panacea for poverty, unemployment, and societal ills. This led to the entrenched belief that constant economic expansion was possible and desirable. However, the consequences of this belief on the environment, the economy, and society have become increasingly evident and alarming. The pursuit of limitless growth has often been at the expense of the planet's finite resources, leading to widespread resource depletion, pollution, and ecological degradation.[37] The unwavering emphasis on economic growth has led to significant disparities in wealth and income, whereby the affluent amass resources to the detriment of the impoverished and marginalized populations.

To tackle the intertwined challenges of climate change and resource depletion, along with the inequitable distribution of resources, it is crucial to question the existing paradigms and investigate alternative economic frameworks that emphasize sustainability and equity. Among these alternatives are the circular economy, de-growth, and steady-state economics.[38] The circular economy model encourages the optimal utilization of resources by designing products and systems that reduce waste while enhancing the opportunities to reuse, repair, and recycle materials.[39] The circular economy seeks to minimize the environmental effects of production and consumption by maintaining resources in a perpetual cycle, thereby promoting economic development and innovation.[40] The de-growth model advocates intentionally downscaling production and consumption to achieve ecological sustainability and social well-being.[41] De-growth challenges the idea that economic growth is the only path to prosperity, instead emphasizing the importance of redistributing wealth, reducing work hours, and prioritizing community engagement and well-being.[42] The steady-state economics model proposes a stable, non-growing economy within the Earth's ecological limits.[43] Steady-state economics seeks to establish a sustainable and resilient society by prioritizing the fair allocation of resources alongside the welfare of individuals and the environment. Transitioning to these alternative models presents both benefits and challenges. The benefits include reduced environmental impact, more significant social equity, and increased resilience to external shocks,

[35] Akram et al. (2024).
[36] Mokyr (2018, 1–127).
[37] Subramanian (2018, 8–19).
[38] Charonis (2021, 75–94).
[39] Tambovceva et al. (2021).
[40] Ogunmakinde et al. (2022).
[41] Smith et al. (2021).
[42] Tomlinson (2022).
[43] Rees (2021a, b, 98–118).

such as climate change and resource scarcity.[44] However, challenges include overcoming ingrained beliefs in limitless growth, altering established economic and political systems, and managing the potential short-term economic disruptions associated with these transitions.

Addressing the interconnected issues of climate change, resource depletion, and inequalities in resource consumption requires a collaborative strategy that engages individuals, communities, and governmental entities.[45] Some practical steps that can be taken at various levels include, at the individual level, adopting sustainable consumption habits by reducing waste,[46] recycling, and purchasing environmentally friendly products[47]; saving energy and water by implementing energy-efficient measures and reducing unnecessary usage[48]; supporting and engaging with local initiatives focused on sustainability, social equity,[49] and environmental conservation; supporting policies and practices that prioritize sustainability and equality in the workplace, local community, and at the governmental level.[50] The establishment of policies that encourage the sustainable utilization of resources, including the principles of a circular economy and incentives for renewable energy[51]; investing in public infrastructure that facilitates sustainable living, including public transportation, green spaces,[52] and waste management systems[53]; developing and enforcing regulations that limit pollution, resource extraction, and deforestation,[54] while protecting biodiversity and ecosystems; fostering international cooperation and partnerships to address global challenges, including climate change, resource depletion, and socio-economic inequality.

The simultaneous challenges of climate change and resource depletion necessitate a comprehensive reassessment of our existing economic frameworks and ideologies.[55] People, communities, and governments must collaborate to address these interconnected challenges and embrace alternative models prioritizing the well-being of both people and the planet.[56]

[44] Nyambuu and Semmler (2023).

[45] Antonio (2023, 464–489).

[46] Suski et al. (2021).

[47] Moser (2016, 389–397).

[48] Cenci et al. (2022).

[49] Prentice et al. (2024, 1–19).

[50] Ciplet and Harrison (2020).

[51] Nunes et al. (2023, 485).

[52] Van Oijstaeijen et al. (2020)

[53] Jouhara et al. (2017, 485–506).

[54] Bebbington et al. (2018, 13164–13173).

[55] Rees (2021a, b, 22–52).

[56] Pereira et al. (2020, 1172–1195).

2.1.1 Natural Disasters Within the Framework of Climate Change

Due to climate change's impacts, natural disasters such as hurricanes, floods, wildfires, and droughts are becoming more severe.[57] These disasters' rising occurrence and severity have catastrophic effects on human communities and natural ecosystems.[58] The interconnection between natural resource depletion and disasters is critical because disasters can exacerbate the scarcity of already stretched resources. At the same time, resource crunches can also make populations more vulnerable to disasters.[59]

When natural disasters strike, they destroy critical infrastructure, water systems, forests, and cropland, leading to immediate resource shortages. For instance, hurricanes can cause freshwater contamination through storm surges and flooding, while wildfires can decimate forests that serve as key resources for local communities. This destruction compounds existing resource scarcity, leading to heightened social pressures. The destruction of natural resources during disasters can further erode environmental sustainability.[60] For example, following extreme droughts or floods, agricultural land becomes less fertile, making it harder to recover food production. In coastal regions, increasing sea levels and storm surges lead to land erosion, while saltwater intrusion jeopardizes freshwater resource availability.[61] These processes often push communities into even more precarious situations, where surviving on limited resources becomes increasingly difficult.

2.1.2 Violence and Conflict Over Scarce Resources

As resources become scarcer due to gradual degradation and sudden shocks from natural disasters, competition for these dwindling assets often intensifies. This competition can escalate into violence, manifesting as localized conflicts, civil wars, or cross-border tensions. Environmental sustainability studies increasingly recognize the "environmental conflict" hypothesis, which suggests that resource scarcity and environmental degradation can trigger or exacerbate violent conflict.[62]

Resource scarcity, particularly for water, land, and food, has long been recognized as a driver of conflict. In areas where essential resources are limited, populations may resort to violence to secure access.[63] The Darfur conflict in Sudan that erupted in the early 2000s is a key example often cited in climate change literature.[64] Although the conflict had multiple causes, competition for water and grazing land, severely

57 Dey and Lewis (2021, 177–193).

58 Upadhyay (2020, 159).

59 Rajan (2021, 1).

60 Montt et al. (2018, vii).

61 Guimond and Michael (2021, 57(2), e2020WR028326).

62 Galgano (2019, 1–17).

63 Adaawen et al. (2019, 2, 15–31).

64 Roettinger (2019, 30529).

strained due to prolonged droughts and desertification, played a significant role in fueling violence between nomadic and agricultural communities.

Climate-induced resource depletion drives mass migration as people search for better living conditions.[65] Migrant populations may then conflict with host communities over access to scarce resources.[66] The large-scale movement of individuals resulting from climate-induced disasters, such as droughts in Sub-Saharan Africa, has contributed to heightened tensions in host regions, potentially leading to outbreaks of violence.[67] The 2011 Syrian civil war is often partially attributed to a prolonged drought from 2006 to 2011, which devastated agriculture and drove rural populations into urban centers, exacerbating existing political and social tensions.[68]

In states with weak governance and institutions, competition for resources often devolves into violence because there are no mechanisms to mediate conflicts peacefully.[69] Climate change exacerbates governance challenges in fragile or failing states, making it difficult for these governments to provide basic services and maintain social order.[70] This fragility often leads to a spiral of violence as various groups vie for control over limited resources. In certain areas of the Sahel region in Africa, the interplay of climate change, limited resources, and ineffective governance has contributed to the emergence of armed factions, resulting in heightened insecurity and violence.[71]

3 Climate Change as a Catalyst for Conflict

Climate change functions as a "threat multiplier" by exacerbating conditions that contribute to conflict, such as resource scarcity, natural disasters, and migration pressures.[72] Climate change, while not a direct instigator of violence, intensifies pre-existing vulnerabilities and tensions, frequently acting as a catalyst for the escalation of conflict dynamics.[73] In certain areas, violent non-state actors, including terrorist organizations and militias, are taking advantage of the impacts of climate change.[74] These groups may manipulate local resource scarcity and grievances to recruit members or gain support. In northern Nigeria and the Lake Chad Basin, the terrorist organization Boko Haram has exploited extreme droughts and water scarcity to enhance its power, providing security or resources in return for allegiance.[75]

65 Kaczan and Orgill-Meyer (2020, 281–300).

66 Oucho and Williams (2019, 1–18).

67 Adaawen et al. (2019, 15–31).

68 Ünver (2019, Political economy of pre-civil war Syria. 3536).

69 Raleigh and Linke (2018, 88–93).

70 Chin-Yee (2019, 2–31).

71 Dieng (2021, 765–779).

72 Mamshai (2023, 1–16).

73 Peters (2021, 102092).

74 Akinyetun (2022, 36–58).

75 Bharkhada (2022).

Beyond immediate conflict, resource scarcity due to climate change can lead to economic decline, reduced agricultural yields, and rising food prices. These factors contribute to widespread social instability and can lead to civil unrest. In many cases, the resulting violence is directed at both state institutions and rival groups as people vie for control of limited resources or express their dissatisfaction with deteriorating living conditions.

4 Environmental Sustainability and Peace Building

Efforts to address the interconnections between resource scarcity, natural disasters, and violence are critical to promoting environmental sustainability and peace.[76] Scholars and policymakers are increasingly focusing on the need for adaptive strategies that build resilience in communities vulnerable to climate change.[77] These strategies include sustainable resource management, disaster preparedness, and conflict resolution mechanisms.

Sustainable management of resources like water, land, and forests is key to reducing competition and fostering cooperation, particularly in regions prone to scarcity.[78] For instance, transboundary water agreements, such as those governing the Nile or the Mekong rivers, can help reduce tensions between neighbouring countries by providing a legal framework for resource sharing.[79] Enhancing disaster preparedness and response systems is crucial for reducing the effects of climate-related disasters on at-risk communities.[80] Developing robust infrastructure and promoting adaptive agricultural methods can enable communities to better endure the impacts of climate change, thereby decreasing the chances of resource-related conflicts.[81] Integrating environmental management into peacebuilding efforts is another approach gaining traction.[82] Environmental peacebuilding emphasizes the role of cooperative natural resource management as a tool for fostering peace and preventing conflicts.[83] In post-conflict settings, restoring ecosystems and rehabilitating resource-based livelihoods can help reduce the risk of violence re-emerging. In places like Colombia, environmental restoration programs have been used as part of broader peacebuilding strategies to help former combatants reintegrate into society through sustainable agricultural practices.[84]

---

[76] Bamidele and Erameh (2023, 103274).

[77] Joakim et al. (2021, 13–31).

[78] Mondal and Palit (2022, 29–59).

[79] Hossen et al. (2023, 1–10).

[80] Banwell et al. (2018, 793).

[81] Sarzana et al. (2022, 7–18).

[82] Hulme (2023, 230–252).

[83] Chavez-Miguel et al. (2022, 36–45).

[84] Schmid and Vlaskamp (2024, 568–598).

5 The Prolonged Health Impacts of Natural Disasters

Residual issues arising from a natural disaster are typically those that result in prolonged repercussions. Many individuals are aware of the immediate actions taken in response to natural disasters, which focus on mitigating the impacts of the event itself. A range of emergency measures is required to stabilize the population that has been most significantly impacted. The concluding stage of recovery is known as the reconstruction phase, which may extend over several months or even years. In this phase, mental health challenges arising from the traumatic event become apparent.[85] Anxiety and depression may arise from various factors during this period, including the pace of recovery, changes in socioeconomic conditions following the disaster, and persistent reminders of the event, such as media reports on the anniversary of the disaster.[86] Natural disasters in low- to middle-income nations frequently exacerbate pre-existing health inequalities.[87] These occurrences may result in an imbalanced allocation of disease burden, disproportionately impacting at-risk populations. Insufficient access to healthcare, clean water, and sanitation intensifies the transmission of diseases, thereby increasing the vulnerability of marginalized communities. Natural disasters often result in widespread destruction, marked by the destruction of infrastructure, large-scale displacement of communities, and a decline in hygiene conditions.[88] These occurrences subject the impacted population to a complex web of public health challenges. A delayed reaction intensifies the crisis, paving the way for significant outbreaks and a dire impact on human life. The complex relationships between disease transmission pathways and their associated risk factors can result in significant alterations in health trends.

Mental health issues are among the most prevalent enduring consequences of natural disasters.[89] The devastation of losing family members, residences, and sources of income can profoundly affect numerous individuals. Following a natural disaster, it is common for individuals to experience emotions of sorrow and disbelief. However, when individuals struggle to regain a sense of normalcy, mental health professionals are tasked with addressing the enduring psychological consequences of such traumatic events. Hygienic issues, coupled with the deterioration of essential infrastructure, elevate the likelihood of the emergence and transmission of communicable diseases. The two primary strategies for addressing and eliminating such health risks are vaccination and the reinstatement of appropriate hygiene practices.[90]

The impacts of natural disasters on mental health introduce a further layer to the pre-existing disparities in health.[91] Individuals belonging to vulnerable populations, especially those with a history of mental health challenges and limited access to

[85] Naushad et al. (2019, 632–643).

[86] Tulane University School of Social Work

[87] Nashwan et al. (2023a, b, c, 1–8).

[88] Kelman (2020, viii).

[89] Zhong et al. (2018, 165–194).

[90] Rodrigues and Plotkin (2020, 1526).

[91] Waramlah et al. (2018, 56–67).

mental health services, face an unequal burden.[92] This unfortunate circumstance is exacerbated in less affluent nations, where women are particularly impacted due to their societal roles and various contributing factors. A study examining research from Asia, Latin America, and the Caribbean indicated a rise in mental health problems among individuals after experiencing climate-related disasters.[93] Research indicates that there is a heightened likelihood of poisoning and a range of infections following a natural disaster.[94] These concerns encompass infections related to the skin, respiratory system, gastrointestinal tract, and zoonotic diseases. The rise in these infections can be attributed to the deterioration of infrastructure and sanitation systems that are usually effective in managing the dissemination of pathogens responsible for infections.[95] Individuals suffering from chronic medical conditions are especially at risk of experiencing further medical complications.[96] Health service infrastructures are vulnerable to substantial harm during natural disasters, and the repercussions can persist long after the immediate crisis has been addressed.[97] The systemic effects may be exacerbated in low- and middle-income nations, thereby highlighting the challenges these countries encounter even under typical conditions.[98]

Social determinants significantly influence the difficulties encountered by health services both during and following natural disasters.[99] Significant populations generally migrate away from the impacted region. As the inhabitants depart, the motivation to rebuild the healthcare service system diminishes. The time required to adequately reconstruct the hospital infrastructure will increase as resources for restoring the general community also decline.

6 Food Scarcity and Insecurity

The issue of food insecurity looms large, affecting around 12% of the world's population, which equates to an alarming total of approximately 924 million individuals, according to data from the World Health Organization.[100] 31.9% of women and 27.6% of men, respectively, experience moderate to severe food insecurity, a worrying trend of ever-widening gender disparity.[101] The growing frequency and intensity of natural disasters, such as droughts, floods, and storms, further complicate the difficulties encountered by the agricultural sectors in low- and middle-income

[92] Mezzina et al. (2022, 894370).

[93] Sharpe and Davison (2021, e051908).

[94] Erickson et al. (2019, 157–171).

[95] Ferreira et al. (2021, 146279).

[96] Caldeira et al. (2021, 102494).

[97] Hidalgo and Baez (2019, 591–607).

[98] Kola et al. (2021, 535–550).

[99] Mao and Agyapong (2021, 658528).

[100] Nashwan et al. (2023a, b, c, 1–7).

[101] Nashwan et al. (2023a, b, c, 1–7).

countries, thereby worsening the problem of food insecurity.[102] This situation establishes obstacles that hinder the relentless endeavors aimed at eradicating hunger and poverty, thereby complicating the journey towards sustainable development. Food scarcity is not solely a consequence of natural disasters that devastate food supplies and infrastructure; it also stems from the erosion of assets and the means of livelihood.[103]

A relationship exists between food scarcity and the prevalence of chronic diseases.[104] This complex issue is initiated by the significant obstacle of food unaffordability, which subsequently leads to malnutrition. The result of this scenario is a heightened probability of developing chronic health issues, which in turn depletes financial resources and significantly affects household budgets.[105] Natural disasters act as a significant catalyst in this precarious environment, intensifying stress levels, worsening malnutrition, and exhausting financial resources, thus sustaining this detrimental cycle.[106]

In Pakistan, more than 3.4 million children are currently suffering from chronic hunger.[107] Recent assessments conducted after the flood indicate a concerning situation, with 76,000 children facing severe food shortages and being at risk of acute malnutrition. The flooding has caused considerable damage to farmland and livestock, resulting in a significant shortfall in food production. As a result, the cost of food commodities has surged dramatically, rendering them unaffordable for families affected by the floods. The number of individuals lacking sufficient access to food has increased from 5.96 million before the floods to a concerning 8.62 million in the aftermath. As winter approaches, the imminent danger to the lives of hundreds of thousands becomes increasingly severe unless immediate actions are taken. In response to this difficult landscape, it is crucial to implement a holistic approach aimed at enhancing agricultural resilience, ensuring equitable food distribution, and developing sufficient healthcare infrastructure and training, particularly in low- and middle-income nations.[108] Engagement in disaster preparedness and response, coupled with the creation of social safety nets and support for the most vulnerable populations, is crucial for effectively tackling the challenges associated with food insecurity and its related concerns.

[102] Sankar-Gorton et al. (2024, 71).
[103] Reddy et al. (2019, 290–330).
[104] Himmelgreen et al. (2022, 113124).
[105] Jacobson (2018, 3–32).
[106] Nashwan et al. (2023a, b, c, 1–7).
[107] Hameed et al. (2023, 2563–2585).
[108] Nashwan et al. (2023a, b, c, 1–7).

7 Resource Scarcity and Conflict Risks

Resource scarcity denotes the restricted access to vital resources, including water, cultivable land, food, energy, and minerals, which are essential for human existence and economic progress.[109] As the world's population continues to increase and environmental deterioration intensifies because of climate change, urban expansion, and unsustainable practices, the issue of resource scarcity has emerged as a critical global challenge. The lack of resources, particularly in vulnerable and underdeveloped areas, may lead to an increase in conflict risks, exacerbating geopolitical tensions and social instability.[110]

As resources become increasingly scarce, competition among nations, regions, or factions intensifies, leading to conflicts and acts of aggression. A relevant illustration of this phenomenon is the dispute over water rights in transboundary river basins, including the Nile and Tigris-Euphrates,[111] as well as disputes over territories abundant in oil, exemplified by the South China Sea.[112] Resource scarcity within nations can intensify inequality and provoke social upheaval. For instance, increases in food prices or shortages of water may incite protests, riots, and insurgencies, as evidenced during the Arab Spring, where the lack of essential resources contributed to widespread public dissatisfaction.[113] Resource scarcity has historically been acknowledged as a significant catalyst for conflict worldwide.[114] As vital resources, including water, food, land, and energy, diminish, the competition for their access and control escalates. The implications of resource scarcity can vary significantly, encompassing everything from localized disputes and social upheaval to extensive armed conflict.[115] Conceptually, resource scarcity arises when the demand for a particular resource surpasses its supply or when access to that resource is limited.[116] This phenomenon can be categorized into three primary types:

- Physical scarcity denotes a condition where a resource is fundamentally restricted or depleted, as illustrated by the limited access to fresh water in dry areas.
- Economic scarcity refers to a situation in which resources are available but remain unattainable due to economic obstacles, inadequate governance, or insufficient infrastructure.
- Political scarcity refers to the limitations imposed on access resulting from factors such as political dominance, corruption, or ineffective governance.

The various manifestations of scarcity, particularly when intensified by elements such as increasing population, climate change, or inadequate resource management,

[109] Wang and Azam (2024, 101757).

[110] Buhaug and Von Uexkull (2021, 545–568).

[111] Gelaw and Debele (2024, 1–13).

[112] Ernst (2022, 8–30).

[113] Barakat and Fakih (2021, 1–15).

[114] Link et al. (2016, 495–515).

[115] Abrahams (2020, 104998).

[116] Damkjaer and Taylor (2017, 513–531).

frequently lead to conflict as communities or nations compete for dominance over dwindling resources.[117]

Numerous theories elucidate the relationship between resource scarcity and conflict:

- Environmental Security Theory posits that environmental degradation and resource scarcity increase the likelihood of violent conflict, particularly in regions that are heavily dependent on natural resources.[118]
- The Malthusian Theory posits that the rate of population increase surpasses the availability of resources, resulting in competition and conflict, which may ultimately culminate in the disintegration of society.[119]
- Resource Curse Theory suggests that countries rich in natural resources but plagued by poor governance are more likely to encounter conflict.[120] This vulnerability arises from issues such as mismanagement, corruption, and social inequality.

Water scarcity has emerged as a critical catalyst for conflict in the Middle East, a region characterized by its arid climate, increasing population pressures, and restricted freshwater availability.[121] The Tigris-Euphrates River basin, which spans Turkey, Syria, and Iraq, has long been a source of conflict.[122] The development of the Atatürk Dam by Turkey, as a component of its Southeastern Anatolia Project (GAP), has resulted in reduced water availability for downstream nations, particularly Syria and Iraq.[123] The current situation has intensified political tensions and given rise to accusations of unfair distribution of water resources. The Jordan River Basin, serving as a vital source of water for Israel, Jordan, Syria, Lebanon, and the Palestinian territories, has also emerged as a significant area of dispute.[124] Conflicts regarding water rights and accessibility have played a significant role in exacerbating the wider Arab–Israeli conflict.[125] Water scarcity exacerbates the instability in the region, particularly during periods of drought, driving both state and non-state actors to compete for this vital resource.[126]

The situation in Darfur, Sudan, often described as the first "climate change war," exemplifies how the depletion of land and water resources can lead to conflict.[127] The region has experienced considerable desertification due to climate change and overgrazing, resulting in reduced availability of arable land and water resources. The

[117] Maja and Ayano (2021, 271–283).

[118] Hardt (2017, xv).

[119] Farasat et al. (2023, 741–751).

[120] Adams et al. (2019, 128–140).

[121] Tinazzi (2024, 1–121).

[122] Bachmann et al. (2019, 1–24).

[123] Jongerden et al. (2021, 383–399).

[124] Zeitoun et al. (2019, 1064–1094).

[125] Farag (2018, 299–317).

[126] Unfried et al. (2022, 102633).

[127] Perkel (2021, 8–20).

combination of environmental degradation and increasing population has heightened the rivalry between Arab nomadic herders and African agricultural communities. The pastoralists, dependent on seasonal movement for grazing, have progressively intruded upon agricultural lands, resulting in violent confrontations. This competition for resources, exacerbated by ethnic divisions and political exploitation, has contributed to one of the most lethal conflicts of the twenty-first century. The situation in Darfur exemplifies how the scarcity of resources, especially in vulnerable states, can lead to widespread violence.

In Tunisia, Egypt, and Syria, increasing food insecurity played a significant role in fostering widespread public discontent, leading to protests and, ultimately, resulting in regime changes or civil conflicts.[128] In Syria, the circumstances were exacerbated by a significant drought that occurred between 2006 and 2010, leading to extensive damage to the agricultural industry and resulting in the displacement of hundreds of thousands of farmers to urban regions.[129] The sudden influx of displaced rural populations exerted considerable pressure on urban resources and intensified the socio-economic difficulties that were instrumental in igniting the Syrian civil war. The Arab Spring serves as a compelling illustration of how the lack of food, frequently intensified by environmental conditions, can trigger social upheaval and lead to political turmoil.

The Niger Delta region of Nigeria exemplifies a notable case of conflict stemming from the scarcity of resources, especially concerning energy resources.[130] Nigeria stands as the foremost oil producer on the African continent, however, the economic gains derived from oil extraction have not translated into tangible benefits for the local communities in the Delta region.[131] This disparity has resulted in pervasive poverty, high levels of unemployment, and significant environmental damage, particularly from oil spills. Local communities, perceiving themselves as marginalized by both the Nigerian government and international oil corporations, have resorted to armed resistance as a form of protest. The Movement for the Emancipation of the Niger Delta (MEND) and various other militant organizations have participated in acts of sabotage, abductions, and assaults on oil facilities in order to advocate for an increased allocation of oil revenues and reparations for ecological harm.[132] This conflict illustrates the ways in which limited economic prospects and environmental deterioration in regions abundant in resources can incite violence.

The South China Sea is an area abundant in resources, featuring significant deposits of oil, natural gas, and marine life.[133] This maritime region is subject to territorial claims, either partially or entirely, by multiple nations, including China, Vietnam, the Philippines, Malaysia, and Brunei.[134] China's proactive stance and

[128] Buck (2016, 13–28).

[129] De Châtel (2019, 286–304).

[130] Shebbs and Njoku (2016, 32–36).

[131] Iheriohanma (2016, 134–135).

[132] Chris et al. (2021, 19–27).

[133] Nguyen (2015, 15–35).

[134] Barinaadaa and Bariledum (2021, 155–162).

actions in the region, such as the development of artificial islands and military installations, have heightened tensions with neighbouring countries.[135] The circumstances in the South China Sea exemplify how the rivalry for energy resources and fishing entitlements, especially in a region of substantial strategic importance, can result in geopolitical tensions.[136]

In recent years, increased concern over climate change and resource depletion has sparked speculation that these elements may elevate the probability or frequency of violent conflict.[137] Changes in our environment, along with the consequent disruptions—ranging from droughts to flooded coastal regions and the degradation of cultivable land—are expected to be significant catalysts for warfare and conflict.[138] Climate change and resource scarcity are rarely, if at all, the sole causes of violent conflict. Instead, they ought to be regarded as "threat multipliers" that engage with a range of other risk factors and diverse sources of vulnerability in real-world scenarios.[139] It is essential to emphasize that although climate change and resource scarcity are not the exclusive causes of violent conflict, they significantly increase the likelihood of such events.

8 Natural Disasters and Human Vulnerability

Natural disasters, including earthquakes, floods, hurricanes, droughts, and volcanic eruptions, have consistently influenced human history, presenting significant risks to human life, property, and ecological systems.[140] The occurrence of these events is an inherent aspect of the Earth's natural processes; however, the degree to which they affect human societies is significantly shaped by human vulnerability. This vulnerability constitutes a multifaceted framework that encompasses social, economic, political, and environmental elements, collectively influencing the extent to which individuals, communities, and nations experience the impacts of disasters. A natural disaster transpires when a natural hazard, including phenomena such as storms, earthquakes, or floods, intersects with susceptible human communities, resulting in extensive damage.[141] A hazard is classified as a disaster when it substantially disturbs human existence, economic stability, and infrastructural integrity. Natural hazards frequently occur without warning and cannot be entirely prevented; however, their effects can be alleviated through effective preparedness and risk management strategies.[142]

135 Turcsányi (2018, 31–60).
136 Fels and Vu (2016, 3–23).
137 Black et al. (2022).
138 Allen et al. (2024, 63).
139 Enemchukwu (2019, 3474844).
140 Cooke et al. (2022, 122–140).
141 Bonati (2021, 659–662).
142 Tabish and Syed (2015, 227–252).

Vulnerability denotes the extent to which a community or individual is at risk of experiencing adverse effects from disasters.[143] Vulnerability is influenced by a variety of elements, such as economic deprivation, insufficient access to essential resources, inadequate infrastructure, ineffective governance, social disparities, environmental deterioration, and a lack of readiness.[144] The extent of vulnerability directly correlates with the severity of the outcomes resulting from a disaster.[145]

Disaster risk refers to the likelihood of experiencing loss or damage resulting from the interplay between hazards and vulnerabilities.[146] Although hazards are predominantly outside the realm of human influence, vulnerability is an aspect that can be mitigated through the implementation of proactive risk reduction strategies.[147]

9 Human Vulnerability: An Intricate Concept with Multiple Dimensions

Social vulnerability is a consequence of social inequality and marginalization, rendering specific populations more prone to the impacts of disasters.[148] Poverty serves as a fundamental catalyst for social vulnerability, as individuals in impoverished conditions frequently reside in high-risk environments, experience inadequate housing and healthcare, and possess limited resources to manage and recuperate from disasters.[149] Marginalized populations, including women, children, the elderly, and individuals with disabilities, encounter elevated levels of risk.

Economic systems are particularly vulnerable to disturbances caused by natural disasters.[150] This is particularly applicable to developing economies, which often rely on agriculture or the extraction of natural resources. Such economies are at heightened risk from climate-related events, including droughts, floods, and hurricanes. Economic vulnerability can also stem from inadequately constructed infrastructure, including roads, bridges, and power grids, which are prone to damage. Additionally, the absence of sufficient insurance coverage can hinder individuals and businesses from effectively recovering from financial losses.

Human activities have played a substantial role in the deterioration of the environment, thereby heightening susceptibility to natural disasters.[151] The ability of the natural environment to alleviate the impacts of disasters, including floods and landslides, is compromised by factors such as deforestation, urban development,

143 Rufat et al. (2015, 470–486).

144 Forbes-Mewett and Nguyen-Trung (2019, 5–27).

145 Benevolenza and DeRigne (2019, 266–281).

146 Jia et al. (2021, 11187).

147 Kitchin and Dodge (2020, 47–65).

148 Carter (2021, 1–26).

149 Voola and Kalyanasundaram (2021, 902–912).

150 Mata-Lima et al. (2013, 45–64).

151 Koubi (2019, 1–55).

and poor land management practices.[152] Climate change intensifies environmental susceptibility by elevating both the frequency and intensity of extreme weather phenomena.

Governance is fundamental in shaping the preparedness and response of societies to natural disasters. Ineffective governance, alongside corruption and insufficient disaster planning, exacerbates political vulnerability.[153] When governmental authorities neglect to implement building regulations, fail to distribute resources for disaster readiness, or do not create early warning mechanisms, communities become increasingly vulnerable to the risks associated with disasters. Additionally, political instability may obstruct efficient disaster response and recovery initiatives.

The earthquake that struck Haiti in 2010 was devastating, leading to more than 200,000 fatalities and extensive devastation. The country's susceptibility to such a disaster was influenced by several interrelated factors, including pervasive poverty, inadequate governance, fragile infrastructure, and insufficient disaster readiness.[154] The collapse of informal housing in densely populated regions, coupled with the absence of a robust emergency response mechanism, exacerbated the humanitarian crisis. The earthquake underscored the susceptibility of a fragile state to be overwhelmed by a natural disaster, particularly in light of its significant social, economic, and political vulnerabilities.[155]

Typhoon Haiyan, acknowledged as one of the most powerful tropical cyclones ever documented, struck the Philippines in 2013, leading to considerable loss of life and widespread destruction of property.[156] Although the Philippines has a long-standing history of experiencing typhoons, the impact of Haiyan was intensified by factors such as widespread poverty, substandard housing conditions, and ineffective governance.[157] Coastal communities residing in temporary shelters experienced severe devastation due to the storm surge, while relief operations faced significant obstacles stemming from logistical difficulties and inadequate infrastructure. This calamity underscored the critical need for enhancing resilience through strategies aimed at poverty alleviation, infrastructure development, and improved disaster preparedness.

Cyclone Idai, which affected Mozambique, Zimbabwe, and Malawi in 2019, is considered one of the most catastrophic tropical storms in the recorded history of Southern Africa.[158] The cyclone caused the deaths of over 1300 people and resulted in the displacement of hundreds of thousands.[159] Mozambique, recognized as one of the most impoverished nations globally, faced heightened susceptibility owing to inadequate infrastructure, ineffective governance, and restricted capacity for disaster

---

152 Kumar et al. (2022, 19 46).

153 Choudhury et al. (2019, 1761–1781).

154 Best and Burke (2019, 1647–1681).

155 Sadiq and Tyler (2017, e 177).

156 Bankoff (2016, 187–207).

157 Bowen (2015, 7482).

158 Mutasa (2022, 175–189).

159 Deprez and Labattut (2020, 14).

response.[160] The cyclone resulted in significant flooding and the devastation of agricultural produce, thereby worsening food insecurity and driving numerous individuals further into poverty. The occurrence emphasized the unequal effects that natural disasters impose on the most impoverished nations, drawing attention to the necessity for global assistance and sustainable development strategies with a view to mitigating vulnerability.

10 The Impact of Climate Change on Heightened Vulnerability

Climate change is increasingly acknowledged as a significant factor contributing to natural disasters and exacerbating human susceptibility. The increase in global temperatures is leading to a higher occurrence and intensity of heatwaves, floods, storms, and droughts, which in turn exacerbates the difficulties encountered by populations that are already vulnerable.[161] The Intergovernmental Panel on Climate Change (IPCC) has warned that the impacts of climate change will disproportionately impact the poorest and most vulnerable regions, particularly in the Global South, where adaptation capabilities are often limited.[162] In Sub-Saharan Africa, extended periods of drought and unpredictable rainfall are jeopardizing food security, heightening the risk of famine, displacement, and conflicts over limited resources.[163]

11 Reducing Vulnerability: Building Resilience to Natural Disasters

Disaster Risk Reduction (DRR) represents a pivotal strategy for diminishing vulnerability to disaster.[164] This holistic framework prioritizes the minimization of disaster risk through a triad of preparedness, prevention, and mitigation efforts. Essential elements of disaster risk reduction include the establishment of robust infrastructure, the implementation of rigorous building codes, the improvement of early warning systems, and the integration of sustainable land-use strategies. Further, DRR underscores the significance of community engagement, ensuring that local populations actively participate in the processes of disaster planning and response.

Climate Adaptation: Addressing the risks linked to climate change requires the adoption of climate adaptation strategies.[165] These strategies may include enhancing

[160] Sunu (2024, 61–65).

[161] Majedul Islam (2022, 21–36).

[162] Butler and Hanigan (2019, 1243–1252).

[163] Ehi et al. (2024, 37–57).

[164] Ishiwatari (2022, 3019–3045).

[165] Grafakos et al. (2019, 87–106).

flood protection systems, optimizing water resource management, and investing in agriculture that is resilient to climate impacts. Adaptation initiatives are particularly essential for regions that are highly susceptible, such as island nations and developing countries, where the impacts of climate change are most evident.

Poverty Reduction and Social Safety Nets: Mitigating poverty is an essential approach to decreasing susceptibility to disasters.[166] Social safety net initiatives, including cash assistance, food security measures, and insurance programs, can facilitate faster recovery for at-risk communities following disasters, enabling them to restore their livelihoods. Additionally, ensuring access to healthcare, education, and secure housing can significantly diminish the enduring effects of disasters.

Enhancing governance and institutional capacity is crucial for the effective management of disasters.[167] It is imperative for governments to formulate and execute national policies aimed at disaster risk reduction, allocate necessary resources for both preparedness and response efforts, and collaborate with international organizations to foster resilience.

International Collaboration: Numerous at-risk nations are deficient in the financial and technical capabilities essential for effectively preparing for and responding to natural disasters.[168] International collaborative initiatives, supported by entities like the United Nations Office for Disaster Risk Reduction (UNDRR), can provide essential assistance through financial resources, expert technical expertise, and the coordination of humanitarian relief efforts.[169]

12 State Fragility and Natural Disasters: A Vicious Cycle

In fragile states, the effects of natural disasters are frequently intensified by inadequate governance and limited institutional capabilities.[170] Natural disasters exacerbate existing tensions in fragile contexts, particularly where governmental structures are insufficiently equipped to handle crises efficiently. The relationship between state fragility and natural disasters establishes a detrimental cycle characterized by increased vulnerability, instability, and conflict.[171]

Fragile states frequently exhibit inadequate governance frameworks and institutions that lack the capacity to effectively address natural disasters.[172] Following such events, these states may face significant challenges in organizing relief operations, securing necessary resources, and reconstructing damaged infrastructure. The absence of accountability and transparency in resource management frequently

166 Akand (2022, 55–64).
167 Albris et al. (2020, 101647).
168 Cui et al. (2021, 216–223).
169 Frennesson et al. (2022, 103410).
170 Tierney (2012, 341–363).
171 Mena et al. (2019, 37).
172 World Bank Group (2018, 33–39).

results in delays and inefficiencies during disaster response and recovery efforts. Fragile states frequently depend on precarious economic sectors, particularly agriculture, which are extremely vulnerable to disasters caused by climate change.[173] Droughts and floods can result in agricultural failures, the death of livestock, and interruptions in food supply networks, thereby exacerbating the poverty of already vulnerable populations. The financial repercussions of natural disasters can result in enduring impacts on development, thereby diminishing the state's ability to deliver essential services.[174]

In numerous vulnerable nations, natural calamities intensify pre-existing social fractures, especially those related to ethnicity, religion, or socioeconomic status. Marginalized groups, such as rural populations, women, and indigenous communities, are frequently the most vulnerable to the effects of disasters and are often excluded from the decision-making processes related to disaster response and recovery initiatives.[175] The absence of inclusion may result in dissatisfaction and friction, thereby heightening the likelihood of social upheaval and discord. The Syrian civil war, which has led to the deaths of hundreds of thousands and the displacement of millions, was preceded by one of the most severe droughts recorded in the nation's history, occurring from 2006 to 2011. The drought led to extensive agricultural failures, compelling rural communities to relocate to cities, where they encountered high levels of unemployment and poverty. The government's failure to adequately address the drought and its consequences significantly contributed to the growing dissatisfaction with the regime, which ultimately culminated in armed conflict. While the drought was not the sole catalyst for the war, it exacerbated existing political, social, and economic tensions.

The Syrian civil war, resulting in the loss of hundreds of thousands of lives and the displacement of millions, was preceded by one of the most significant droughts in the country's history, which took place from 2006 to 2011.[176] The drought led to extensive agricultural failures, compelling rural communities to relocate to cities, where they encountered high levels of unemployment and poverty. Although the drought was not the exclusive trigger for the war, it intensified pre-existing political, social, and economic strains. The Sahel region of Africa, which includes countries such as Mali, Niger, and Burkina Faso, is one of the most vulnerable areas to the effects of climate change.[177] Increasing temperatures and extended periods of drought have resulted in desertification, diminished agricultural output, and a scarcity of water resources. Environmental challenges have significantly contributed to the tensions between pastoralist and agricultural communities while simultaneously enabling the rise of extremist groups. The failure of regional governments to effectively tackle these issues has intensified state fragility, resulting in persistent conflict and displacement.

173 Peters et al. (2020, 13–30).
174 Botzen et al. (2019, 167–184).
175 Grow et al. (2022, 957).
176 Selby et al. (2017, 232–244).
177 Mbaye (2020, 12).

13 Tackling Vulnerability Stemming from Climate Change

Fragile states require enhanced governance frameworks capable of addressing the challenges associated with climate change and natural disasters.[178] This necessitates the advancement of disaster preparedness measures, the allocation of resources towards infrastructure resilience, and the fortification of institutions that can facilitate coordinated and effective responses.

Developing resilience to climate change is crucial for mitigating the susceptibility of vulnerable nations. This procedure encompasses the promotion of sustainable land management practices, the allocation of resources towards renewable energy initiatives, and the enhancement of water management systems. International assistance, encompassing climate finance and technical support, is essential for enabling vulnerable nations to adjust to the impacts of climate change.

Addressing the root causes of conflict, such as resource scarcity and social inequality, is essential to prevent conflicts that are intensified by climate change. Engaging local communities in mediation and peacebuilding efforts can significantly alleviate tensions and foster social unity in vulnerable nations.[179]

Climate change and natural disasters represent significant global challenges that necessitate collaborative efforts on an international scale.[180] Fragile states, often marked by insufficient resources to respond effectively, necessitate support from the international community to strengthen their resilience and mitigate the effects of climate change. This encompasses the delivery of humanitarian aid, the advancement of sustainable development initiatives, and the tackling of the fundamental factors contributing to state fragility.

14 The Interconnection of Migration, Displacement, and Resource Crunch

The interplay among migration, displacement, and resource pressure is intricate and layered. It is becoming increasingly acknowledged that the scarcity of resources and the degradation of the environment serve as significant catalysts for both voluntary migration and involuntary displacement. As vital resources become increasingly scarce or are distributed inequitably, individuals frequently find themselves compelled to relocate in pursuit of improved living conditions. This movement of populations can subsequently impose further burdens on the regions or nations that receive them, intensifying pre-existing resource challenges and introducing new difficulties.

[178] Peters et al. (2020, 13–30).

[179] Smidt (2020, 344–372).

[180] Seddon et al. (2020, 20190120).

Environmental degradation and climate change are significant catalysts for migration and displacement today. Rising temperatures, changes in rainfall patterns, desertification, deforestation, and the depletion of freshwater resources make certain regions increasingly uninhabitable.[181] The phenomenon of migration induced by climate change is anticipated to increase significantly in the forthcoming decades as more regions worldwide face heightened susceptibility to severe weather events, rising sea levels, and diminished agricultural output.

The rapid increase in population, especially in developing nations, exerts considerable stress on natural resources, including land and water. As rural areas face increasing population density and a decline in available resources, a significant portion of the population migrates to urban areas in search of employment prospects and enhanced access to various services. Nevertheless, this sudden surge of residents into cities can overwhelm infrastructure, housing, and public services, thereby presenting new difficulties for resource management and maintaining social cohesion.

In urban centers such as Lagos[182] (Nigeria), and Dhaka (Bangladesh),[183] rapid urbanization, propelled by internal migration and displacement, has resulted in significant overcrowding, housing scarcity, and a heightened demand for essential services such as water, energy, and sanitation. The swift expansion of urban slums in these locations illustrates the failure of urban infrastructure to accommodate the surge of incoming migrants, exacerbating competition for resources and intensifying social conflicts.

Resource pressure is also created by a decline in agricultural productivity, often affected by environmental degradation, soil erosion, and insufficient water resources. Numerous rural communities dependent on subsistence agriculture are compelled to relocate when they can no longer maintain their livelihoods because of crop failures, drought conditions, or the degradation of arable land. Food insecurity serves as both a catalyst and a result of migration, prompting individuals to depart from regions characterized by scarcity in pursuit of more reliable food supplies. Water scarcity has emerged as a significant catalyst for migration and displacement, especially in areas where clean water is crucial for agricultural practices, industrial activities, and everyday living.[184] Competition for scarce water resources can result in conflict, prompting individuals to migrate or become displaced.[185] In certain instances, the displacement caused by water shortages may generate tensions between established communities and newcomers as both populations vie for access to the same limited resources.

[181] Tinazzi (2024, 1–21).

[182] Omilusi (2020, 35–56).

[183] Vučković and Adams (2022, 397–429).

[184] Bertassello et al. (2023, 2706).

[185] Niyitunga (2019, 95–117).

15 Social, Economic, and Political Implications

The convergence of migration, displacement, and resource constraints carries profound social, economic, and political consequences.[186] When populations relocate due to a lack of resources, they frequently encounter difficulties concerning integration, job opportunities, and access to essential services.[187] The influx of significant numbers of migrants or displaced individuals can place considerable pressure on the resources and infrastructure of host communities, potentially resulting in social discord, competition for resources, and, in certain instances, conflict.[188] The influx of migrants or displaced individuals often results in social friction between established communities and newcomers, mainly where essential resources like housing, water, and food are scarce. Such frictions can intensify into conflict, especially when residents perceive that migrants are usurping employment opportunities or are being granted preferential access to assistance and services.

Migration can exert a dual influence on the economic landscape of host communities. On one side, migrants may enhance the labour pool, drive economic expansion, and introduce diverse skills and expertise.[189] Conversely, significant migration flows can significantly overwhelm existing infrastructure, social services, and labour markets when the number of incoming individuals surpasses the local economy's ability to integrate them effectively.

In areas where migration and displacement are influenced by resource scarcity, political instability may emerge as governments face difficulties addressing the influx of individuals and the related resource challenges.[190] In certain instances, this instability can give rise to populist or nationalist movements that promote stringent immigration policies or blame migrants for existing economic and social issues. Venezuela's economy's deterioration, intensified by ineffective political governance and a decrease in oil income, has resulted in significant shortages of food and medical supplies, driving millions of Venezuelans to seek refuge in adjacent nations.[191] The significant influx of migrants has placed considerable pressure on the resources of host nations such as Colombia, Brazil, and Peru, where the rapid arrival of individuals has exceeded the capacity of social services and infrastructure.

[186] Schierup et al. (2018, 733–745).

[187] Cotroneo (2017, 283–318).

[188] Itumo and Nwefuru (2016, 24–38).

[189] Brunow et al. (2015, 1027–1075).

[190] Adaawen et al. (2019, 15–31).

[191] Tapia et al. (2022, 4–6).

16 Policy Implications

Addressing the challenges posed by migration, displacement, and resource pressure requires comprehensive policy responses that consider the interconnected nature of these phenomena. Some key strategies include:

- Enhancing resilience to climate change via sustainable resource management, adopting advanced agricultural techniques, and developing infrastructure can mitigate the necessity for migration by tackling the fundamental causes of resource strain.
- Nations that host significant populations of migrants or displaced individuals require assistance from the global community to address the resulting resource challenges effectively. This support encompasses financial resources, humanitarian aid, and the exchange of effective strategies for integrating migrants into local communities.
- Establishing legal frameworks that acknowledge the rights of migrants and individuals who have been displaced is crucial for safeguarding their protection and facilitating their access to essential resources. This process encompasses the creation of secure migration routes and the provision of legal recognition and employment opportunities for those who have been displaced.

17 Findings

The absolute scarcity of physical resources does not solely determine conflict potential. Instead, the critical elements are the inequality in distribution and access to these resources. Many natural resources exhibit significant geographical and market concentration, with deposits frequently in politically unstable regions. This situation heightens the likelihood of conflict.

Countries exhibit varying degrees of susceptibility to shortages and price volatility. Mutual interdependencies can incentivize collaborative efforts, whereas unilateral dependency may exacerbate tensions and conflict.

Conflicts regarding resources are seldom motivated solely by economic competition. Nevertheless, the likelihood of escalation rises when competition for resources intertwines with additional factors, such as internal power struggles among governing elites or aspirations for regional dominance.

The relationship between perception and "securitization" is complex. The likelihood of conflict arising from competition over a resource is influenced not solely by tangible issues related to access and distribution. The way political entities and media portray the situation, along with the subjective perceptions of various stakeholders, plays a crucial role. Heightened rhetoric and the propagation of fear exacerbate the sense of threat, resulting in the "securitization" of economic conflicts.

The problem-solving abilities and willingness of the participating individuals and institutions ultimately determine the extent to which competition results in either latent or overt conflict. Numerous tools are available to mitigate conflict risks and address ongoing disputes. These tools encompass national resource management, resource governance, and dispute resolution mechanisms. Although specific instruments have proven their efficacy, others are still under development.

18 Conclusion

Resource conflicts have occurred historically, are currently ongoing, and are expected to persist. Such conflicts may become increasingly common in the years ahead as the demand for resources rises while their availability declines. Situations where resources are a contributing factor present distinct challenges and opportunities for individuals and organizations focused on mitigating the risk of violence. To effectively prevent violent conflicts, it is essential to have some form of early warning system; resource scarcity can be a significant indicator. When analyzed within a historical and contextual framework, this scarcity can aid in forecasting the potential emergence of violent conflict.

Effective resource management is essential to mitigate the risk of violent conflict and to avert the resurgence of violence in conflicts that have already been resolved. To prevent the emergence of violent conflict, it may be most beneficial to implement dual strategies that promote fair and equitable distribution of resources while also managing them to reduce poverty. In instances where conflicts have previously occurred, renewed violence can be prevented if peace agreements incorporate provisions for natural resource management, establishing mechanisms within the agreements to address this critical issue.

References

Abdallah S, Akayim AR, Amedor EN, Afetsu JY (2023) Climate change: Who is responsible? Open J Appl Sci 13(12):2431–2446

Abrahams D (2020) Conflict in abundance and peacebuilding in scarcity: challenges and opportunities in addressing climate change and conflict. World Dev 132:104998

Adaawen S, Rademacher-Schulz C, Schraven B, Segadlo N (2019) Drought, migration, and conflict in sub-Saharan Africa: What are the links and policy options? Curr Direct Water Scarcity Res 2:15–31

Adams D, Adams K, Ullah S, Ullah F (2019) Globalization, governance, accountability and the natural resource 'curse': implications for socio-economic growth of oil-rich developing countries. Resour Policy 61:128–140

Akand MAH (2022) The role of social safety nets in poverty alleviation of rural poor in Bangladesh: a social survey. Doctoral dissertation, © University of Dhaka

Akinyetun TS (2022) The proliferation of armed non-state actors in the Sahel: drivers, ramifications and the way forward. J Central Eastern Eur Afr Stud 2(2):36–58

Akram R, Ai F, Srivastava M, Sharma R (2024) Considering natural gas rents, mineral rents, mineral depletion, and natural resources depletion as new determinants of sustainable development. Resour Policy 96:105200

Albris K, Lauta KC, Raju E (2020) Strengthening governance for disaster prevention: the enhancing risk management capabilities guidelines. Int J Dis Risk Reduct 47:101647

Allen E, Kevany S, Sitaraman S (2024) Everything everywhere all at once: wars, climate change, natural disasters, coups, and economic collapse. J Indo-Pacific Aff 7(3):1

Antonio RJ (2023) Social theory and climate change in the interregnum. In: Handbook on inequality and the environment. Edward Elgar Publishing, pp 464–489

Avruch K (2022) Culture and conflict resolution. In: The Palgrave encyclopedia of peace and conflict studies. Springer, Cham, pp 254–259

Bachmann A, Tice V, Al-Obeidi LA, Kılıç DT (2019) Tigris-Euphrates river ecosystem: a status report. In: Mesopotamia water forum

Bamidele S, Erameh NI (2023) Environmental degradation and sustainable peace dialogue in the Niger Delta region of Nigeria. Resour Policy 80:103274

Bankoff G (2016) Wind, water, and risk: shaping a transnational history of the western North Pacific. Trans Trans Regional Natl Stud Southeast Asia 4(1):187–207

Banwell N, Rutherford S, Mackey B, Chu C (2018) Towards improved linkage of disaster risk reduction and climate change adaptation in health: a review. Int J Environ Res Public Health 15(4):793

Barakat Z, Fakih A (2021) Determinants of the Arab spring protests in Tunisia, Egypt, and Libya: What have we learned? Soc Sci 10(8):282

Barinaadaa N, Bariledum K (2021) The united nations convention on the law of the sea and international maritime disputes: a case study of the south China Sea. NIU J Soc Sci 7(3):155–162

Bebbington AJ, Humphreys Bebbington D, Sauls LA, Rogan J, Agrawal S, Gamboa C, Verdum R (2018) Resource extraction and infrastructure threaten forest cover and community rights. Proc Natl Acad Sci 115(52):13164–13173

Benevolenza MA, DeRigne L (2019) The impact of climate change and natural disasters on vulnerable populations: a systematic literature review. J Hum Behav Soc Environ 29(2):266–281

Bertassello L, Müller MF, Wiechman A, Penny G, Tuninetti M, Müller-Itten MC (2023) Food demand displaced by global refugee migration influences water use in already water-stressed countries. Nat Commun 14(1):2706

Best R, Burke PJ (2019) Macroeconomic impacts of the 2010 earthquake in Haiti. Empir Econ 56:1647–1681

Bharkhada J (2022) Identifying the causal link between the desiccation of Lake Chad and Boko-Haram recruitment in the Sahel region

Black R, Busby J, Dabelko GD, De Coning C, Maalim H, McAllister C,... Staudenmann JA (2022) Environment of peace: security in a new era of risk. Stockholm International Peace Research Institute.

Bonati S (2021) Natural hazards. In: Encyclopedia of security and emergency management. Springer, Cham, pp 659–662

Botzen WW, Deschenes O, Sanders M (2019) The economic impacts of natural disasters: a review of models and empirical studies. Rev Environ Econ Policy 2019:1

Bowen T (2015) Social protection and disaster risk management in the Philippines: the case of typhoon Yolanda (Haiyan). World bank policy research working paper, 7482.

Brunow S, Nijkamp P, Poot J (2015) The impact of international migration on economic growth in the global economy. Handb Econ Int Migr 1:1027–1075

Buck EA (2016) Navigating the Arab spring: the power of food prices and the stability of monarchies. Oxford University Press

Buhaug H, Von Uexkull N (2021) Vicious circles: violence, vulnerability, and climate change. Annu Rev Environ Resour 46(1):545–568

Burger A, Oz T, Kennedy WG, Crooks AT (2019) Computational social science of disasters: opportunities and challenges. Fut Int 11(5):103

Butler CD, Hanigan IC (2019) Anthropogenic climate change and health in the Global South. Int J Tuberc Lung Dis 23(12):1243–1252
Byaro M, Nkonoki J, Mafwolo G (2023) Exploring the nexus between natural resource depletion, renewable energy use, and environmental degradation in sub-Saharan Africa. Environ Sci Pollut Res 30(8):19931–19945
Caldeira C, Gui X, Reynolds TL, Bietz M, Chen Y (2021) Managing healthcare conflicts when living with multiple chronic conditions. Int J Hum Comput Stud 145:102494
Carter B (2021) Impact of social inequalities and discrimination on vulnerability to crises
Cenci MP, Scarazzato T, Munchen DD, Dartora PC, Veit HM, Bernardes AM, Dias PR (2022) Eco-friendly electronics—a comprehensive review. Adv Mater Technol 7(2):2001263
Challenges and Formulating Resilient Strategies. Int J Environ Sci 7(3):57–68
Charonis GK (2021) Degrowth, steady state and circular economies: alternative discourses to economic growth. Soc Regist 5(3):75–94
Chaudhary MT, Piracha A (2021) Natural disasters—origins, impacts, management. Encyclopedia 1(4):1101–1131. https://doi.org/10.3390/encyclopedia1040084
Chavez-Miguel G, Bonatti M, Ácevedo-Osorio Á, Sieber S, Löhr K (2022) Agroecology as a grassroots approach for environmental peacebuilding: strengthening social cohesion and resilience in post-conflict settings with community-based natural resource management. GAIA Ecol Perspect Sci Soc 31(1):36–45
Chin-Yee S (2019) Climate change and human security: case studies linking vulnerable populations to increased security risks in the face of the global climate challenge. King's College London: EUCERS Strategy Paper, 2-31.
Choudhury MUI, Uddin MS, Haque CE (2019) "Nature brings us extreme events; some people cause us prolonged sufferings": the role of good governance in building community resilience to natural disasters in Bangladesh. J Environ Plan Manag 62(10):1761–1781
Chris OI, Victor O, Felicia OC (2021) Militancy and oil wars in Africa: the Niger Delta experience in Nigeria. Natl J Multidiscip Res Dev 6(2):19–27
Ciplet D, Harrison JL (2020) Transition tensions: mapping conflicts in movements for a just and sustainable transition. Environ Polit 29:435–456
Cooke SJ, Galassi DM, Gillanders BM, Landsman SJ, Hammerschlag N, Gallagher AJ, Lennox RJ (2022) Consequences of "natural" disasters on aquatic life and habitats. Environ Rev 31(1):122–140
Cotronco A (2017) Specificitics and challenges of responding to internal displacement in urban settings. Int Rev Red Cross 99(904):283–318
Cui P, Peng J, Shi P, Tang H, Ouyang C, Zou Q, Lei Y (2021) Scientific challenges of research on natural hazards and disaster risk. Geography Sustain 2(3):216–223
Damkjaer S, Taylor R (2017) The measurement of water scarcity: defining a meaningful indicator. Ambio 46(5):513–531
De Châtel F (2019) The role of drought and climate change in the Syrian uprising: untangling the triggers of the revolution. In: Green planet blues. Routledge, London, pp 286–304
De Coning C (2018) Sustaining peace: Can a new approach change the UN?
Deprez S, Labattut E (2020) Lessons learnt from CARE's shelter responses to cyclone Idai in Malawi, Mozambique and Zimbabwe
Dey R, Lewis SC (2021) Natural disasters linked to climate change. In: The impacts of climate change. Elsevier, London, pp 177–193
Dieng A (2021) The Sahel: challenges and opportunities. Int Rev Red Cross 103(918):765–779
Ehi OE, Emegha M, Awakessien S, Asomah JK (2024) Climate change and food security in sub-Saharan Africa: The development of African-rooted adaptation strategies. J Geography Environ Manag 10(3):37–57
Enemchukwu N (2019) The chain of climate change, scarcity and conflict in global politics. Scarcity and Conflict in Global Politics, June 24, 2019

Erickson TB, Brooks J, Nilles EJ, Pham PN, Vinck P (2019) Environmental health effects are attributed to toxic and infectious agents following hurricanes, cyclones, flash floods, and significant hydrometeorological events. J Toxicol Environ Health Part B 22(5–6):157–171

Ernst M (2022) Control through cooperation? Assessing China's economic and military-strategic interests in the south China Sea. J Territ Mar Stud 9:7

Farag M (2018) Can water mitigate the Palestinian-Israeli conflict? The case for environmental peacemaking. Arab World Geographer 21(4):299–317

Farasat A, Siddiqui S, Ghous M, Shafqat A, Hanif A, Sohail M, Jabbar M (2023) Exploring the paradox of population growth and food resources in Punjab–Pakistan: reassessment in the light of Malthusian theory. Pak J Sci 75(04):741–751

Fels E, Vu TM (2016) Introduction: understanding the importance of the disputes in the south China Sea. In: Power politics in Asia's contested waters: territorial disputes in the South China Sea, pp 3–23

Ferreira DC, Graziele I, Marques RC, Gonçalves J (2021) Investment in drinking water and sanitation infrastructure and its impact on waterborne diseases dissemination: the Brazilian case. Sci Total Environ 779:146279

Forbes-Mewett H, Nguyen-Trung K (2019) Defining vulnerability. In: Vulnerability in a mobile world. Emerald Publishing Limited, pp 5–27

Forzieri G, Bianchi A, e Silva FB, Herrera MAM, Leblois A, Lavalle C, Feyen L (2018) Escalating impacts of climate extremes on critical infrastructures in Europe. Global Environ Change 48:97–107

Frennesson L, Kembro J, de Vries H, Jahre M, Van Wassenhove L (2022) International humanitarian organizations' perspectives on localization efforts. Int J Dis Risk Reduct 83:103410

Gahlawat IN, Lakra P (2020) Global climate change and its effects. Integr J Soc Sci 7(1):14–23

Gaillard JC, Cadag JRD, Rampengan MM (2019) People's capacities in facing hazards and disasters: an overview. Nat Hazards 95:863–876

Galgano FA (2019) The environment–conflict nexus. In: The environment-conflict nexus: climate change and the emergent national security landscape, pp 1–17

Gelaw A, Debele M (2024) International water conventions, the exploitation of transboundary rivers and the Nile water dispute among Ethiopia, Sudan and Egypt. Heliyon 2024:1

Grafakos S, Trigg K, Landauer M, Chelleri L, Dhakal S (2019) An analytical framework to evaluate the level of integration of climate adaptation and mitigation in cities. Clim Change 154:87–106

Grow L, Daniels B, Spencer D, Sloan C, Blades N, Gómez MT, Christensen SR (2022) Disaster vulnerability. BCL Rev 63:957

Guimond JA, Michael HA (2021) Effects of marsh migration on flooding, saltwater intrusion, and crop yield in coastal agricultural land subject to storm surge inundation. Water Resour Res 57(2):e2020WR028326

Guo R (2018) Chapter 2, globalization, natural resources and borders, cross-border resources management, 3rd edn, pp 23–29

Hameed A, Padda IUH, Karim S (2023) Spatial analysis of food and nutrition security in Pakistan: a holistic pathway towards zero hunger policies. GeoJournal 88(3):2563–2585

Hardt JN (2017) Environmental security in the Anthropocene: assessing theory and practice. Routledge, London

Hidalgo J, Baez AA (2019) Natural disasters. Crit Care Clin 35(4):591–607

Himmelgreen D, Romero-Daza N, Heuer J, Lucas W, Salinas-Miranda AA, Stoddard T (2022) Using syndemic theory to understand food insecurity and diet-related chronic diseases. Soc Sci Med 295:113124

Hossen MA, Connor J, Ahammed F (2023) How to resolve transboundary river water sharing disputes. Water 15(14):2630

Hulme K (2023) Assessing the contribution of human rights actors to environmental peacebuilding. In: Research handbook on international law and environmental peacebuilding. Edward Elgar Publishing, London, pp 230–252

Ide T (2023) Rise or recede? How climate disasters affect armed conflict intensity. Int Secur 47(4):50–78
Ide T, Scheffran J (2014) On climate, conflict and cumulation: suggestions for integrative cumulation of knowledge in the research on climate change and violent conflict. Global Change Peace Sec 26(3):263–279
Iheriohanma VI (2016) Environmental impact assessment of oil and gas industry in Niger Delta, Nigeria: a critical environmental and legal framework assessment
Ishiwatari M (2022) Disaster risk reduction. In: Handbook of climate change mitigation and adaptation. Springer, Cham, pp 3019–3045
Itumo A, Nwefuru NH (2016) Nigerian state and responses to plights of persons internally displaced by Boko Haram insurgents: Implications for socio-economic and political development. Res Humanit Soc Sci 6(15):24–38
Jacobson JL (2018) Women's health: the price of poverty. In: The health of women. Routledge, London, pp 3–32
Jia H, Chen F, Du E (2021) Adaptation to disaster risk—an overview. Int J Environ Res Public Health 18(21):11187
Joakim EP, Mortsch L, Oulahen G (2021) Using vulnerability and resilience concepts to advance climate change adaptation. In: Environmental hazards and resilience. Routledge, London, pp 13–31
Jongerden J, Akıncı ZS, Ayboğa E (2021) Water, politics and dams in the Mesopotamia basin of the northern middle east: how Turkey instrumentalises the South-Eastern Anatolia Project for Political, Military and strategic interests. In: Tigris and Euphrates rivers: their environment from headwaters to mouth, pp 383–399
Jouhara H, Czajczyńska D, Ghazal H, Krzyżyńska R, Anguilano L, Reynolds AJ, Spencer N (2017) Municipal waste management systems for domestic use. Energy 139:485–506
Judge E (2014) The natural environment: global warming, pollution, resource depletion, and sustainable development. The Business Environment: Themes and Issues in a Globalizing World, 199
Kaczan DJ, Orgill-Meyer J (2020) The impact of climate change on migration: a synthesis of recent empirical insights. Clim Change 158(3):281–300
Kelman I (2020) Disaster by choice: how our actions turn natural hazards into catastrophes. Oxford University Press, Oxford
Khan SAR, Razzaq A, Yu Z, Shah A, Sharif A, Janjua L (2022) Disruption in food supply chain and undernourishment challenges: an empirical study in the context of Asian countries. Socioecon Plann Sci 82:101033
Khan T, Samiullah M, Rouf II, Sultana S, Rahman S, Rahman MB, Khanum R (2024) The nexus of water scarcity and climate change: understanding interconnected challenges and formulating resilient strategies. Int J Environ Sci 7(3):57–68
Kitchin R, Dodge M (2020) The (in) security of smart cities: vulnerabilities, risks, mitigation, and prevention. In: Smart cities and innovative urban technologies. Routledge, London, pp 47–65
Kiyala JCK (2022) Civil society, climate change, security risks management and peacebuilding in the Anthropocene. In: Civil society and peacebuilding in sub-Saharan Africa in the Anthropocene: an overview. Springer, Cham, pp 67–93
Kola L, Kohrt BA, Hanlon C, Naslund JA, Sikander S, Balaji M, Patel V (2021) COVID-19 mental health impact and responses in low-income and middle-income countries: reimagining global mental health. Lancet Psychiatry 8(6):535–550
Koubi V (2019) Sustainable development impacts of climate change and natural disaster. In: Background paper prepared for sustainable development outlook, pp 1–55
Kumar R, Kumar A, Saikia P (2022) Deforestation and forests degradation impacts on the environment. In: Environmental degradation: challenges and strategies for mitigation. Springer, Cham, pp 19–46

Link PM, Scheffran J, Ide T (2016) Conflict and cooperation in the water-security nexus: a global comparative analysis of river basins under climate change. Wiley Interdiscip Rev Water 3(4):495–515

Lu C, Wang K (2023) Natural resource conservation outpaces and climate change: roles of reforestation, mineral extraction, and natural resources depletion: resources. Policy 86:104159

Maja MM, Ayano SF (2021) The impact of population growth on natural resources and farmers' capacity to adapt to climate change in low-income countries. Earth Syst Environ 5(2):271–283

Majedul Islam MM (2022) Threats to humanity from climate change. In: Climate change: the social and scientific construct. Springer, Cham, pp 21–36

Maltarich MA, Kukenberger M, Reilly G, Mathieu J (2018) Conflict in teams: modeling early and late conflict states and the interactive effects of conflict processes. Group Org Manag 43(1):6–37

Mamshai FHA (2023) "Climate change as a threat multiplier": Security and communal implications for Iraq. Commun Change 4(2):1

Mao W, Agyapong VI (2021) The role of social determinants in mental health and resilience after disasters: implications for public health policy and practice. Front Public Health 9:658528

Martin JL, Maris V, Simberloff DS (2016) The need to respect nature and its limits challenges society and conservation science. Proc Natl Acad Sci 113(22):6105–6112

Mata-Lima H, Alvino-Borba A, Pinheiro A, Mata-Lima A, Almeida JA (2013) Impacts of natural disasters on environmental and socio-economic systems: What makes the difference? Ambiente Soc 16:45–64

Mbaye AA (2020) Climate change, livelihoods, and conflict in the Sahel. Geo J Int'l Aff 21:12

Mena R, Hilhorst D, Peters K (2019) Disaster risk reduction and protracted violent conflict. Overseas Development Institute, London

Mezzina R, Gopikumar V, Jenkins J, Saraceno B, Sashidharan SP (2022) Social vulnerability and mental health inequalities in the "Syndemic": call for action. Front Psych 13:894370

Mokyr J (2018) Editor's introduction: the new economic history and the industrial revolution. In: The British industrial revolution. Routledge, London, pp 1–127

Mondal S, Palit D (2022) Challenges in natural resource management for ecological sustainability. In: Natural resources conservation and advances for sustainability. Elsevier, London, pp 29–59

Montt G, Fraga F, Harsdorff M (2018) The future of work in a changing natural environment: climate change, degradation and sustainability. International Labour Organization, London

Moser AK (2016) Consumers' purchasing decisions regarding environmentally friendly products: an empirical analysis of German consumers. J Retail Consum Serv 31:389–397

Mutasa C (2022) Revisiting the impacts of tropical cyclone Idai in Southern Africa. In: Climate impacts on extreme weather. Elsevier, Berlin, pp 175–189

Nashwan AJ, Ahmed SH, Shaikh TG, Waseem S (2023a) Impact of natural disasters on health disparities in low-to-middle-income countries. Discover Health Syst 2(1):23

Nashwan AJ, Ahmed SH, Shaikh TG, Waseem S (2023b) Impact of natural disasters on health disparities in low-to middle-income countries. Discover Health Syst 2(1):23

Nashwan AJ, Ahmed SH, Shaikh TG, Waseem S (2023c) Discover health systems

Naushad VA, Bierens JJ, Nishan KP, Firjeeth CP, Mohammad OH, Maliyakkal AM, Schreiber MD (2019) A systematic review of the impact of disaster on the mental health of medical responders. Prehosp Disaster Med 34(6):632–643

Nguyen TLA (2015). Origins of the south China sea dispute. In: Territorial disputes in the South China Sea: navigating rough waters. Palgrave Macmillan, London, pp 15–35

Niyitunga EB (2019) A conceptual analysis for understanding water scarcity and its threats to international peace and security. Afr J Public Aff 11(3):95–117

Nunes AMM, Coelho Junior LM, Abrahão R, Santos Júnior EP, Simioni FJ, Rotella Junior P, Rocha LCS (2023) Public policies for renewable energy: a review of the perspectives for a circular economy. Energies 16(1):485

Nyambuu U, Semmler W (2023) Sustainable macroeconomics, climate risks and energy transitions. Contrib Econ 2023:1

Ogunmakinde OE, Egbelakin T, Sher W (2022) Contributions of the circular economy to the UN sustainable development goals through sustainable construction. Resour Conserv Recycl 178:106023

Omilusi MO (2020) Thinking global through consensual and reformist approaches: considered governance innovations for Lagos mega-city status. Bangl J Sociol 17(1):1

Oucho LA, Williams MDA (2019) Challenges and solutions to migrant integration, diversity and social cohesion in Africa. UNU-WIDER available: https://www.wider.unu.edu/sites/default/files/OUCHO,%20Linda%20Adhiambo_paper.Pdf

Pereira LM, Davies KK, den Belder E, Ferrier S, Karlsson-Vinkhuyzen S, Kim H, Lundquist CJ (2020) Developing multiscale and integrative nature–people scenarios using the Nature Futures Framework. People Nat 2(4):1172–1195

Perkel RD (2021) Climate changing minds: reintroducing climate change into national security

Peters LE (2021) Beyond disaster vulnerabilities: an empirical investigation of the causal pathways linking conflict to disaster risks. Int J Dis Risk Reduct 55:102092

Peters K, Dupar M, Opitz-Stapleton S, Lovell E, Budimir M, Brown S, Cao Y (2020) Climate change, conflict and fragility: information and analysis to support programme design scoping for the climate and Resilience Framework Programme (CLARE)

Prentice V, Tao J, Bamanie M, Humphrey D, Joseph-Stanislaus L (2024) Municipal sustainability plans and the inclusion of social equity initiatives. J Soc Equity Public Admin 2(2):1–19

Rajan R (2021) Disaster management, sustainability, and globalization: an integrative planetary approach. Int J Interdiscip Global Stud 16(2):1

Raleigh C, Linke A (2018) Subnational governance and conflict: an introduction to a special issue on governance and conflict. Polit Geogr 63:88–93

Reddy V, Devi MJ, Anbumozhi V (2019) Ensuring food and nutritional security in the face of disasters and climate change: what is the adaptive solution. Towards Resilient ASEAN 1:290–330

Rees WE (2021a) Growth through contraction: conceiving an eco-economy. Real World Econ Rev 96:98–118

Rees WE (2021b) Achieving sustainability: Reform or transformation? In: The Earthscan reader in sustainable cities. Routledge, London, pp 22–52

Rivera FI, Kapucu N (2015) Disaster vulnerability, hazards and resilience. Springer, Berlin

Robins NS, Fergusson J (2014) Groundwater scarcity and conflict–managing hotspots. Earth Perspect 1:1–9

Rodrigues CM, Plotkin SA (2020) Impact of vaccines, health, economic and social perspectives. Front Microbiol 11:1526

Roettinger J (2019) A critical assessment of the link between climate change and violent conflict in sub-Saharan Africa: The case of Darfur

Rufat S, Tate E, Burton CG, Maroof AS (2015) Social vulnerability to floods: review of case studies and implications for measurement. Int J Dis Risk Reduct 14:470–486

Sadiq AA, Tyler J (2017) Agency coordination and cross-sector collaboration in fragile states. In: Oxford research encyclopedia of natural hazard science

Sankar-Gorton L, Lubeck-Schricker M, Hossain R, Iwu CD, Hess J, Levin C (2024) The impact of climate change on public health in humanitarian settings. Gates Open Res 8(71):71

Sarzana C, Melgar A, Meddings G, Laderach P, Pacillo G (2022) Evidence and processes for mainstreaming peacebuilding in climate adaptation efforts. A framework and a safeguard approach for peace-responsive climate action: the climate security sensitiveness tool (CSST). AICCRA working paper

Schierup CU, Likić-Brborić B, Wise RD, Toksöz G (2018) Migration, civil society and global governance: an introduction to the special issue. Globalizations 15(6):733–745

Schmid D, Vlaskamp MC (2024) Public preferences of environmental peacebuilding: the case of forest conservation projects in Colombia. Environ Sec 2(4):568–598

Seddon N, Chausson A, Berry P, Girardin CA, Smith A, Turner B (2020) Understanding the value and limits of nature-based solutions to climate change and other global challenges. Philos Trans R Soc B 375(1794):20190120
Selby J, Dahi OS, Fröhlich C, Hulme M (2017) Climate change and the Syrian civil war revisited. Polit Geogr 60:232–244
Sharpe I, Davison CM (2021) Climate change, climate-related disasters and mental disorder in low-and middle-income countries: a scoping review. BMJ Open 11(10):e051908
Shebbs EU, Njoku R (2016) Resource control in Nigeria-issues of politics, conflict and legality as challenge to development of the Niger delta region. J Good Gover Sustain Develop Afr (JGGSDA) 3(3):32–36
Shivanna KR (2022) Climate change and its impact on biodiversity and human welfare. Proc Indian Natl Sci Acad 88(2):160–171
Smidt HM (2020) United Nations peacekeeping locally: enabling conflict resolution, reducing communal violence. J Conflict Resolut 64(2–3):344–372
Smith TS, Baranowski M, Schmid B (2021) Intentional degrowth and its unintended consequences: uneven journeys towards post-growth transformations. Ecol Econ 190:107215
Subramanian KR (2018) The crisis of consumption of natural resources. Int J Recent Innov Acad Res 2(4):8–19
Sunu NE (2024) Assessing differences in household food insecurity vulnerabilities post-cyclone Idai in Beira, Mozambique
Surjan A, Kudo S, Uitto JI (2016) Risk and vulnerability. In: Sustainable development and disaster risk reduction, pp 37–55
Suski P, Speck M, Liedtke C (2021) Promoting sustainable consumption with LCA-A social practice-based perspective. J Clean Prod 283:125234
Tabish SA, Syed N (2015) Disaster preparedness: current trends and future directions. Int J Sci Res 4(6):227–252
Tagarev T, Ratchev V (2020) A taxonomy of crisis management functions. Sustainability 12(12):5147
Tambovceva TT, Melnyk LH, Dehtyarova IB, Nikolaev SO (2021) Circular economy: tendencies and development perspectives
Tapia MS, San-Blas G, Machado-Allison CE, De Jimenez ML, Carmona AJ (eds) (2022) Food security and food safety challenges in Venezuela. Lausanne: Frontiers Media SA
Tierney K (2012) Disaster governance: social, political, and economic dimensions. Annu Rev Environ Resour 37(1):341–363
Tinazzi I (2024) Water scarcity, migrations and climate change: an assessment of their nexus
Tomlinson C (2022) Green growth or degrowth? The economics of climate change
Tröger D, Braun AC (2024) "Industry impacts more than nature"—Risk perception of natural hazards in more-than-human-worlds. Int J Dis Risk Reduct 2024:104568
Tulane University School of Social Work, Health Effects of Natural Disasters: Safety and Preparedness Resources, March 15, 2018. https://socialwork.tulane.edu/blog/health-consequences-natural-disasters/
Turcsányi RQ (2018) Chinese assertiveness in the south China sea. Power sources, domestic politics, and reactive foreign policy. Springer, Cham
Unfried K, Kis-Katos K, Poser T (2022) Water scarcity and social conflict. J Environ Econ Manag 113:102633
Ünver S (2019) Political economy of pre-civil war Syria
Upadhyay RK (2020) Markers for global climate change and its impact on social, biological and ecological systems: a review. Am J Clim Chang 9(03):159
Van Oijstaeijen W, Van Passel S, Cools J (2020) Urban green infrastructure: a review on valuation toolkits from an urban planning perspective. J Environ Manag 267:110603
Verma S, Rawat A, Awasthi K, Ramakant Mishra A (2024) Assessing interdependency of climate change and food security: an overview. In: Sustainability and health informatics: a systems approach to address the climate action induced global challenge, pp 211–221

Voola SI, Kalyanasundaram PI (2021) Rural poverty and sustainable development goals. In: No poverty. Springer, Cham, pp 902–912

Vučković M, Adams AM (2022) Population and health policies in urban areas. In: International handbook of population policies. Springer, Cham, pp 397–429

Wallace D (2018) Introduction: security and global climate change. In: Climate change, policy and security. Routledge, London, pp 1–40

Wang J, Azam W (2024) Natural resource scarcity, fossil fuel energy consumption, and total greenhouse gas emissions in top emitting countries. Geosci Front 15(2):101757

Waramlah R, Puganeswary T, Navindrakumar K, Rosliza AM, Juni MH (2018) Public health impact of natural disaster: socio-ecological perspective. Int J Public Health Clin Sci 5(2):56–67

World Bank Group (2018) Democratic republic of Congo systematic country diagnostic: policy priorities for poverty reduction and shared prosperity in a post-conflict country and fragile state. World Bank

Yildiz V, Hatipoglu MA, Kumcu SY (2022) Climate change impacts on water resources. In: Water and wastewater management: global problems and measures. Springer, Cham, pp 17–25

Zeitoun M, Abdallah C, Dajani M, Khresat SE, Elaydi H, Alfarra A (2019) The Yarmouk tributary to the Jordan River I: agreements impeding equitable transboundary water arrangements

Zhang Y, Khan I, Zafar MW (2022) Assessing environmental quality through natural resources, energy resources, and tax revenues. Environ Sci Pollut Res 29(59):89029–89044

Zhong S, Yang L, Toloo S, Wang Z, Tong S, Sun X, Huang C (2018) The long-term physical and psychological health impacts of flooding: a systematic mapping. Sci Total Environ 626:165–194

Application of Free Trade Pact for Climate Action: The Agreement Establishing the African Continental Free Trade Area

Semia Sama

Abstract African states continue to face the dual challenges of lifting millions of Africans out of extreme poverty and raising incomes through sustained economic growth. At the same time, these countries are among the most vulnerable to the devastating impacts of climate change, as global warming, rising sea levels, flooding, and desertification threaten lives and livelihoods across the continent. Amid these challenges, Member States of the African Union have negotiated a groundbreaking free trade agreement. The *Agreement Establishing the African Continental Free Trade Area* to facilitate the free movement of people, goods, and services. Climate change is a complex multidimensional problem that demands collective solutions from diverse disciplines and actors at the global, regional, and local levels. The international trade regime can play a vital role in advancing environmental sustainability. Many trade agreements (global, regional, and bilateral) now include substantive environmental chapters that commit parties to addressing issues such as climate change, sustainable forest management, meaningful public engagement, marine pollution, and sustainable fisheries management. This chapter examines weather the *Agreement Establishing the African Continental Free Trade Area* serves as an effective instrument for achieving environmental objectives, including climate change mitigation and adaptation. It concludes by proposing a climate-friendly path forward for the *Agreement Establishing the African Continental Free Trade Area* as its parties continue to deposit their instruments of ratification.

1 Introduction

African states continue to face the pressing challenge of lifting millions of Africans out of extreme poverty and raising incomes through sustained economic growth. At the same time, these countries are among the most vulnerable to the devastating impacts of climate change, as global warming, rising sea levels, flooding,

S. Sama (✉)
Faculty of Law, Lakehead University, Thunder Bay, ON, Canada
e-mail: ssama@lakeheadu.ca

D. Oghenerobor Agelebe (ed.), *Climate Ethics in Africa*, Global Perspectives on Law and the Environment 1, https://doi.org/10.1007/978-3-032-08757-7_5

and desertification threaten lives and livelihoods across the continent existence.[1] Amid these challenges, Member States of the African Union (AU) have negotiated a groundbreaking free trade agreement, the *Agreement Establishing the African Continental Free Trade Area* (AfCFTA). Signed in March 2018 and entering into force in May 2019 following the deposit of the 24th instrument of ratification within the African Commission, the AfCFTA seeks to create a single market for goods and services across 55 African countries. By progressively eliminating tariffs and non-tariff barriers, the agreement aims to deepen regional economic integration and promote sustainable development.[2] If fully implemented, this agreement has the potential to create an integrated market of millions of people and foster inclusive, long-term economic growth across Africa.

Many global, regional, and bilateral trade agreements have included clauses to ensure they do not hinder climate change mitigation, adaptation, and sustainable development efforts.[3] While the literature on the AfCFTA and development/trade abound,[4] no formal studies on the environmental implications of AfCFTA in Africa have been undertaken thus far. This paper aims to fill this gap by examining AfCFTA's role in combating climate change—reducing the sources of greenhouse gases (GHG) (that are primarily driving climate change) and enhancing their sinks (such as forests) to promote environmental sustainability in Africa. This chapter questions whether AfCFTA can be enlisted as an effective economic tool to achieve the African Union's Agenda 2063 goals for sustainable development, good governance, respect for human rights, peace and security, people-driven development, food security and climate-resilient economies.[5] The outcome of this study will provide valuable insights for informing and guiding African governments on how to incorporate substantive environmental provisions in AfCFTA to stimulate collaboration in combatting climate change in Africa.

The paper is organized into five sections. Following this introduction, Section 2 provides background information on Africa's experience with climate change. Section 3 presents an overview of the AfCFTA and how it may either strengthen or undermine the global climate and sustainability agenda. Section 4 offers guidance on how the AfCFTA can be leveraged to promote climate change mitigation and adaptation, and sustainable development across Africa. The central argument is that leveraging the AfCFTA could serve as an effective mechanism for reducing emissions of greenhouse gases and achieving environmental objectives on the African continent. The findings aim to help identify climate-friendly pathways as African Union Members continue to deposit their instruments of ratification, thereby

[1] World Meteorological Organization (2020a, 197).

[2] African Union, *The Agreement Establishing the African Continental Free Trade Area* (AfCFTA).

[3] Protocol Replacing the NAFTA with the USMCA; GATT, Marrakesh Agreement Establishing the World Trade Organization (1994) at 154.

[4] Thuo (2019) 1028 at 1083; Adewopo et al. (2022) 33 at 62; Forere (2018) 43 at 73; (2020) 773 at 792.

[5] AU *Agenda 2063* (2015) at 2.

contributing to the broader goal of developing a more effective global climate change regime. This possibility, therefore, warrants closer examination.

2 Background on Africa's Experience with Climate Change

Why should climate change and sustainable development be essential considerations in the AfCFTA? The science is clear: the impacts of climate change, including extreme weather events, are taking a significant toll on Africa--home to billions of people, including youths, women, older persons, Indigenous peoples, persons with disabilities, and low-income communities—many of whom are disproportionately vulnerable to climate-related risks. The World Meteorological Organization (WMO) studied Africa's State of the Climate in 2019. It concluded that the continent experienced (1) rising temperatures at a rate that is faster than global mean surface temperature; (2) low annual precipitation totals due to continued rainfall deficit and flooding; (3) sea-level increase; (4) tropical cyclones; (5) droughts; (6) flooding; and (7) extreme heat, with temperatures exceeding 45 °C in parts of Mozambique, South Africa, and Zimbabwe.[6] As the report highlights, these climatic changes have severely affected key sectors such as agriculture, food security, and water resources, while also creating favorable conditions for spreading vector-borne diseases, including dengue, malaria, and yellow fever.[7]

In 2018, the Intergovernmental Panel on Climate Change (IPCC) projected that climate-related risks to human health and food security, livelihoods, water supply, and economic growth will increase with global warming of 1.5 °C and surge further to 2 °C. While Sub-Saharan Africa has experienced dramatic consequences of climate extremes, temperature increases are projected to be higher in this region than in the global mean temperature increase.[8] At global warming of 2 °C and 1.5 °C, West and Central Africa (identified as climate-change hotspots with negative impacts from climate change on crop yields and production) are projected to display substantial increases in the number of hot days. According to the report, Southern Africa, especially South Africa, Namibia, and Botswana, are expected to experience the most significant temperature increases. Over Eastern Africa, annual rainfall projections show a robust rise in precipitation, especially in Somalia.[9] Overall, millions of people in Africa will experience the most significant reductions in economic growth at global warming of 2 °C.[10]

Changes in climatic conditions, such as erratic precipitation and rainfall patterns, inexplicable wind speed, a general decrease in annual rainfall volumes, a novel weather pattern with seasonal weather variations, and extreme weather conditions

[6] WMO (n2) 5 at 17.

[7] WMO (n3) at 2.

[8] MR Allen et al. (2018) at 39.

[9] IPCC Special Report (n2) at 197.

[10] IPCC Special Report (n3) 39.

and temperature values, have been experienced in the Eastern Cape in South Africa.[11] Such changes have negatively affected local agriculture, ecosystems, human settlement and sanitation, and livelihoods.[12] If the current climate variability trends persist, South African climate variables will continue to oscillate in seasonal and annual averages and reach extremes by 2100.[13] Although varying future predictions of the monsoonal rainfall (indispensable for rain-fed agriculture) were noted by Sultan and Gaetani in 2016, robust evidence of yield loss, driven by global warming, has emerged in West Africa. They argue in this context that better knowledge of how climate will change in this subregion and how these changes will affect crop productivity is vital to inform policies that may counteract the harsh effects. From their perspectives, further efforts are required to improve the modelling of the monsoon system and best quantify the ambiguity in its changes under a warmer climate.[14]

Agriculture is the pillar of Africa's economy. As specified by the World Bank, the agricultural economy employs 65–70% of Africans and accounts for 30–40% of Africa's GDP.[15] Given its moisture deficits, low capacity to adapt to climate change impacts, comparatively high dependence on rainfed crops, and the fact that about 70% of the continent's poor live in rural areas, with agriculture their most significant economic activity, consequently, renders Africa a "hot spot" for climate change impacts. Projections under the IPCC's Representative Concentration Pathway 8.5 suggest that warming scenarios will disturb agricultural yields and food security.[16] Effectual solar-powered micro-irrigation systems, for instance, can surge farm-level incomes by 5–10 times, enhance yields by up to 300% and cut water usage by up to 90% while offsetting emissions of greenhouse gases by generating up to 250 kilowatts of clean energy.[17] Therefore, hunger and poverty reduction efforts in the agricultural sector are an optimistic approach to reducing the undesirable impacts of climate change and promoting economic growth in Africa.

With the decline in economic growth and welfare losses, Southern Africa, accompanied by sub-Saharan Africa, North Africa, and East Africa, in sequence, will experience the most considerable impacts from climate change.[18] Therefore, concerted efforts at adaptation are needed. These should include introducing temperature-sensitive varieties, diversifying livelihoods and production systems, shifting to sustainable agriculture, including irrigation agriculture, and tackling institutional challenges (poor social and physical infrastructure, absence of access to credit and crop insurance, market imperfections, etc.). Over time, prominence should be given to spreading from agriculture to industry and services.[19]

---

[11] Apraku et al. (2019) 295 at 308.

[12] Apraku 295 at 308.

[13] Apraku 295 at 308.

[14] Sultan and Gaetani (2016, 1262).

[15] The World Bank (2013) at 13.

[16] WMO (2020b) at 3.

[17] AS above.

[18] J Asafu-Adjaye (2014, 49).

[19] As above 23:2 ii at 17–ii 49.

While reviewing climate change impacts and policy developments for West African livestock, fishery, and crop production sectors, Robert Zougmoré et al.[20] If appropriate mitigation and adaptive measures are not implemented, cereal production will reduce by 40% in 2050, future rainfall will decline by 5%, and droughts will increase aridity by 5–8%. They will modify the distribution of fish species.[21] These scenarios will consequently impact millions of people already facing water scarcity, food insecurity, and unhealthy diet crises.[22] With about 411 million people needing a basic drinking water service,[23] Climate change could further degrade the environment and hamper the attainment of the 2015 Sustainable Development Goals by 2030 and the Agenda 2063 (Africa We Want). Hence, developing or reinforcing adaptive mechanisms to address the harmful effects of climate change must be a high priority, with promoting and facilitating trade in environmental goods and services earmarked as an essential opportunity in global and regional trade agreements.

3 Agreement Establishing the African Continental Free Trade Area

Considering the aspirations of Agenda 2063 and conscious of the need to create a secure and expanded market for Africa's goods and services and promote its sustainable economic development, the member States of the African Union established the *Agreement Establishing the African Continental Free Trade Area* (AfCFTA) on March 21, 2018. The AfCFTA came into effect on 30 May 2019, after the 24th instrument of ratification was deposited with the African Union Commission.

This Agreement was supposed to be negotiated and implemented in three phases: Phase I negotiations covered trade in goods and services; Phase II negotiations (2021) addressed intellectual property rights, investment, and competition policy; Phase III negotiations, still to be negotiated, will cover e-commerce. So far, these negotiations have culminated in many protocols outlining the continent's objectives, guiding principles, and modalities for trade liberalisation. The Protocol on Trade in Goods, Protocol on Trade in Services, and Protocol on Rules and Procedures for the Settlement of Disputes are the milestones of Phase I negotiations.[24] AfCFTA envisages the establishment of the African Continental Free Trade Area (a free trade zone in Africa where goods and services move) and a continental customs union in the future. It aims to[25]

---

[20] Zougmoré et al. (2016)

[21] Zougmoré (n2).

[22] FAO; IFAD, UNICEF; World Food Programme and World Health Organization (*2021*, 27).

[23] United Nations Children's Fund and World Health Organization (2022) at 1.

[24] AfCFTA Article 7.

[25] AfCFTA Article 3.

- Create a solo market for goods and services, enabled by the movement of persons to expand the economic integration of Africa and following the Pan African Vision of "An integrated, prosperous and peaceful Africa" enshrined in Agenda 2063;
- Fashion a liberalised market for goods and services via continuous rounds of negotiations;
- Foster and attain gender equality, inclusive and sustainable socio-economic development, and structural transformation of the participating states;
- Improve the competitiveness of the economies of all parties within Africa and the global market;
- Boost industrial development via diversification, agricultural and regional value chain development, and food security;
- Settle the challenges of numerous and overlapping memberships and accelerate the continental and regional integration processes.

To accomplish the objectives delineated above, State Parties have agreed to the following:

- Gradually remove tariffs and non-tariff barriers to trade in goods and open trade in services;
- Cooperate on all trade-related areas, including investment, intellectual property rights, competition policy, and customs matters;
- Establish a dispute settlement mechanism concerning their rights and an institutional framework for implementing and administering the AfCFTA.[26]

Regarding principles, the AfCFTA shall be governed by Most-Favoured-Nation Treatment, National Treatment, reciprocity, transparency and disclosure of information, preservation of the acquis, substantial liberalisation, consensus in decision-making, and variable geometry principles.[27] The AfCFTA builds on and brings together regional economic communities (REC) in the continent, such as the Economic Community of West African States, the Arab Maghreb Union, the Common Market for Eastern and Southern Africa, the Southern African Development Community, the East African Community, the Economic Community of Central African States, and the Sahel Saharan States Community.[28] These RECs constitute the cornerstones to facilitate economic integration between bloc members. According to the United Nations, RECs have a specific role to play in integrating the United Nations Sustainable Development Goals (SDGs) and Africa's Agenda 2063 following their convening power, universal coverage, intergovernmental strength and nature, broad-based cross-sectoral mandate, and experience in mobilising regional consensus on vital international agreements.[29]

The institutional framework for implementing, administering, facilitating, monitoring, and evaluating AfCFTA includes the Assembly, the Council of Ministers, the

---

[26] AfCFTA Article 4.

[27] AfCFTA Article 5.

[28] Government of Cameroon *National Development Strategy 2020–2030* (2022) at 60.

[29] UN SDG *and AU Agenda 2063* (2017) at 6.

Committee of Senior Trade Officials, and the Secretariat. The Assembly, the highest decision-making organ of the African Union, has the exclusive authority to provide strategic guidance on AfCFTA and adopt interpretations of it on the recommendation of the Council of Ministers.[30] This Council, which shall consist of the Ministers for Trade or officials duly designated by a State Party, shall meet twice a year to decide measures needed to promote the AfCFTA's objectives, collaborate with the relevant organs and institutions of the African Union, and oversee the work of all working groups and committees established under the AfCFTA. The decisions of the Council of Ministers, within its mandate, shall be incumbent on member states.[31]

The Committee of Senior Trade Officials (CSTO) shall comprise the Principal or Permanent Secretaries or other officials selected by each Member State. They shall implement the decisions of the Council of Ministers as directed, develop action plans and programs for implementing the Agreement, and ensure the proper functioning of the AfCFTA. In overseeing the implementation of this Agreement, the CSTO may request a Technical Committee to investigate any matter, instruct the Secretariat to assume specific assignments, and undertake any other function that is consonant with this Agreement or as demanded by the Council of Ministers. The CSTO shall meet at least twice yearly and submit its report/recommendations to the Council of Ministers.[32] The Secretariat (temporarily the Commission) shall be an operationally independent institutional body within the African Union system with an autonomous legal personality. Its funds shall come from the overall annual budgets of the African Union, and the Council of Ministers of Trade shall determine the roles and responsibilities of the Secretariat.[33]

As some researchers claim, the African continental free trade area is bound to face insurmountable challenges, including the preponderance of neo-protectionist tendencies, conflicts, economic stagnation, and limited political will to cooperate.[34] Others have maintained that the AfCFTA can improve Africans' livelihoods by increasing intra-African trade and spawning new employment opportunities in an integrated African labour market.[35] Hence, Mude has introduced the concept of "transformative integration." It denotes a more collaborative and accommodative integration designed to transform trade, cooperation, and the socio-economic dynamics of life in the African continent.[36]

The AfCFTA is mandated to intensify the economic integration of the African continent in line with the Pan-Africanist vision of a prosperous, integrated, and peaceful Africa.[37] Yet, questions have been raised regarding the extent to which this Agreement is a purely African project. Given the long-standing capitalist orientation

[30] AfCFTA Article 10.

[31] AfCFTA Article 11.

[32] AfCFTA Article 12.

[33] AfCFTA Article 13.

[34] Nwangwu et al. (2019) 111 at 132.

[35] Lungu (2019) 1.

[36] Mude (2020) 41 at 53, at 42.

[37] AfCFTA Article 3.

of the global political economy—partly characterised by Africa's structural dependence on the developed world,[38] Mude argues that external interference in the trajectory of the economic procedures of the AfCFTA cannot easily be dismissed.[39] While this is a legitimate concern, it must be emphasised that Article 3 (a) of the Agreement, which sets as a core objective the creation of a single market for goods and services to attain an integrated, peaceful and prosperous Africa, underscores a distinctly African character and Agenda. The AfCFTA represents a significant milestone in Africa's integration project [40] It provides an opportunity to leverage the combined economic size of Africa, foster export diversification and industrialisation, coordinate African trade policy and accelerate the digital and green transitions vital to building a more sustainable economy.[41] Described as the latest vehicle through which Africa can galvanise and strengthen economic integration, food security, industrialisation, and international investment and intellectual property regimes, a fully operational AfCFTA will "pitch Africa's new economic order as a competitor, purportedly on equal footing, against other multilateral trading regimes."[42]

In practical terms, the AfCFTA is designed to help curb the growing winds of protectionism and promote the welfare of Africa—the world's poorest and least integrated continent—through trade liberalisation and facilitation, tackling intellectual property rights issues and settling trade disputes.[43] The AfCFTA is comprehensive in its effort to eliminate barriers (tariff and non-tariff) to trade in goods and services in Africa, thereby establishing "a stepping stone to an African customs union and, later, a fully-fledged African economic community."[44] The need to improve the transport infrastructure in Africa to facilitate the ACFTA suggests that the security of such infrastructure (like ports) and the promotion of the security interests of State Parties are very significant aspects of the Agreement.[45]

Regarding concrete benefits, implementing the AfCFTA will usher in deep reforms needed to cut red tape and simplify customs procedures, thereby shaping the African economy and lifting 68 million people from moderate poverty and an additional 30 million from severe poverty.[46] According to the World Bank, actual income gains from AfCFTA's full operationalisation could grow 7% by 2035 (almost US$450 billion).[47] Gathii believes that if all African Union member states ratify, the AfCFTA will oversee "a market of 1.2 billion people and a Gross Domestic Product" of US $2.5 trillion.[48]

[38] Mude (n2).

[39] Mude (n3) 41 at 53, at 43.

[40] Woolfrey et al. (2019) 5.

[41] Songwe et al. (2021) at 13.

[42] Akinkugbe (2020) 141.

[43] Berahab and Dadush (2018, 1).

[44] United Nations Conference on Trade and Development (2016) at 8.

[45] Berahab and Dadush (n2).

[46] World Bank (2020, 5).

[47] World Bank (n2) at 3.

[48] Gathii (2019, 1028).

Seven (Cameroon, Ghana, Rwanda, Kenya, Mauritius, Egypt, and Tanzania) of the 44 countries State Parties to AfCFTA have already started preferential trade on products for which rules of origin have been decided. According to Business in Cameroon, Cameroon made its first shipment (consisting of dried safou, dried pineapple, and ginger tea) under the preferential rate of AfCFTA to Ghana. This was realised after the customs issued the first AfCFTA certificate of origin for its products to Ghana on 6 October 2022.[49] The World Bank estimates Cameroon's income gains due to AfCFTA in 2035 at 8%,[50] with the most significant expected improvements from implementing this initiative anticipated in Cameroon, the Democratic Republic of Congo, Egypt, Nigeria, and Tanzania, with a decline in trade costs of 10 percentage points.[51]

While seen as the vehicle through which the African Union anticipates achieving sustainable economic development,[52] whether AfCFTA is enlisted as an effective tool to achieve Africa's Agenda 2063 goals remains. Can/will this Agreement, through its objectives, institutions, ability to monitor potential trade protectionism, enforcement mechanism, and toolbox of rules, promote greater access to environmental goods and services at a cheaper cost and combat climate change?

In the preamble, AU Member States reaffirm their right to regulate within their territories and their flexibility to achieve legitimate policy objectives in the environment, among other areas. In the context of the environment, "flexibility" denotes the "degree of latitude possible within a given governance structure, its substantive rules, standards, and norms, and its procedural requirements without fundamentally displacing or breaching the governance systems as a whole.[53] This suggests that the AU Member States can modify the content of their environmental obligations under AfCFTA, and regarding the question of whether states should be allowed increased flexibility to achieve ecological goals, Åkermark and Mårsäter say that the "more treaty flexibility is available, the more important it is to have institutionalised mechanisms, bilateral or multilateral, for a continuous reevaluation of the flexibility devices."[54]

In the preamble, Member States acknowledge their existing rights and responsibilities under international agreements they have signed and ratified. African Union Member States have ratified numerous multilateral environmental agreements (MEAs), including global climate agreements with near-universal membership.[55] Member States also reached a deal, under Article 3, to create a single market to

[49] Business in Cameroon *Cameroon now exports goods to Ghana per AfCFTA tariffs*.

[50] World Bank at 4.

[51] World Bank at 37.

[52] UNCTAD (2016) at 1.

[53] Craig et al. (2017, 3).

[54] Åkermark et al. (2005, 509).

[55] United Nations Treaty Collection (1992) at 107 (1992); Adoption of the Paris Agreement1 at CP.21, the Convention on Biological Diversity (1760) UNTS at 79; Vienna Convention for the Protection of the Ozone Layer, 1513 UNTS at 293.

deepen Africa's economic integration, following the Pan-African Vision of "an integrated, prosperous and peaceful Africa" protected in Agenda 2063. Agenda 2063 is Africa's shared strategic framework for sustainable development and inclusive growth, with its Aspiration 1—"a prosperous Africa based on inclusive growth and sustainable development."[56]

AU Member states aspire that, among others, Africa's ecosystems are healthy, protected, and valued, with climate-resilient communities and economies.[57] As can be seen, Article 3 of the Agreement aims to ensure that, among others, Aspiration 1 of Agenda 2063 is realistically achieved in Africa. According to Article 3, Members must ensure that environmental sustainability and trade liberalization goals are at the same level and cooperate to promote an open regional trading system that would result in sustainable development in all African countries to address the problem of climate change better.

Besides these preambular provisions, Article 26 (Part VIII) of AfCFTA on "Exceptions" outlines specific instances in which Member States may be exempted from AfCFTA rules. Some exceptions provide space for shielding environmental and sustainability measures from the stringency of AfCFTA and its Protocols only if certain conditions are met. First, it stipulates that "nothing in this Protocol shall be construed as preventing the adoption or enforcement of measures by any State Party necessary to protect human, animal or plant life or health." Despite the failure to explicitly mention the environment, sustainability, or climate change mitigation, the Article 26 (Part VIII) exception can help protect fauna, flora, and environmental media, including air, land, and water.

Similarly, Article 26 (Part VIII) of the Agreement applies to measures concerning the conservation of non-renewable natural resources if made effective with restrictions on internal production or consumption. Similar language is found in GATT Article XX (b) and (g), and the World Trade Organization's Appellate Body has made it clear that exhaustible natural resources include living and non-living resources.[58] Some researchers have defined "exhaustible natural resources" as natural materials with an exceptionally slow growth rate relative to their consumption rate, which puts them at risk for depletion. Metals, coal, oil, and natural gas are exhaustible natural resources.[59] They are synonymous with non-renewable resources. Consequently, decisions about their exploration and exploitation will influence the well-being of current and future generations.

Besides general exceptions and preambular language, AfCFTA missed many opportunities to copy progressive sustainability measures in regional and global trading systems. AfCFTA does not refer to the enforcement of MEAs, including provisions committing Parties to adopt and implement laws and other necessary

---

[56] African Union Goals & Priority Areas of Agenda 2063.

[57] African Union Commission *Agenda 2063: The Africa We Want* (2015) at 2.

[58] World Trade Organization 'United States—Import Prohibition of Certain Shrimp and Shrimp Products' *Report of the Appellate Body* 127 at 131.

[59] Abu-Kalla et al. (2020) at 20, at 2.

measures to fulfill their respective obligations under MEAs and to essential principles found in other RTAs such as environmental impact assessment, environmental democracy, environmental cooperation, and sustainable development.[60]

The Agreement notably does not explicitly refer to sustainable development and climate change, as opposed to regional trade agreements concluded by the EU in 2019–2020.[61] The AfCFTA does not include an environment chapter or provisions stating commitments to strengthen government capacity to promote sustainable forest management, such as legally harvested timber and forest products, even though this is found in the USMCA agreement.[62] Considering the significant role of forests in climate change mitigation and the rapid degradation of forests in Africa and other tropical nations,[63] it is disappointing that the AfCFTA failed to duplicate provisions from other regional trade agreements.

While the Paris Agreement's heart is the nationally determined contribution (NDC), its achievement by State Parties is not a legally binding commitment.[64] This means that Parties are not legally obligated to pursue measures to achieve their NDCs. Because Parties do not have binding obligations, a party could fall short of its NDC without breaching any international legal duty.[65] The African Union failed to address this weakness—it could not include an obligation in the AfCFTA to "effectively implement" NDCs, including engaging in best efforts to realize the goals meted out in their NDCs even though this is found in most EU's free trade agreements. Regarding climate change provisions, about 64 RTAs include one or more provisions referring explicitly to global warming, climate change, creating a low-emission economy or reducing greenhouse gases (GHG).[66] So, how can AfCFTA be amended to promote climate change mitigation and sustainable development in the African continent?

4 Recommendations

There has been a discernible upward trend in the incorporation of substantive environmental provisions in global and regional trade agreements, such as those that refer to ecological cooperation, public participation, and dispute settlement. Indeed, approximately 97% of all regional trade agreements notified to the World Trade Organization (WTO) contain at least an environmental provision.[67] Against this backdrop,

[60] Article 24.8 AfCFTA.

[61] Monteiro et al. (2021, 8).

[62] Article 24.23 AfCFTA.

[63] Semie Sama (2021, 2).

[64] Paris Agreement Article 4 (2).

[65] Rajamani and J Brunnee (2017, 542).

[66] JA Monteiro et al. at 9.

[67] World Trade Organization, 1.

the AfCFTA, as the analysis indicates, stands out for its lack of explicit environmental or sustainability provisions within its core framework.

Article 27 of the Agreement provides that Member States of the AU shall review the AfCFTA every 5 years following its entry into force. The purpose of this periodic review is to ensure its continued effectiveness, promote deeper integration, and enable adaptation to advance regional and international development. Complementing this review mechanism, Article 28 permits any State Party to propose amendments to the Agreement. Such proposals must be based on the experience acquired and progress achieved during the implementation of AfCFTA. Amendment proposals are to be made to the Depositary which circulates them to all State Parties. Thereafter, the relevant committees and sub-committees shall present the proposed amendments to the Council of Ministers for consideration, and the Assembly shall adopt the proposed amendments.

To begin with, I propose that the African Union strengthen the environmental and sustainability standards embedded within the AfCFTA. If trade liberalization under the AfCFTA is to enhance, rather than undermine, Africa's ecological integrity, State Parties should commit to trade-related environmental provisions in relevant sections of the Agreement. As explained by Van and Signe, the extent to which a trade agreement, bilateral, regional, or global, positively affects environmental sustainability partly depends on the specific provisions set out in the agreement.[68] Several multilateral trade agreements, including WTO agreements, the EU FLEGT Plan, and the Canada-United States-Mexico Agreement, contain chapters or clauses dedicated to sustainable development. These agreements comprise several environmental provisions with standards that seek to reinforce the environment-climate nexus. As the following section will demonstrate, trade agreements, through their institutions, can serve as a critical enabling framework for environmental protection at bot regional and global levels..

4.1 The World Trade Organization Agreements

The WTO agreements are also the Final Act of the 1986–1994 Uruguay Round of trade negotiations. They cover goods, services, and intellectual property and outline the liberalization principles and exceptions, including the General Agreement on Tariffs and Trade (GATT) (for goods), the General Agreement on Trade in Services (GATS) (for services), and the Trade-Related Aspects of Intellectual Property (TRIPS) (for intellectual property).[69] To help achieve environmental protection goals, GATT has included Article XX on enshrined General Exceptions, which seeks to ensure, among other things, that environmental measures are not applied arbitrarily and are not used as disguised protectionism.[70]

---

[68] van der Ven and Landry (2021)

[69] GATT.

[70] GATT.

Additionally, the WTO Agreement on Technical Barriers to Trade and the WTO Agreement on the Application of Sanitary and Phytosanitary Measures provides an opportunity for WTO members to adopt regulatory measures to protect the environment and advance a green economy while imposing disciplines to ensure such measures are not unnecessary restrictions on international trade.[71] Rules such as the WTO Agreement on Subsidies and Countervailing Measures leave member states with policy space for providing subsidies that distort global trade. At the same time, TRIPS offers a framework for applying the intellectual property system to promote access to and dissemination of green technologies, provides policy space to encourage public interest in sectors of significance to socio-economic and technological development, and specific incentives for technology transfer and exclusions of environmentally damaging technologies from intellectual property (IP) protection.[72]

To reiterate Principle 12 of the Rio Declaration,[73] WTO member states further recognized in the 2001 WTO Doha Ministerial Declaration that while promoting the liberalization of trade policies, State Parties should be allowed to take environmental measures, such as actions to protect humans and biodiversity at appropriate levels. However, they agreed that such measures should not be applied in ways that would promote unwarrantable discrimination between member states where the exact requirements prevail or a hidden restriction on international trade and are otherwise following the provisions of the WTO Agreements.[74] Therefore, on the one hand, WTO agreements offer policy space for Member States to adopt trade-related measures at the regional and national levels to advance genuine environmental protection goals while ensuring that such measures are not arbitrarily applied.

4.2 *The European Union FLEGT*

In 2003, to help increase sustainable and biodiverse forest coverage worldwide, European Union (EU) Member States adopted the Forest Law Enforcement, Governance, and Trade (FLEGT) Action Plan and the European Union Communication on Stepping up EU Action to Protect and Restore the World's Forests, as one of five priorities of the EU to encourage products from deforestation-free supply chains,[75] Including countries with the highest global deforestation rates, the FLEGT aims to address illegal logging by supporting timber-producing countries, promoting/ supporting trade in legal timber products, public procurement policies, private-sector initiatives, investment and finance, using legislation, and addressing conflict.[76]

[71] GATT

[72] GATT.

[73] Rio Declaration on Environment and Development (1992).

[74] GATT.

[75] European Commission *EU Communication (2019).*

[76] European Commission (n2).

Among other measures to fight illegal trade in forest products was the introduction, under the FLEGT, of the EU Timber Regulation and Voluntary Partnership Agreements (VPA).[77] VPAs define a Timber Legality Assurance System (TLAS), which, once wholly operational, permits the issuance of FLEGT licenses for timber consignments, offering green lane access to the EU market.[78] The EU Timber Regulation obliges forest-exporting nations that place timber, including timber products, on the EU market to exercise due diligence. This includes adopting rules to curtail the risk of importing unlawfully harvested timber into the EU.[79] The overarching goal of the FLEGT Action Plan is to obligate countries that have signed a VPA with the EU to ensure that only legally harvested timber and timber products enter the EU.[80]

Regarding its effectiveness, a 2016 evaluation of the FLEGT Action Plan indicates that it is relevant to addressing illegal logging, raising awareness, contributing to forest governance globally, and reducing the demand for unlawful timber in the EU. Its legally binding nature differentiates this action plan from other global forest agreements that provide UN Member States with a framework for fostering sustainable forest management. Unlike the Forest Instrument[81] and the Forest Principles,[82] The VPA, a key component of the FLEGT Action plan, is a legally binding trade agreement between the European Union and a timber-producing country outside the EU.[83] The VPAs are lawfully binding upon ratification.

The EU FLEGT is a regional agreement that has been leveraged to promote the expansion of trade and foster sustainable development in the world. For example, it was reported in 2021 that this action plan would have positive environmental and socio-cultural impacts, including promoting environmental democracy, a combination of three rights: the right to access environmental information, the right to participate in decision-making, and the right to access justice). Environmental democracy is crucial in addressing global environmental problems.[84] According to the United Nations, data are essential for designing short-term and long-term responses and recovery measures, particularly for the sustainable management of forests.[85]

To further prevent the importation of deforestation into the EU and improve forest governance and sustainable forest management, which are crucial for reversing biodiversity declines and enhancing resilience to climate change,[86] The EU adopted a

---

77 Forest Law Enforcement.

78 FLEGT.

79 EU Regulation No. 995 of the European Parliament and of the Council of 20 October 2010.

80 Sama 1 at 26, at 13.

81 UNGAResolution adopted by the General Assembly on 22 December 2015.

82 UN at 26 vol. III.

83 FLEGT.

84 Sama above.

85 United Nations *The Global Forest Goals Report 2021: An overview of progress* (2021) at 4.

86 United Nations *Making Good on the Glasgow Climate Pact: A Call to action achieve one gigatone of emissions reductions from forests by 2025* (2022) at 5.

FLEGT Work Plan 2018–2022.[87] Even so, a few researchers have challenged some of the positive assessments of the EU FLEGT Action Plan. For instance, it is argued that the FLEGT Action Plan is an EU initiative that is transported to other regions, which could enhance the likeliness of different interpretations.[88] And that VPA implementation still needs to secure sustainable forest management or promote social equity in forest-dependent communities in Ghana.[89]

4.3 Canada–United States–Mexico Agreement

The North American Free Trade Agreement, which entered into force on January 1, 1994, was amended according to Article 2202, resulting in the Agreement between the United States of America, the United Mexican States and Canada (USMCA). The Protocol replacing NAFTA was signed on November 30, 2018, and became effective on July 1, 2020. The central aim of Chapter 24 (Environment) is to encourage mutually supportive trade and environmental policies and practices, high levels of environmental protection, effective environmental law enforcement, and the capacities of State Parties to address trade-related environmental problems in the furtherance of sustainable development.[90]

Chapter 24 includes provisions on enforcing environmental laws, protecting the ozone layer, the marine environment from pollution, air quality, trade, terrestrial and aquatic biodiversity, including their conservation and sustainable management, and ecosystem services. Under paragraph 24, for instance, Parties acknowledge the significance of trade and investment in clean technologies, among other environmental goods and services, in advancing economic and environmental performance, contributing to green jobs and growth, and promoting sustainable development.[91] Parties have further agreed to cooperate in facilitating and liberalizing global trade in environmental goods and services.

Conscious of forests' role in carbon storage, soil stabilization, and water purification, State Parties commit to fighting illegal logging, trading in illegally harvested timber, promoting trade in legally harvested timber and timber products, and strengthening capacity and frameworks to enhance sustainable forest management. Given their socioeconomic and environmental values, Parties also commit to combat illegal take and trade in fauna and flora resources by exchanging information and experiences on related issues and implementing resolutions to fight their unlawful taking and trading.[92] This also includes measures to eliminate subsidies that can promote

[87] European Commission *EU Communication (2019).*

[88] Heeswijk and Tunhout (2013, 7).

[89] Hansen et al. (2018, 82).

[90] Agreement between the United States of America, the United Mexican States, and Canada 7/1/20 Article 24.2.

[91] USMCA Article 24.24.

[92] USMCA Article 24.22.

overfishing and working in other trade forums to strengthen international rules on providing fishing subsidies and enhancing transparency.[93]

The United Nations reported 2016 that trade policy could impact ecosystem protection, ensure satisfactory nutrition, and enable access to essential treatments. Accordingly, environmental provisions in trade agreements are a pertinent example of how sustainable development can be integrated into trade policy at the global and regional levels.[94] Another study by the World Trade Organization revealed that international trade, including trade in sustainable goods, can create significant economic opportunities and make available the resources needed to implement other development targets in the environmental and social sphere.[95] Both findings suggest that international trade can contribute to the realization of SDG 13 (take urgent action to combat climate change and its impacts). Thus, it can serve as a groundwork for building regional and global policies for climate change and sustainable development. However, for state parties to regional and international trade agreements to fully reap trade benefits, adopting measures aiming to mainstream environmental and sustainable development strategies into trade agreements is essential.

In addition to dedicating a standalone chapter to climate change, environmental protection, and sustainable development, the AU could mainstream environmental provisions in other chapters of the Agreement, such as the Sanitary and Phyto-Sanitary (SPS) Measures chapter. One of the AU's SPS Policy Framework objectives is to increase investment in SPS systems by enhancing opportunities to expand the intra-African trade of animals, plants, and food through reinforced public–private cooperation, resource mobilization, and awareness raising.[96] In the SPS Annex of AfCFTA, Article 4 affirms that State Parties shall facilitate trade while safeguarding animal plants, human life, or health in the territory of State Parties and ensuring that the implementation of SPS measures does not become unjustifiable barriers to trade.[97] To ensure that trade liberalisation advances a sustainable agricultural system, the AU could further refine these provisions to explicitly safeguard each member's right to protect against risks to public health, food safety, or animal or plant life. At the same time, such measures should be required to remain transparent, science-based, and non-discriminatory, so as to prevent the creation of redundant or unjustifiable sanitary and phytosanitary trade restrictions. Embedding sustainable principles within the SPS framework would thus align agriculture with long-term ecological resilience and food security objectives.

---

[93] USMCA Article 24.21.

[94] United Nations Conference on Trade and Development (2016) at 6.

[95] World Trade Organization *Mainstreaming trade to attain the Sustainable Development Goals 2.*

[96] African Union *Sanitary and Phytosanitary (SPS) Policy Framework for Africa* (2019) at 16.

[97] African Union *Annex 7 Sanitary and Phytosanitary measures.*

5 Conclusion

In summary, this chapter examined whether the Agreement Establishing the African Continental Free Trade Area (AfCFTA) can effectively advance sustainability objectives, including climate change mitigation and adaptation. The African Union has lauded the AfCFTA as a landmark initiative capable of accelerating intra-African trade and boosting Africa's position in the global economy by strengthening the continent's collective voice and policy space in global trade negotiations. Although the AFCFTA represents a significant step toward the long-sought liberalization and expansion of trade and development in Africa, its contribution to environmental protection and sustainability remains limited. The AfCFTA missed key opportunities to incorporate progressive sustainability measures commonly found in other regional and global trading frameworks, as it failed to make explicit reference to climate change and sustainable development explicitly.

It therefore remains uncertain whether African leaders will take concrete steps to include more substantive environmental provisions in the AfCFTA, such as those addressing climate change, sustainable forest management, sustainable fisheries management, environmental goods and services, environmental cooperation, access to environmental information and public participation, and dispute settlement. Leveraging the AfCFTA to promote collaboration on climate change mitigation, adaptation, and sustainable development in Africa, while simultaneously strengthening regional and global trade, represents both a plausible and strategically sound approach.

References

Abu-Kalla H et al (2020) Hoard or exploit? Intergenerational allocation of exhaustible natural resources. Energies 2020:13

Adewopo A, Oriakhogba D, Okorie C (2022) Negotiating the intellectual property protocol under the agreement establishing the African continental free trade area: priorities and opportunities for Nigeria. Law Develop Rev 15:1

African Union, Annex 7 sanitary and phytosanitary measures. https://au-afcfta.org/wp-content/uploads/2022/01/Annex-7-to-the-CFTA-Agreement.pdf

African Union, Goals and priority areas of agenda 2063. https://au.int/en/agenda2063/goals

African Union, Sanitary and phytosanitary (SPS) policy framework for Africa (2019) 16. https://au.int/sites/default/files/documents/40036-doc-au_sps_policy_framework_-_full_document.pdf

African Union, The agreement establishing the African continental free trade area, Kigali, March 2018

African Union Commission (2015) Agenda 2063: the Africa we want. African union commission

Agreement Between the United States of America, the United Mexican States, and Canada 7/1/20 (USMCA)

Åkermark Sia, Spiliopoulou and Mårsäter, Olle, "Treaties and the Limits of Flexibility" (2005) 74 Nordic Journal of International Law

Akinkugbe OD (2020) Dispute settlement under the African continental free trade area agreement: a preliminary assessment. Afr J Int Compar Law 28:138

Allen MR et al (2018) 'Framing and Context' in Masson-Delmotte and others, Global Warming of 1.5 °C. An IPCC Special Report on the impacts of global warming of 1.5 °C above pre-industrial levels and related global greenhouse gas emission pathways in the context of strengthening the global response to the threat of climate change, sustainable development, and efforts to eradicate poverty (Summary for policymakers). IPCC, Tokyo

Apraku A, Moyo P, Akpan W (2019) Coping with climate change in Africa: an analysis of local interpretations in Eastern Cape, SA. Develop Southern Afr 36:3

Asafu-Adjaye J (2014) The economic impacts of climate change on agriculture in Africa. J Afr Econ 23:2

Berahab R, Dadush U (2018) Will the African free trade agreement succeed? OCP Policy Center, Policy Brief: 18/10, April

Business in Cameroon "Cameroon now exports goods to Ghana, per AfCFTA tariffs" Cameroon now exports goods to Ghana, per AfCFTA tariffs—Business in Cameroon

Craig RK et al (2017) Balancing stability and flexibility in adaptive governance: an analysis of tools available in US environmental law. Ecol Soc 22:2

European Commission, EU Communication (2019) on stepping up EU action to protect and restore the world's forests. http://ec.europa.eu/environment/forests/eu_comm_2019.html

FAO, IFAD, UNICEF, WFP and WHO (2021) The state of food security and nutrition in the world 2021: transforming food systems for food security, improved nutrition, and affordable healthy diets for all. FAO and UNICEF, Rome

FAO-EU FLEGT Programme, FLEGT action plan and EU deforestation regulation. https://www.fao.org/in-action/eu-fao-flegt-programme/background/context/ru//

Forere MA (2018) The agreement establishing the African continental free trade area: Will it spur foreign direct investment in Africa? Acta Jurid 2018(1):43–73

Gathii JT (2019) Agreement establishing the African continental free trade area. Int Legal Mater 58:5

General Agreement on Trade and Tariffs, Marrakesh Agreement Establishing the World Trade Organization, April 15, 1994, 1867 UNTS

Government of Cameroon (2022) NDS30: National Development Strategy 2020–2030: For structural transformation and inclusive development Government of Cameroon. Government of Cameroon, Yaoundé

Hansen CP, Rutt R, Acheampong E (2018) Experimental or business as usual? Implementing the European Union Forest Law Enforcement, Governance and Trade (FLEGT) Voluntary Partnership Agreement in Ghana. For Policy Econ 2018:96

Lungu I (2019) A fresh chance for Africa's youth: market efforts of the African continental free trade area (AfCFTA). GIZ-AU briefing paper

Monteiro J-A, Chobanova S, Ramos D (2021) Trade and climate change: climate change in regional trade agreements, information brief no. 2, virtual trade dialogues with business—trade 4 climate, October 2021: Background Note 8

Mude T (2020) The African continental free trade agreement and intra-African trade. Afr Insight 50:1

Ndonga D, Laryea E, Chaponda M (2020) Assessing the potential impact of the African continental free trade area on least developed countries: a case study of Malawi. J Southern Afr Stud 46(4):773–792

Non-Legally Binding Authoritative Statement of Principles for a Global Consensus on the Management, Conservation and Sustainable Development of All Types of Forests UN Doc. A/CONF/151/26 (vol. III)

Nwangwu C et al (2019) Regionalism in world politics: interrogating the relevance of the Economic Community of West African states in global political economy. Central Eur J Int Sec Stud 13:2

Protocol Replacing the North American Free Trade Agreement with the Agreement Between the United States of America, the United Mexican States and Canada

Rajamani L, Brunnee J (2017) The legality of downgrading nationally determined contributions under the Paris agreement: lessons from the US disagreement. J Environ Law 29:3

Regulation (EU) No 995/2010 of the European Parliament and of the Council of 20 October 2010 laying
Restore the World's Forests. https://ec.europa.eu/environment/forests/eu_comm_2019.htm
Rio Declaration on Environment and Development, UN Doc. A/CONF.151/26. Rev.1 13 June, 1992
Sama SM (2021) Strengthening the role of forests in climate change mitigation through the European union forest law enforcement, governance and trade action plan. J Environ Law Policy 1:3
Songwe V, Macleod JA, Karingi S (2021) The African continental free trade area: a historical moment for development in Africa. J Afr Trade 8:12–23
Sultan B, Gaetani M (2016) Agriculture in west Africa in the twenty-first century: climate change and impacts scenarios, and potential for adaptation. Front Plant Sci 7:1262
The World Bank (2013) Unlocking Africa's agricultural potential, Washington, DC
UNCTAD (2016) African continental free trade area: advancing pan-African integration, Some Considerations. United Nations, Geneva
United Nations (2016) Trading into sustainable development: trade, market access, and the sustainable development goals. UNCTAD, New York
United Nations (2017) The sustainable development goals and agenda 2063. UN Women, Nairobi
United Nations (2021) The global forest goals report 2021: an overview of progress. United Nations Forum on Forests Secretariat, New York
United Nations (2022) Making good on the Glasgow climate pact: a call to action, aims to achieve one gigaton of emissions reductions from forests by 2025. UN Environment
United Nations Children's Fund and World Health Organization (2022) Progress on drinking water, sanitation, and hygiene in Africa 2000–2020: Five years into the SDGs. UNICEF and WHO, New York
United Nations Framework Convention on Climate Change 1771 UNTS 107, 9 May 1992; Adoption of the Paris Agreement1/CP.21, FC/CP/2015/L.9/Rev.1 Annex; the Convention on Biological Diversity (CBD) 1760 UNTS 79, 5 June 1992; Vienna Convention for the Protection of the Ozone Layer, 1513 UNTS 293, 22 March 1985
United Nations General Assembly (2015) Resolution adopted by the general assembly on 22 December 2015 United Nations Forest instrument, 70/199, A/RES/70/199, Seventieth session Agenda item 20
van der Ven C, Signe L (2021) Greening the AfCFTA: it is not too late. Africa Growth Initiative, Policy Brief
van Heeswijk L, Tunhout E (2013) The discursive structure of FLEGT (Forest Law Enforcement, Governance and Trade): the negotiation and interpretation of legality in the EUA and Indonesia. For Policy Econ 32:6–13
Woolfrey S, Apiko P, Pharatlhatlhe K (2019) Nigeria and South Africa: shaping prospects for the African continental free trade area. Political Economy Dynamics of Regional Organisations in Africa. Discussion paper no. 242
World Bank (2020) The African continental free trade area: economic and distributional effects. The World Bank, Washington, DC
World Meteorological Organization (2020a) State of the climate in Africa 2019. WMO, Geneva
World Meteorological Organization (2020b) State of the climate in Africa 2019. WMO, Geneva, 5–17; IPCC Report, 2018
World Trade Organization, United States—import prohibition of certain shrimp and shrimp products, report of the appellate body, WT/DS58/AB/R
World Trade Organization, Trade and climate change, information brief no. 2, p 1. https://www.wto.org/english/news_e/news21_e/clim_03nov21-2_e.pdf
World Trade Organization, Mainstreaming trade to attain the sustainable development goals, 2. https://www.wto.org/english/res_e/booksp_e/sdg_e.pdf
Zougmoré RB et al (2016) Toward climate-smart agriculture in West Africa: a review of climate change impacts, adaptation strategies and policy developments for the livestock, fishery, and crop production sectors. Agric Food Sec 5:26

AfCFTA: A Catalyst for Green Tourism in Africa

Collins Odote and Elizabeth Gitaru Mitaru

Abstract Green Tourism and Sustainable Development have recently become buzzwords in national and international forums. The African Continental Free Trade Area (AfCFTA) is poised to reshape Africa's economic landscape, fostering regional integration and boosting intra-African trade. However, to truly harness the potential of this historic agreement, it is essential to consider the role of green tourism as a catalyst for sustainable development. This article explores the synergies between the AfCFTA and green tourism, highlighting their interconnectedness and advocating for the integration of sustainable practices within the trade framework. The article advocates for the inclusion of green tourism within the AfCFTA, highlighting its potential to drive responsible and inclusive economic growth, create employment opportunities, and empower local communities. By prioritizing green tourism, African nations can address the neglected aspects of sustainable development, reinforcing solidarity and demonstrating a commitment to a sustainable and environmentally conscious future. This support is crucial to ensure that the agreement's potential benefits are fully realized, contributing to the development and growth of green tourism on the continent. The AfCFTA agreement has significant implications for green tourism in Africa. The chapter emphasizes the necessity of proactive policy formulation, practical strategies, and the diligent implementation of these measures by governments to maximize the positive impact of the AfCFTA on the tourism sector. It acknowledges that the AfCFTA is an ongoing project that requires sustained commitment and collaboration to achieve its full potential in Africa.

C. Odote (✉) · E. G. Mitaru
Faculty of Law, University of Nairobi, Nairobi, Kenya
e-mail: Collins.odote@gmail.com

D. Oghenerobor Agelebe (ed.), *Climate Ethics in Africa*, Global Perspectives on Law and the Environment 1, https://doi.org/10.1007/978-3-032-08757-7_6

1 Introduction

International economic law, a branch of Public International Law, typically emphasizes International Trade and International Investment law as its primary pillars.[1] The pivotal role that International Environmental law plays has often been overlooked. Sustainable development and tourism are now the new kid on the block, and scholars, practitioners, and students are usually oblivious that a golden thread runs between trade and the environment. The twenty-first century has placed many nations between a rock and a hard place as many countries and people realize that talking about the environment is the missing piece to the trade puzzle. At this point, we must fathom why deep and informative discussions on the environment are still not driving the point home in the twenty-first century.

The history of international trade in Africa is a tale of exploitation and resilience. From the ancient trade routes to the challenges of colonization and the pursuit of economic integration, Africa has endured numerous transformations.[2] While facing persistent obstacles, the continent's commitment to regional cooperation, infrastructure development, and diversification of its trade has paved the way for a brighter future.[3]

Africa's engagement in international trade can be traced back to ancient times when the continent's geographical position facilitated trade with other regions. The trans-Saharan trade routes, established as early as the ninth century, linked North Africa to the Sahel and Sub-Saharan regions, facilitating the exchange of goods, ideas, and cultural influences.[4] Precious commodities such as gold, salt, ivory, and enslaved people were transported across the vast desert, fostering economic growth and cross-cultural connections.[5]

The nineteenth century witnessed the scramble for Africa by European powers, leading to the colonization of vast territories.[6] This period profoundly impacted international trade in Africa, as colonial powers exploited the continent's resources for their economic gains.[7] European nations established trading posts, monopolized key industries, and imposed trade policies favouring their interests. The extraction of raw materials and the introduction of cash-crop agriculture transformed Africa into a supplier of primary goods to the global market.[8]

Following the wave of decolonization in the mid-twentieth century, African nations regained their political independence. However, they faced significant challenges in reshaping their economies and redefining their roles in international

[1] Herdegen (2016, 3).

[2] Mazrui (1986), Gordon (2008), Mbembe (2014) and Ngwena (2018).

[3] Luke (2019).

[4] Gordon (1997, 98) and Hegel (1956).

[5] JFA. Ajayi '30.

[6] Scholvin (2015, 59).

[7] As above.

[8] Smith (1987) at 1–4.

trade.[9] Many countries promoted industrialization and diversified their trade portfolios to achieve economic self-reliance and reduce dependence on primary exports. Regional economic integration became a key strategy to enhance trade within Africa.[10] Organizations such as the Economic Community of West African States (ECOWAS), the Common Market for Eastern and Southern Africa (COMESA), and the Southern African Development Community (SADC) were established to promote intra-regional trade, harmonize policies, and foster economic cooperation. These efforts aimed to create more extensive and more competitive markets, attract foreign investment, and facilitate the movement of goods and services across borders.

From this rich history, we realize that much-concerted effort placed on international trade is an offspring of the injustices the Africans suffered. This injustice and classification of Africa as a dark continent has made the peoples of Africa more concerned with making profits as opposed to preaching the importance of Sustainable Development. More effort needs to be put into the area of green tourism. Green tourism can be defined as a form of tourism that considers the triple bottom line of sustainability—environmental, socio-cultural, and economic aspects. It aims to minimize negative impacts on natural resources, promote biodiversity conservation, preserve cultural heritage, and contribute to local community development. By adopting sustainable practices, green tourism seeks to balance visitor satisfaction and host destinations' long-term well-being.

The African continent has witnessed remarkable progress in international trade in recent years. The establishment of the African Continental Free Trade Area (AfCFTA) is a testament to this advancement. The AfCFTA, driven by the desire to enhance intra-African trade, holds great promise for economic integration, development, and prosperity across the continent.[11] This ambitious initiative seeks to create a single market encompassing 55 African countries, facilitating the movement of goods, services, and people and fostering deeper regional cooperation.

The rationale behind the AfCFTA is multi-faceted. Firstly, it is envisioned as a catalyst for economic growth, industrialization, and diversification. By promoting trade and investment, the AfCFTA aims to unlock Africa's potential, stimulate innovation, and create employment opportunities.[12] It is expected to bolster intra-African trade, which historically has been relatively low compared to other regions. By removing barriers, harmonizing trade rules, and streamlining customs procedures, the AfCFTA seeks to increase trade volumes, expand markets, and attract foreign direct investment.[13]

Furthermore, the AfCFTA has made efforts to create a critical link between trade, sustainable development and environmental governance. With its three pillars of economic growth, social inclusion, and environmental protection, sustainable development is central to the AfCFTA's agenda. The AfCFTA aims to ensure that

9 Dokotum (2020).

10 Ababio (2020, 9).

11 The World Bank (ed) (26 January 2023).

12 Olontoruba (2023).

13 United Nations (2012) at 29–37.

trade contributes to inclusive and environmentally responsible growth by integrating sustainable development principles into trade policies and practices. This includes promoting sustainable production and consumption patterns, enhancing resource efficiency, and protecting biodiversity and ecosystems.

The AfCFTA also recognizes the importance of environmental governance in achieving sustainable development objectives. Environmental governance entails the establishment of effective policies, regulations, and institutions to manage natural resources, prevent pollution, and address environmental challenges. The AfCFTA allows African countries to strengthen their ecological governance frameworks, foster cooperation on environmental issues, and promote adoption of sustainable practices across sectors. Above all, African nations need to do a lot in terms of green tourism and push for more binding resolutions and policies.

This article exhaustively addresses the topic of the African Continental Free Trade Area as a catalyst for green tourism from an environmental lens. The article follows the proceedings before forming the African Continental Free Trade Area and provides that synergies exist between it and Green Tourism.

The article then shifts gears and examines Green Tourism and Trade status during the pandemic by examining countries such as Rwanda and South Africa. It then argues and justifies why green tourism should be an enigma in the African Continental Free Trade Area. The article concludes that if Green Tourism is sufficiently considered, it will tighten a few loose nuts that the leaders in Kigali forgot. Then, it would have been the best hallmark for solidarity in the economic arena.

2 The African Continental Free Trade Area

In March 2018, all roads leading to Kigali, Rwanda, were imprinted on the minds of Africans for a historic moment. The African States, 54 out of 55, agreed to sign the agreement Establishing the African Continental Free Trade Area. As much as it seems to be the first 'solidarity' ever witnessed in the African area, much faith needs to be given to the history that shaped and will shape what Africans will be enjoying in the following decades. The primordial stages of the African Continental Free Trade Area began in the early 1990s when the Abuja Treaty Establishing the African Economic Community (Abuja Treaty)[14] and the 2000 Constitutive Act of the African Union[15] Established the legal basis for a pan-African trade pact, building upon the Organization of African Unity (OAU) established under the 1980 Lagos Plan of Action.[16]

[14] Treaty Establishing the African Economic Community (Abuja Treaty) (adopted on 3 June 1991, entered into Force on 12 May 1994).

[15] The Constitutive Act of the African Union 2000.

[16] Organisation of the African Unity 'Lagos plan of action for the economic development of Africa 1980–2000'.

In 2012, African nations, building upon the 2000 Constitutive Act of the African Union, decided to expedite establishing a continental free trade agreement, with the idea now being actualized. The AfCFTA combines elements of Africa's innovative trade agreement model and the multilateral legal framework of the World Trade Organization (WTO). It places significant emphasis on development, aiming to foster economic and social progress while promoting legal harmonization.[17] The Agreement aligns with the African Union's Agenda 2063, which outlines the Africa we want, prioritizes inclusive socioeconomic development, and connects Africa's growth and integration with the Sustainable Development Goals (SDGs).[18] The African Continental Free Trade Area (AfCFTA) is poised to be a game-changer for Africa's economic and social development. By creating a single market for goods and services across the continent, AfCFTA can increase intra-African trade, promote industrialization, and foster economic growth and job creation.

The culmination of this new instrument for solidarity has ensured that the area of International Environmental Law is not overlooked, ensuring that all African states, in their bid to create a new block, sing the anthem of sustainable development and green tourism daily.

2.1 *Synergies Between the African Continental Free Trade Area Agreement, Its Protocols and Green Tourism*

As much as the agreement Establishing the African Continental Free Trade Area seems to be oscillating between two poles of Trade in Goods and Trade in Services, the area on sustainable development, mainly green tourism, seems to have been captured and enshrined in the agreement, giving it such a push. This commendable stride is in line with the continent's strategic framework that aims to deliver on its goal for inclusive and sustainable development and is a concrete manifestation of the pan-African drive for unity, self-determination, freedom, progress, and collective prosperity pursued under Pan-Africanism and African Renaissance,[19]

The pivotal role that Agenda 2063 has played in shaping green tourism is praiseworthy. Aspiration 1 of Agenda 2063 outlines that in the next 40 years, we envision an Africa with a blue economy, a climate change litigator, and a continent that produces fewer carbon emissions.[20] By Agenda 2063 enshrining sustainable development principles, green tourism as a sector can be seen as a priority that deserves keen attention by nations, especially regarding the economic value it can garner.

[17] Agreement Establishing the African Continental Free Trade Area (adopted 21 March 2018, entered into force 30 May 2019), art 3.

[18] African Union 'Agenda 2063: The Africa We Want'.

[19] African Union 'Agenda 2063: The Africa We Want'.

[20] African Union, *Agenda 2063: The Africa We Want* (African Union Commission 2015), aspiration 1.

When we examine the preamble, the core of the agreement, we see that it captures the fact that states, in their hot pursuit of creating national policy objectives and introducing new regulations, should ensure primarily that the environment is protected.[21] However, these preambular references do not set out legally binding obligations. They cannot, therefore, be interpreted as reflecting a strong commitment to the environment on the part of AfCFTA State Parties. There has been much animosity among nationals of various states that much protection is being given to investors who, to the detriment of the community and the environment, create more carbon emissions to the African Continent. If, as a continent, we are aiming for the Africa we want, such actions should be strongly guarded against, particularly by member states. On the flip side, this particular provision requires nations to take individual responsibility for formulating progressive laws that allow for integrating AfCFTA into their policy objectives. This enables the creation of sustainable related policies that focus mainly on the conservation of the environment that engulfs green tourism, especially when it comes to its close ties to trade. The mandate to introduce the new regulations, therefore, brings about the introduction of green economies' sustainable development, allowing for active and progressive action towards environmental care.

A glimpse of Article 3 of the Agreement Establishing the African Continental Free Trade Area reveals two things.[22] One, the area of sustainable development is known. However, it has not been given as much consideration as it should be. Two, the scope of sustainable development is too narrow even to capture the essential area of green tourism.

Although the agreement is extensive in its need to include sustainable development principles, its provisions solely around tourism have failed to be captured. This article aims to be a wake-up call to a completely forgotten area. Any amendment to the agreement and its protocol needs to be considered.

The Protocol on Trade in Goods[23] and the Protocol on Trade in Services[24] has created general exceptions that states may resort to when it is necessary to adopt specific measures of their own will to protect various things. Among them is the protection of animal or plant life. As much as this provision is an echo of the General Agreement on Trade and Tariffs (GATT, 1994),[25] it is a good move. However, the drafters of the AfCFTA Agreement should have done more for Africans other than just transplanting laws. This particular article lays down a foundation for exploitation from companies and government agencies that seek to harness the power to interpret it with a selfish connotation in mind. While the agreement can remain eager to promote sustainable green economies, wrongful interpretation can dismantle any

[21] Agreement Establishing the African Continental Free Trade Area (adopted 21 March 2018, entered into force 30 May 2019), preamble.

[22] Agreement Establishing the African Continental Free Trade Area (adopted 21 March 2018, entered into force 30 May 2019), art 3(e).

[23] Protocol on Trade in Goods (adopted 21 March 2018, entered into force 30 May 2019), art 26(b).

[24] Protocol on Trade in Services (adopted 21 March 2018, entered into force 30 May 2019), art 15(b).

[25] General Agreement on Tariffs and Trade (GATT 1994), art XX(b).

action toward progressive growth in the same light. It is therefore worth noting that there is a great need to include restrictions and binding measures around the circumstance in the article, which later ensures the safeguard of the green tourism sector overall.

Since AfCFTA objectives are based mainly on promoting interregional integration through trade, sustainable growth can reduce pressure on natural resources by targeting product design, promoting circular economy processes, and encouraging sustainable consumption.[26] The effects can be predicted to have impeccable benefits for the environment, which seeks to offer relief after having experienced profound climate changes that have negatively impacted it. This leads to the greenhouse effect affecting nations all over the world. Therefore, the environment improves significantly, and tourism also reaps the benefits.

In furtherance, the AfCFTA agreement shall have to formulate trade agreements that can positively and negatively impact the environment. On the one hand, trade agreements seek to increase trade and enhance economic growth, which tends to accelerate climate change by, among other things, generating emissions related to transportation and deforestation. In addition, many trade agreements restrict a country's ability to adopt specific environmental policies. For instance, trade agreements often restrict the types of subsidies countries can provide, limiting the options for subsidizing renewable energy. Conversely, trade agreements can advance environmental objectives. By creating new markets, trade agreements present an opportunity to create economies of scale and share know-how, promoting economic efficiency and innovation and supporting ecological outcomes. Indeed, the AfCFTA could accelerate the development of sustainable technologies and infrastructure.[27]

It is therefore worth noting that the potential impact AfCFTA has on Africa is that it has the exponential ability to circumvent nations into tremendous growth only if strategically implemented. It can not only ensure that environmental rights are enhanced but also introduce necessary chapters on trade and sustainable development, which requires trade partners, among other things, to follow international environmental standards, effectively enforce environmental laws, trade sustainably in natural resources, combat illegal trade in threatened and endangered species, encourage trade that supports tackling climate change, and promote practices such as corporate social responsibility.

3 Trade and Tourism: The Status of Intra-African Tourism During the Pandemic

Implementing the agreement will reap juicier and riper fruits for the African Continent. Liberalizing trade in goods will result in a 50% reduction in actionable trade barriers in the five priority service sectors: tourism, transportation, communication,

[26] Sommer and MacLeod (2019) at 71; Hoekman (2005).

[27] Carlos Lopes (4 January 2021).

financial services, and business services.[28] However, as with trade, Africa's global share of the tourism industry is relatively small, which could be attributed to a similarly low level of intra-regional trade and tourism.

Prevailing data and studies suggest a relationship between tourism and trade. A 10% increase in intra-African tourism boosts continental exports by 1.4 and 3.2%.[29] These results enormously posit that boosting regional tourism could catalyze intra-African trade, especially within the African Continental Free Trade Area context. During the initial 3 months of the pandemic, the African Union approximated that Africa suffered a loss of approximately $55 billion in revenue from travel and tourism and a decline of two million job opportunities.[30] According to projections by the International Monetary Fund, African nations heavily reliant on tourism experienced a contraction of 12% of their real GDP in 2020.[31]

The pandemic exposed Africa's nudity to its dependence on foreign travellers to drive its economy. This is especially true for countries in Eastern and Southern Africa, which have developed significant leisure and safari-oriented facilities that appeal to European, American, Asian, and other visitors. A survey conducted in December 2020 by the International Finance Corporation (IFC) and Dalberg Advisors focused on tourism companies in Tanzania, Uganda, and Zambia. The findings indicated that these companies were bracing for a substantial decline in revenue during the 2020–2021 season. Specifically, they anticipated a staggering two-thirds reduction in their earnings from international tourism receipts.[32]

This realization was a wake-up call for many African Countries, who decided to retool and focus on local tourism. To the south, South Africa was facing the worst hit it has ever faced with its tourism sector. National Parks and museums were similar to morgues, as no one was coming to visit except for a few.[33] The country welcomed 10 million international tourists in 2019; however, in 2020, it was limping along to check the revenue it had garnered. This prompted it to make local tourism a focus of its recovery plan.[34]

This move took shape in various countries. For instance, with advisory support from IFC, Rwanda is developing a recovery strategy that includes developing local and regional markets. Meanwhile, in Kenya, entry fees to all game parks and reserves have been cut for one year to attract more local tourists.[35] In West Africa, some hotels in Ghana have enjoyed their second-best month ever in December 2020—thanks to the power of domestic tourism. Under AfCFTA, 97% of commodities and goods are tariff-free. There will be a subsequent and directly connected increase in regional

[28] Allali Sara (Policy Brief 2021).

[29] Manyara and Naliaka (2021, 793).

[30] A Behsudi (December 2020).

[31] O Monnier (*International Finance Corporation*, May 2021).

[32] As above.

[33] W Roelf 31 August 2020.

[34] As above.

[35] Kenya Wildlife Services (6 July 2020).

trade. Since trade can promote tourism in Africa, it implies that regional integration is necessary to foster regional tourism.

Intra-African travel has, to date, received little empirical investigation, although most African outbound travel is to other African countries. The continent is, therefore, the leading destination for African tourists. It is clear that the reasons international tourists travel to Africa—for safari, culture, sun, sea and sand—are not the main driving forces of intra-African tourism. The results indicate that African tourists travel more to more prominent, developed African countries close to their country (geographically and culturally). Therefore, the notion that a large portion of African travel is for shopping purposes is supported. This holds important implications for policymakers as well as tourism destination managers. Policies to promote inter-regional travel and movement of goods (i.e., regional economic integration) would facilitate tourism growth. This implies improved accessibility regarding lower visa restrictions and better transport infrastructure. It is also essential for tourism destination managers to recognize that African tourists travelling to their countries are from countries with similar tourism resources (protected and marine areas) and would, therefore, not necessarily be interested in these tourism products. The preference might, therefore, be for more urban tourism products.

While research explored intra-African tourism, future research should expand the model to include a more comprehensive set of countries, including non-African countries, and distinguish between different travel reasons, such as visiting friends and family, business or holiday. Furthermore, it would be worthwhile to determine the effect of other trade agreements in Africa on bilateral tourism flows.

4 The Seat of Green Tourism at the Intra-African Trade Table

We start with the fundamental premise that the environment, tourism and its relationship with trade have been discussed in isolation for many decades, yet they all work synchronously. The 'trio' all remain affected by the migration of people. From a reading of Agenda 2063, the close relationship between trade and the environment can be witnessed. The document repeatedly echoes the importance of Sustainable Consumption and production patterns, water security, climate resilience, natural disaster preparedness and prevention while incorporating renewable energy, which can demonstrate the close relationship between the environment and trade.[36] State agencies should issue a tremendous need for support to help propel environmental policies that promote trade as an economic activity and as a recreational activity that allows for the promotion of tourism.

The tourism industry, especially green tourism, has become a key economic sector in Africa. The sector's embrace could be attributed to the failure of traditional sectors (such as agriculture and the extractives industries, which have hitherto dominated the

[36] African Union, *Agenda 2063*.

continent's economic space) to bring about meaningful economic growth and development.[37] Countries have taken considerable measures to develop their respective tourism sectors to attract tourists all year round rather than the seasonable travelling most Western countries are geared towards.[38]

At the continental level, the African Union has identified the tourism industry in its Agenda 2063 as a pathway through which structural transformation marked by a shift from agrarian-based economies to those driven by manufacturing and services sectors can be achieved. The African Union has established ambitious goals for the tourism sector, aiming to achieve 100% growth in its contribution to the gross domestic product (GDP) and a doubling of intra-Africa tourism within a decade (2013–2023).[39]

When we look at tourism from a statistical lens, tourism contributes and generates much revenue. With a projected annual growth rate of 5%, the tourism industry currently accounts for 8.5% of the continent's GDP and 6% of total investments valued at USD 29 billion. It employs 24.3 million people, or 6.4% of the total workforce.[40] In 2018, the continent welcomed 78 million visitors, generating revenue of USD 45 billion.[41] This translated to a global share of 5% and 3% in arrivals and receipts, respectively, and an average spend per arrival of USD 600 compared to the worldwide average of USD 1010. The continent's low share could, among others, be attributed to the comparatively low levels of intra-regional tourism, which currently stands at just 46% compared to the global average of 80%. Several factors have been identified as obstacles to regional tourism in Africa, including the high costs of regional travel, particularly air transport, exacerbated by poor connectivity, unfavourable visa regimes that restrict the movement of Africans, and the lack of appropriate tourism products.[42]

Given Africa's statistical position compared to the world, tourism has dramatically influenced interregional trade, as the case has been in Western countries. Africa has yet to exhaust and extensively maximize its potential in the tourism sector, especially with the emphasis on including more green alternatives that attract investors into their countries. It is prudent to conclude that Africa is set to benefit tremendously when both trade and green tourism sectors are addressed collectively to have both sectors merged and working concurrently.

As a result, green tourism is a sustainable approach to combating and mitigating climatic changes that have negatively impacted the environment. Climatic changes have led to wildfires in the grasslands, reduced rainfall, depleting rivers, and the involuntary movement of wildlife, negatively affecting tourism.[43] By adopting green

---

[37] Manyara and Naliaka (n 33) 793.

[38] International Labour Organisation 1 GB.341/POL/PV) para 72(d).

[39] Manyara and Naliaka (n 33) 793.

[40] World Trade and Tourism Council 'Travel and Tourism Impact 2021' (*Global Economic Impacts and Trends*, June 2021).

[41] Landry Signe and Chelsea Johnson (*Africa Growth Initiative* 2018).

[42] Kearney (1994, 21(1) 153).

[43] Lumsdon and Owen (2004) at 157–169.

tourism, African countries can explore modern technology to conserve and protect wildlife and the environment. It aids in further promoting agricultural production, eventually bringing more traction to trade in the area and opening up countries to more trade and industrialization in line with AfCFTA. These technologies can positively impact the agricultural and industrial sectors, enabling economic actors to withstand climate change and pursue industrialization in a way that is less damaging to the environment.[44]

Tourism has great potential to accelerate progress across the Sustainable Development Goals (SDGs). If well managed, the sector can generate quality jobs for durable growth, reduce poverty, and provide other incentives for environmental conservation—a triple win to help countries transition towards more inclusive, resilient economies. However, without proper safeguards and investments, expanding the tourism market will increase pressure on biodiversity and the ecosystems on which the livelihoods of so many people and local communities depend.

Further, well-designed and managed tourism can help preserve natural and cultural heritage assets, empower host communities, generate trade opportunities, and foster peace and intercultural understanding.

Nevertheless, the growing number of people travelling internationally and domestically each year increases the probability of producing pressures regarding greenhouse gas emissions, economic leakages, resource management, and impact on local communities and cultural assets. It is, therefore, of great importance to harness tourism's positive contribution to sustainable development and mitigate the sector's adverse effects, which calls for strong partnerships and decisive action by all tourism stakeholders.

By encouraging intra-African trade, the continent opens up opportunities that encourage green tourism. This, upon analysis, contributes to cross-cultural understanding, migration, and increased population, which inevitably translates to more consumer production innovation and industrialization in the trade industry

Through the interaction of communities through green tourism, people are encouraged to come together, which reduces prejudices while raising awareness on crucial issues around social justice, equality, and the need to implement human rights. It further enhances the need for fewer travel restrictions around obtaining visa documentation, which allows for more free movement in and out of countries in a regulated manner. In addition, the transportation industry has exponentially developed its infrastructure to facilitate people's migration and encourage the enhancement of individual nations' existing cultural and environmental attractions. Therefore, green tourism's importance can be justified, leading to the formulation of clear legislation supporting the sector.

---

[44] Albert (2019); D Harary '24 October 2018.

5 The Current Pitfalls Being Faced by Green Tourism

The conceptual model of green practices encompasses two vital components: green branding and green growth initiatives.[45] However, the current state of green tourism reveals several pronounced challenges that demand attention. A pivotal obstacle encountered within green tourism pertains to the pervasive issue of greenwashing. Greenwashing denotes the deceptive or inaccurate portrayal of a destination or enterprise as environmentally conscientious despite its failure to implement substantive sustainable practices. Numerous establishments assert their eco-friendliness without substantiating these claims through tangible actions. Consequently, the edibility of green tourism is jeopardized, giving rise to perplexity among travellers. Effectively addressing greenwashing necessitates the establishment of standardized certifications and transparent reporting mechanisms, which can serve to authenticate the sustainability assertions made by tourism providers. By implementing such measures, the integrity of green tourism can be fortified, empowering tourists to make informed choices and engendering trust in the industry's commitment to genuine environmental stewardship.[46]

The influx of foreign individuals from diverse cultural backgrounds has significantly contributed to the erosion of indigenous cultures in many countries. The continuous intermingling of cultures poses a considerable risk to preserving a country's distinctive cultural heritage, leading to its gradual extinction. While acknowledging the potential positive aspects of cultural interaction, it is essential to recognize that a region's inherent uniqueness and essence may be at stake due to ongoing cultural blending. Green tourism, undoubtedly, aims to foster an appreciation for a country's natural resources, culture, and people. However, without vigilant regulation, this well-intentioned endeavour can result in detrimental long-term consequences that may be irreversible.[47] Therefore, it becomes the responsibility of a nation to safeguard its culture and people by implementing robust and sustainable laws. Such laws should preserve the cultural identity and protect the local population's livelihoods. By enacting well-structured legislation and regulations, governments can balance promoting tourism and safeguarding the authenticity and integrity of their cultural heritage. These measures should address the potential negative impacts of tourism, such as commodification, exploitation, and cultural homogenization. By doing so, countries can foster sustainable tourism practices that allow for the appreciation and celebration of diverse cultures while preserving their distinctiveness for future generations.[48]

[45] Chivandi et al. (2023) 10(2) 99.

[46] Clorfene (1994) 17.

[47] International Labour Organisation (1 GB.341/POL/PV) para 72(d).

[48] Pintassligo (2016, 406).

To add insult to injury, the deficiency in awareness and comprehension among tourists and industry stakeholders represents a significant impediment to the advancement of green tourism. Numerous travellers remain uninformed about the environmental ramifications of their choices and are often oblivious to sustainable alternatives.[49] Simultaneously, tourism industry businesses may lack the knowledge and understanding to implement sustainable practices effectively. Addressing this issue necessitates proactive measures to enhance awareness and understanding. Educational campaigns, travel advisories, and training programs can serve as invaluable tools in closing this gap and fostering a culture of responsible tourism. By disseminating information about the environmental impacts of various tourism activities and promoting sustainable alternatives, tourists can make more informed decisions that align with ecological conservation goals. Moreover, by providing comprehensive training to industry stakeholders, businesses can acquire the necessary skills and expertise to integrate sustainable practices into their operations effectively. By actively engaging tourists and industry stakeholders through awareness-building initiatives, the foundation for responsible tourism practices can be solidified. This, in turn, will contribute to the preservation of natural resources, the promotion of sustainable development, and the long-term viability of green tourism.

The implementation of sustainable tourism practices necessitates substantial investments in infrastructure and resources. However, numerous destinations, particularly those in developing countries, encounter challenges from limited financial resources and inadequate infrastructure to support green tourism initiatives. Building eco-friendly accommodations, waste management systems, and renewable energy infrastructure often incur significant costs and require considerable time for development. To address these constraints, fostering public–private partnerships is instrumental in mobilizing financial resources and expertise. Collaborative efforts between governments, private sector entities, and local communities can facilitate the pooling of funds, knowledge, and resources. This collective approach allows for a more efficient allocation of resources and can help overcome financial barriers that hinder the adoption of sustainable practices.[50]

Additionally, providing financial incentives can incentivize investments in sustainable tourism infrastructure. Governments can offer tax incentives, grants, or subsidies to encourage businesses to adopt eco-friendly practices and develop sustainable infrastructure. Such measures alleviate the financial burden on businesses and create a favourable environment for green investments.

International organizations and development agencies can also play a crucial role by providing financial support and technical assistance to needy destinations. Funding programs and capacity-building initiatives can help bridge the gap between limited resources and the demand for sustainable infrastructure.

[49] Dewi et al. (2022, 68).

[50] Tandon et al. (2023, 98).

Governments and relevant stakeholders must recognize the long-term benefits of investing in sustainable tourism infrastructure. These investments contribute to environmental conservation and the preservation of cultural heritage and foster economic growth, job creation, and improved quality of life for local communities.

Another challenge is finding the right balance between conservation efforts and economic growth. Green tourism aims to protect natural resources and preserve cultural heritage, but it must also contribute to the economic development of local communities. Ensuring the local population benefits from tourism while safeguarding the environment requires careful planning and stakeholder collaboration. Engaging local communities, promoting community-based tourism initiatives, and involving them in decision-making can help strike this delicate balance.

6 Crossing the Bridge Between the Environment and Intra-African Trade

The projected increase in intra-African trade facilitated by implementing the African Continental Free Trade Area (AfCFTA) will bring opportunities and challenges. However, it is crucial to anticipate and address potential adversities to mitigate adverse environmental impacts that directly affect the tourism sector. Binding policies and collaborative efforts among member states are essential for sustainable development within the AfCFTA framework.[51]

Firstly, the anticipated rise in trade in agricultural products within Africa may lead to increased farm production and subsequent land use expansion. This could result in deforestation, reducing the absorption of CO_2 and negatively impacting climate change mitigation efforts. Small-scale agriculture, predominant in Africa, has been identified as a significant driver of deforestation globally. Promoting integrated land use approaches and enhancing synergies between agriculture and forest conservation can help mitigate these adverse effects.

Secondly, increased intra-African trade can drive economic diversification and industrialization among member states. However, if industrialization is pursued without adopting climate-friendly technologies, it can contribute to the acceleration of climate change. Urbanisation, which often accompanies industrialization, is also associated with increased greenhouse gas emissions. The extent to which trade-induced industrialization impacts climate change depends on the chosen industries and the technologies member countries adopt. Focusing on sustainable industrial activities and adopting climate-friendly technologies is vital for minimizing adverse environmental consequences.

Expanding trade between African countries can challenge import-competing producers, notably Micro, Small, and Medium Enterprises (MSMEs) and small-scale farmers. These actors may face higher trading costs than larger companies, making them more vulnerable to climate change impacts. Climate change already poses

[51] Ezendu and Syvertsen (2013) at 232.

significant risks to agricultural-dependent economies. When coupled with increased competition from the AfCFTA, these actors may face more significant difficulties adapting to climate change. Supporting and empowering these vulnerable economic actors becomes crucial for enhancing resilience and minimizing the adverse effects of climate change.

Addressing these challenges requires close collaboration among national governments in formulating policies and implementing sustainable practices. It is essential to prioritize the mitigation of environmental effects that can lead to irreversible climatic changes. Through policy creation, coordinated efforts can be undertaken to promote sustainable land use, climate-friendly industrialization, and support mechanisms for vulnerable economic actors. By integrating environmental considerations into trade and development strategies, the AfCFTA can contribute to sustainable economic growth while safeguarding the environment for future generations.

7 Conclusion

The African Continental Free Trade Area (AfCFTA) is poised to reshape Africa's economic landscape, fostering regional integration and boosting intra-African trade. However, to truly harness the potential of this historic agreement, it is essential to consider the role of green tourism as a catalyst for sustainable development. This article explores the synergies between the AfCFTA and green tourism, highlighting their interconnectedness and advocating for the integration of sustainable practices within the trade framework. Before delving into the potential of green tourism within the AfCFTA, the article looks at the journey to establishing this landmark trade agreement. The article provides an overview of the proceedings that unfolded before the formation of the AfCFTA, emphasizing the shared vision of economic prosperity and cooperation among African nations.

The AfCFTA shows that green tourism and the trade agreement share common goals when viewed from an environmental lens. Both strive for sustainable development, recognizing the importance of preserving natural resources, protecting biodiversity, and mitigating the environmental impact of economic activities. Integrating green tourism practices within the AfCFTA framework can catalyze environmentally responsible trade and tourism.

The article argues and justifies why green tourism should be integral to the AfCFTA. By incorporating sustainable tourism practices into the trade agreement, African nations can tap into the vast potential of green tourism, promoting responsible and inclusive economic growth. Furthermore, green tourism can contribute to job creation, poverty alleviation, and the empowerment of local communities, thus enhancing the continent's overall social and economic well being.

The article concludes that by embracing and prioritizing green tourism within the AfCFTA, African leaders have the opportunity to address the neglected aspects of sustainable development. It highlights that if green tourism is sufficiently considered, it can be a driving force in tightening loose ends and solidifying the foundations of solidarity within the economic arena. Integrating green tourism into the AfCFTA would boost intra-African trade and serve as a hallmark of Africa's commitment to sustainable development and environmental stewardship.

References

Ababio MO (2020) Unpacking Africa as a dynamic continent: insights from contemporary development issues in Ghana. Ogua J Soc Sci 9(1):1

African Union, Agenda 2063: the Africa we want. https://au.int/en/agenda2063/overview

Ajayi JFA (1989) Africa at the beginning of the nineteenth century: issues and prospects. In: Ajayi JFA (ed) General history of Africa VI: Africa in the nineteenth century until the 1880s. University of California Press, California

Albert T (2019) The African continental free trade agreement: opportunities and challenges. CUTS International Geneva, Geneva

Behsudi A (2020) Tourism-dependent economies are among those harmed the most by the pandemic. International Monetary Fund (December 2020). https://www.imf.org/en/Publications/fandd/issues/2020/12/impact-of-the-pandemic-on-tourism-behsudi

Chivandi A, Sikhauli M, Mlili T (2023) Green innovation sustainability and green practice behaviours in tourism and hospitality. Athens J Tour 10(2):99

Clorfene L (1994) Green washing. Bus Ethics 8(6):17

Dewi KI et al (2022) Implementation of eco-friendly behaviour by front office employees to support green hotel at the Ritz–Carlton Bali. Int J Green Tour Appl 4(2):68

Dokotum OO (2020) Hollywood and Africa: recycling the "dark continent" myth from 1908–2020. NISC (Pty) Ltd

Editor (2022) A ticket to recovery: reinventing Africa's tourism industry. The point (Gambia) 6 October 2022. https://thepoint.gm/africa/gambia/editorial/a-ticket-to-recovery-reinventing-africas-tourism-industry

European Commission, Circular economy action plan: for a cleaner and more competitive Europe. https://ec.europa.eu/environment/pdf/circular-economy/new_circular_economy_action_plan.pdf

Ezendu A, Syvertsen CM (2013) Green digital enterprise and chaos theory in tourism development: the case of eco-tourism in South Africa. In: Ganesh K, Anbuudayasankar SP (eds) International and interdisciplinary studies in green computing. IGI Global, London, p 232

Gordon LR (1997) African philosophy's search for identity: existential considerations of a recent effort. CLR James J 5:98

Gordon LR (2008) Introduction to Africana philosophy. CUP, London

Harary D (2018) African free trade could increase resilience to climate change and conflict. NewsSecurityBeat (Wilson Center) 24 October 2018. https://www.newsecuritybeat.org/2018/10/african-free-trade-increase-resilience-climate-change-conflict/

Hegel GWF (1956) The philosophy of history. Dover Publications, London

Herdegen M (2016) Principles of international economic law, 2nd edn. OUP, Oxford

Hoekman B (2005) Operationalising the concept of policy space in the WTO: beyond special and differential treatment. J Int Econ Law 405

International Labour Organisation (2022) The future of work in the tourism sector: sustainable and safe recovery and decent work in the context of the COVID-19 pandemic (report for the technical meeting on COVID-19 and sustainable recovery in the tourism sector, 1 GB.341/POL/PV)

Kearney A (1994) Green tourism development in Scotland. Ann Tour Res 21(1):153

Kenya Wildlife Services (2020) Reduction of Kenya wildlife services park entry fees. Kenya Wildlife Service, 6 July 2020. https://kws.go.ke/content/reduction-kenya-wildlife-service-park-entry-fees

Lake Nakuru National Park, Extension of Kenya wildlife services conservation fees. https://www.lakenakurukenya.com/extension-of-kenya-wildlife-services-conservation-fees/
Lopes C (2021) Europe and Africa, needs to see eye to eye on climate change. OECD Development Matters, 4 January 2021. https://oecd-development-matters.org/2021/01/04/europe-and-africa-need-to-see-eye-to-eye-on-climate-change/
Luke D (2019) Making the case for the African continental free trade area, 1st edn. Routledge, London
Lumsdon L, Owen E (2004) Tourism and transport: the green key initiative. In: Les L, Stephen J (eds) Advances in tourism research, tourism and transport. Pergamon, Berlin
Manyara G, Naliaka L (2021) Tourism and trade nexus in Africa: a gravity model approach. Afr J Hosp Tour Leisure 10(3):793
Mazrui AA (1986) The Africans: a triple heritage. Little Brown & Co (T), London
Mbembe A (2014) Out of the dark night: essays on decolonization. Columbia UP, New York
Monnier O (2021) A ticket to recovery: reinventing Africa's tourism industry. International Finance Corporation, May 2021. https://www.ifc.org/wps/wcm/connect/news_ext_content/ifc_external_corporate_site/news+and+events/news/reinventing-africa-tourism#:~:text=In%20July%202020%2C%20the%20African,by%2012%20percent%20in%202020
Mwiti L, 20 staggering facts about tourism in Sub-Saharan Africa, including some real star performers. https://panafricanvisions.com/2014/08/20-staggering-facts-tourism-sub-saharan-africa-including-real-star-performers/
Ngwena C (2018) What is Africanness? Contesting nativism in race, culture, and sexualities. Pretoria University Law, New York
Olontoruba SO (2023) The African continental free trade area and EU–ACP economic partnership agreements contentions and synergies for development. In: Samuel OO, Christopher CN, Leon MT (eds) Africa-EU relations and the African continental free trade area, 1st edn. Routledge, London
Organization of the African Unity, Lagos plan of action for Africa's economic development 1980–2000. https://www.resakss.org/node/6653
Pintassligo P (2016) Green toursim. Encylopedia of tourism 406
Roelf W (2020) With borders closed, South Africa pins hopes on cash-strapped local tourists. Reuters (London) 31 August 2020. https://www.reuters.com/article/us-health-coronavirus-safrica-tourism-idUSKBN25R1LI
Scholvin S (2015) A new scramble for Africa: the rush for energy resources in sub Saharan Africa, 1st edn. Routledge, Berlin
Smith RB (1987) The scramble for Africa. In: Smith RB (ed) The scramble for Africa: documents and debates. Palgrave, London
Sommer L, MacLeod J (2019) How important is special and differential treatment for an inclusive AfCFTA? In: Luke D, MacLeod J (eds) Inclusive trade in Africa: the African continental free trade agreement in comparative perspective. Routledge, Berlin
Tandon A et al (2023) Green and non-green outcomes of green human resource management (GHRM) in the tourism context. Tour Manag 2023:98
United Nations (2012) The theory of free trade areas: the case for an African continental free trade area. In: United Nations (ed) Assessing regional trade in Africa
United Nations Economic Commission for Africa (2021) New assessment of the economic impacts of the agreement establishing the African continental free trade area on Africa. Policy Brief 2021. https://repository.uneca.org/bitstream/handle/10855/46750/%20b11999160.pdf?sequence=1&isAllowed=y
World Trade and Tourism Council (June 2021) Travel and tourism impact 2021. Global economic impacts and trends

Climate Change and Nigeria's Narrative: Impacts, Policies, and Challenges

Olayinka Oluwamuyiwa Ojo, Oluwaseyi Adurapemi Olasoji, and Dennis Oghenerobor Agelebe

Abstract Climate change is the greatest threat to human existence in recent times and perhaps the main threat to public health. To tackle this threat, laws, policies, agencies, and institutions have been enacted to mitigate and adapt to climate change. As part of the global community, Nigeria has experienced many climate-related issues, proving that climate change is no longer a myth but a reality in Nigeria. Nigeria has signed several international agreements and passed many domestic laws and policies to mitigate and adapt to climate change. This article will focus on the impact of climate change in Nigeria, the existing laws, policies, regulations, and institutions tackling climate change in Nigeria, and their effectiveness.

1 Introduction

Climate change is the defining challenge of the present generation.[1] Climate change is perhaps the most critical sustainability issue facing today's generation.[2] Climate change can and will eventually affect other global problems, such as biodiversity, freshwater, marine and forest conservation, poverty, health, and food security. From the industrial revolution to date, there has been a drastic increase in greenhouse gas (GHG), particularly carbon dioxide (CO_2), in the atmosphere than the natural sinks

[1] United Nations Environment Programme on Climate Change (2021).

[2] European Environment Agency 'Climate Change Mitigation' (2021).

O. O. Ojo (✉)
Centre for International Programmes, NEWTON University, Prague, Czech Republic
e-mail: olayinka.ojo@newton.university

O. A. Olasoji
Department of Economics and Development, Faculty of Tropical AgriSciences, Czech University of Life Sciences, Prague, Czech Republic
e-mail: Olasoji@ftz.czu.cz

D. O. Agelebe
Environmental Law Center, University of Cologne, Cologne, Germany
e-mail: dageleb1@uni-koeln.de

D. Oghenerobor Agelebe (ed.), *Climate Ethics in Africa*, Global Perspectives on Law and the Environment 1, https://doi.org/10.1007/978-3-032-08757-7_7

can absorb.[3] The burning of fossil fuels (coal, oil, oil shale, tar sands, natural gas, and peat) has increased the concentration of CO_2 in the atmosphere more than ever, and there are significant variations from year to year.[4] The atmospheric concentration of CO_2 has risen from 280 parts per million (ppm) to 422 ppm between 1750 to Se ptember 2025.[5] If current trends in fossil fuel use and other GHG continue, the concentration of CO_2 could reach 600–700 ppm by the end of the twenty-first century.[6]

Due to the GHG concentrations, temperatures have been rising, resulting in different warming patterns on the earth's surface, generally called global warming.[7] Climate change is not only leading to warmer temperatures but also causing heat waves, storms, melting ice in the Arctic, Greenland, and Antarctic, disappearing glaciers, melting permafrost, rising sea levels, disruptive floods, acidification of the oceans, droughts, desertification, and species extinction.[8] Climate change has been reported to have already claimed more than 150,000 lives and over 5 million illnesses per year.[9] The World Health Organisation projects this figure may double by 2030 if proper checks are not implemented.[10] Unfortunately, stakeholders are slow to take proactive actions in the face of these existential threats.[11]

The potential threats posed by climate change demand an appropriate response that will address areas as diverse as transportation, deforestation, power generation, control over natural resources, industrial and economic growth, and personal liberties.[12] It has been speculated that it will be difficult to reverse the impacts of climate change; however, efforts can be made to mitigate and adapt these impacts. The core of mitigation involves controlling climate change by cutting greenhouse gas (GHG) emissions. Adaptation is taking the proper measures to reduce the harmful effects of climate change (or exploit the positive ones) by making the appropriate adjustments and changes. For humanity to curb climate change and its damaging effects, several conventions, treaties, protocols, and agreements have been enacted at various levels. Some of these conventions, treaties, protocols and agreements have birthed

[3] CD Thomas et al. 'Extinction Risk from Climate Change' (2004, 145).

[4] M Le Page (8 March 2016).

[5] Co2.earth (2025).

[6] S Leahy (31 August 2007).

[7] J Edmonds et al. (2004) at 77.

[8] Thomas (n 4).

[9] J Patz et al. (2005).

[10] AJ McMichael et al. (WHO, 2003) 322.

[11] GF Nemet et al. (2017).

[12] S Bell et al, *Environmental Law* (OUP 2008).

significant climate change measures, such as the United Nations Framework Convention on Climate Change (UNFCCC) of 1992,[13] the Kyoto Protocol of 1997,[14] the Copenhagen Accord of 2009,[15] and the Paris Agreement of 2016.[16]

Although climate change is a global phenomenon, its negative impacts are more felt by poor and developing countries. Nigeria is one of the developing countries affected by climate change. Nigeria is vulnerable because of its high dependence on natural resources, poor resource management, insufficient finance, poverty, illiteracy, corruption, and limited experts and professionals. This paper seeks to evaluate the impact of climate change in Nigeria, the legal frameworks available to address them, and areas of improvement. Specifically, the paper addresses the following questions: What are the impacts of climate change in Nigeria? What laws, strategies and policies have been initiated in Nigeria to tackle climate change? What is Nigeria's capacity to deal with these impacts and challenges?

2 Climate Change and Nigeria's Contribution

Nigeria is a significant player in the oil and gas industry, producing about two million barrels of crude oil daily.[17] It has about 36.89 billion barrels of crude oil as of 2021 and about 200.4 trillion cubic feet of natural gas in reserve.[18] All these data place Nigeria as one of the biggest petroleum producers in Africa and the 13th largest globally.[19] Nigeria's oil and gas industry generated about 70% of the government's total revenue and about 90% of export earnings.[20] By the end of the third quarter of 2023, the oil and gas industry's contribution to Nigeria's gross domestic product (GDP) was approximately 5.5%.[21]

Oil and gas are significant sources of GHGs. Although currently essential to the global economy, they are highly hazardous substances that cause several environmental troubles.[22] The various stages of oil and gas production, including oil exploration, site access, site preparation, drilling, production, pipeline installation, spill control, clean-up, site closure, as well as the final consumption via engines, cars, and

[13] United Nations Framework Convention on Climate Change (9 May 1992) S. Treaty Doc No. 102-38 (1992).

[14] Kyoto Protocol, Decision 4/CMP.1, Guidance Relating to the Clean Development Mechanism, FCCC/KP/CMP/2005/8/Add.1 (2005).

[15] Copenhagen Accord, Report of the Conference of the Parties on its Fifteenth Session, Decision 2/CP.15, U.N. Doc. FCCC/CP/2009/11/Add.1 (2010).

[16] Paris Agreement *in* UNFCCC, COP Report No. 21, Addendum, at 21, U.N. Doc. FCCC/CP/2015/10/Add, 1 (2016).

[17] Central Bank of Nigeria (2025).

[18] EIA 'Country Analysis Executive Summary: Nigeria' *EIA*, (25 June 2020).

[19] GFP 'Oil Production by Country' *Global Fire Power* (2020).

[20] Shell In Nigeria Briefing Notes April 2020.

[21] NBS 'Contribution of Oil Sector to GDP in Nigeria Q3, 2023.

[22] K Kingsley Ezeibe '(2011).

other equipment, have resulted in the emission of GHG, poisonous, and carcinogenic chemicals, pollution of all kinds, destruction of the fauna and flora generally through oil spills, gas flaring, and other oil drilling and handling activities.[23] Gas flaring is a process used to protect against the dangers of overpressuring industrial plant equipment during oil exploration, production, and processing. Flared gases accompany the oil during the extraction and production stages.[24] The gases can be re-injected by properly channelling them into reservoirs, used for other purposes, or flared into the atmosphere.

Gas flaring is one of the ways carbons are emitted into the atmosphere,[25] Which causes global warming and contributes to climate change.[26] According to the World Bank, more than 400 million tons of carbon are emitted globally through gas flaring annually.[27] Gasses have been flared almost daily in Nigeria for 35–40 years. According to satellite data from the Federal Ministry of Environment, over 313 million standard cubic feet of gases are flared annually in Nigeria, emitting 16.5 million tons of carbon. According to the World Bank's Global Gas Flaring Reduction Partnership (GGFRP) report of 2023, Nigeria ranked as the 3rd largest gas-flaring country in the world.[28] There are about 58 locations in Nigeria where gases are flared almost daily.[29] The oil and gas industry is not the only sector in Nigeria that contributes to GHG emissions. Agriculture, forest, other land use (AFOLU), waste, transportation, industrial processes, and product use contribute significantly to GHG emissions.

> The third national communication submitted by the Federal Government of Nigeria to the UNFCCC in March 2020 gave a general outlook on the sectors emitting GHG in Nigeria. The AFOLU sector emitted the largest GHG with 366,733.9 gigagram carbon dioxide equivalent (Gg CO_2-eq), which is 60.1% of the GHG emitted, followed by the energy sector with 206,452 Gg CO_2-eq (33.9%). The waste sector emitted 23,330.3 Gg CO_2-eq (3.8%), and the remaining sectors, transportation, industrial processes and product use, emitted 13,267.1 Gg CO_2-eq (2.2%). CO_2 was responsible for 72.9% of the emissions, CH4 for 21.2%, and N2O for 5.8%. The AFOLU sector was the primary sector emitting CO_2 and N2O with 69.1% and 70.3%, respectively, while the energy sector recorded the highest CH4 emission with 61.0%. The energy sector emitted 62.6% NOX, 74% CO, 53.2% SO_2, and 77.8% NMVOCs.[30] From these data, it can be seen that Nigeria has been one of GHG's major contributors globally. Nigeria is not only contributing to GHG emissions but also having its share of the impact of climate change on our planet.

---

23 JY Ko et al. (2013).

24 S Azaiki (2007, 280).

25 A Hassan & R Kouhy (2013) at 124.

26 WB Kindzierski (2000) at 41.

27 A Rozowska & C Murphy-McGreevey (2020)

28 World Bank (2025).

29 NACGOND (2025).

30 Federal Republic of Nigeria '(2020) at 7.

3 Impact of Climate Change in Nigeria

The constant increase in temperature, rising sea level, erosion, flooding, land degradation, drought and desertification, variable rainfall, loss of biodiversity, extremely harsh weather, and a host of others that are a result of climate change have proven that climate change is no longer a myth but a reality in Nigeria.[31] The temperature has sporadically risen since the 1980s; an increase of 0.2–0.3 per decade has been recorded nationwide.[32] The Federal Ministry of Environment predicts that climate change may cause a further temperature increase of 0.4–1 °C between 2020 and 2050 and an increase of 3.2 °C by 2050.[33] Climatic change experiences and impacts across Nigeria are not the same. The southern part of Nigeria's experience differs entirely from the northern part.[34] This is because Nigeria has a tropical climate with two precipitation regimes: high precipitation in the south and low precipitation in the north.[35] Due to the high precipitation in the south, there have been rising sea levels, floods, and erosions. The south-south part of the country, also known as the Niger Delta region, is the most vulnerable and will experience the highest impact of climate change.[36] Due to the low precipitation in the north, there have been droughts, desertification, aridity, climate migration, and famine. The northwest and northeast are most vulnerable and will experience the highest impact of climate change.[37]

There has been an increase in rainfall in the south and a reduction in rainfall in the north. According to the research carried out by Building Nigeria's Response to Climate Change (BNRCC), there has been an increase of 2–4 mm in annual rainfall in the south, whereas most of the country, particularly the north, have experienced decreased annual rainfall by 2–8 mm between the 1940s–2000s.[38] Furthermore, the sea level in Nigeria[39] has risen between 1990 and 2020[40] by 0.1 m.[41] The Federal Ministry of Environment postulates a rise of 0.2 m in the sea level by 2050 at a low estimate and 1 m by 2050 at a high estimate; it must, however, be noted that a rise in sea level of 1 m could result in a loss of 75% of Nigeria's Niger Delta and some parts of Lagos.[42] Due to these increased rainfalls, rising sea levels, and high precipitation,

---

[31] I Elisha, BA Sawa & EU Lawrence (2017).

[32] I Enete (2014).

[33] Federal Republic of Nigeria 'Nigeria's Second National Communication under the UNFCCC' *Federal Republic of Nigeria* (February 2014)

[34] NE Ebele & NV Emodi (2016) at 1.

[35] H Haider (2019).

[36] Federal Republic of Nigeria 'National Environmental Economic and Development Study (NEEDS) for Climate Change in Nigeria' (FRN, September 2010).

[37] Federal Republic of Nigeria (n 35).

[38] BNRCC *National Adaptation Strategy and Plan of Action on Climate Change for Nigeria (NASPA-CCN)* (BNRCC, November 2011).

[39] RN Amanchukwu, TG Amadi-Ali & N Prince Ololube (2015).

[40] A Akande et al. (2015).

[41] Federal Republic of Nigeria (n 35).

[42] Federal Republic of Nigeria (n 35).

floods, massive runoff, and erosions have occurred.[43] These rainfalls cause buildings to collapse, destroy crops and trees, kill and displace many people, and submerge several villages in Lagos and the Niger Delta.[44] States in Nigeria that have recently experienced heavy rainfalls and floods[45] between June to August 2020 were Lagos,[46] Kwara,[47] Abuja,[48] Niger,[49] Akwa Ibom,[50] and Borno.[51] In September 2019, several states in Nigeria experienced floods that displaced over 20,000 and killed about 12 people.[52] The Nigerian Red Cross reported that between June and September 2019, torrential rainfalls and flash floods hit 124 Local Government Areas within 36 states and the Federal Capital Territory—Abuja (FCT). This affected a total number of 210,117 people, with 171 casualties recorded in hospitals and about 130,610 people reported to be displaced.[53] In 2010, over 2 million people were also affected by a flood in Jigawa State.[54] There are several other undocumented floods and storms across the country. According to Ibukun Odusote, a permanent secretary at the Federal Ministry of Environment, floods that occurred between 2012 and 2018 resulted in the deaths of about 363 Nigerians and the displacement of 3,871,500 people living within the River Niger basin.[55]

Due to heavy rainfalls, several towns and villages in the southeastern states of Abia, Anambra, Enugu, Ebonyi, and Imo[56] are battling severe gully erosions that have created massive gullies.[57] Some gullies' are as deep as 22–150 m with widths of 0.4–5.6 m and lengths of 0.7–2.5 km. Most of them lie along a linear zone of weakness.[58] Due to these gully erosions, vast areas of agricultural land are no longer suitable for cultivation, affecting agricultural productivity.[59] Also, people living in these affected areas regularly relocate to higher grounds from time to time.[60] According to Boniface, Abia has about 300 gullies, Anambra about 700 gullies, Ebonyi about 250 gullies, Enugu about 600 gullies, and Imo about 450 gullies.[61]

---

[43] Enete (n 34).

[44] B Anabaraonye, CJ Okafor & OJ Ikuelogbon (2019).

[45] Sahara Reporters (19 June 2020).

[46] Floodlist News (19 June 2020).

[47] Floodlist News (16 June 2020).

[48] Floodlist News (25 July 2020).

[49] As above.

[50] Floodlist News (14 June 2020).

[51] Floodlist News '(11 June 2020).

[52] Floodlist News (9 October 2019).

[53] Nigerian Red Cross Society (07 October 2019).

[54] Elisha (n 29).

[55] C Ochayi (18 June 2019).

[56] GT Amangabara (2014) at 1.

[57] C Nwankwo & N Ho (2018a).

[58] NN Onu & AI Opara (2012).

[59] BCE Egboka et al. (1990).

[60] BCE Egboka, AE Orji & HO Nwankwoala (2019) at 1.

[61] As above.

In northern Nigeria, due to fewer rainfalls, precipitation has declined consistently, as has temperature and heat. Droughts and desert encroachment have also been consistent. Several wetlands are drying up, and surface water is rapidly reducing.[62] For instance, the southern part of Lake Chad, the lake sections in Nigerian territory, has been drying up. History has it that Lake Chad covered an estimated 25,000 square kilometres in 1963, but it now covers about 1200 square, kilometres and is at risk of disappearing.[63] This occurrence implies that the land is laid to wastage by rising temperatures, leading to the Sahara desert's southward expansion. Hence, farmlands and surrounding villages have become barren due to the overpowering nature of advancing desertification, whose ripple effects have led to the massive migration of people searching for fertile lands.[64] The Sahel creeps south by about 3625 square kilometres annually. Within 20 years, dunes in the north have increased by 400%.[65] Many agrarian and arable lands are no longer suitable for cultivation because dune buries them. It has been postulated that with the constant decline in precipitation and desert encroachment, two-thirds of Bauchi, Borno, Gombe, Jigawa, Kano, Kaduna, Katsina, Kebbi, Sokoto, Yobe, and Zamfara states could become desert or semi-desert by the end of the twenty-first century.[66]

In the same vein, climate migration, although not often discussed in Nigeria, has been the source of many communal clashes and conflicts. For example, the farmer-herder clashes in Nigeria have led to the death of thousands of people.[67] It must be noted that the Fulani herdsmen play a vital role in the Nigerian economy. They contribute about 3.2% to Nigeria's Gross Domestic Product (GDP),[68] Mainly through the meat and milk industry, with an estimated 19.5 million cattle, 975,000 donkeys, 28,000 camels, 72.5 million goats, and 41.3 million sheep.[69] They predominantly reside in Bauchi, Borno, Gombe, Jigawa, Kano, Kaduna, Katsina, Kebbi, Sokoto, Yobe, and Zamfara states. Because arable lands in these states are becoming deserts, the Fulani herders migrate south toward Nigeria's Middle Belt in search of productive land and pastures for their herd. This southern migration was initially seasonal, from December to April, with herders returning to the north.[70] However, recently, due to the unavailability of pastures caused by droughts, low rainfall, low precipitation, and heat in the north, herders have been staying longer, and some are staying permanently in the south. This has continuously triggered disputes, conflicts, and many deaths by destroying farms, crops, and plantations of several farmers in those areas.[71] Plateau, Benue, Kaduna, Adamawa, Taraba, and Kogi are the most affected states. The above

[62] Akande (n 42).

[63] World Bank (2023).

[64] O Akinyemi, A Ogundipe & P Alege (2012).

[65] A Sayne (June 2011).

[66] As above.

[67] Floodlist News (n 49)

[68] IA Johnson and BT Okunola (2017).

[69] International Crisis Group (2017).

[70] As above.

[71] Premium Times (13 May 2012).

means that if climate change is not addressed enthusiastically (through mitigating and adaptation), the effects will threaten Nigeria's existence. Several agencies have been established, and laws, policies, and plans have been proposed and implemented to address, adapt to, and mitigate climate change impacts in Nigeria.

4 Implementation of International Climate Change Agreements in Nigeria

Nigeria joined other nations of the world to mitigate and adapt to climate change by signing the UNFCCC of 1994,[72] The Kyoto Protocol of 1997 and the Paris Agreement of 2017. Nigeria is a non-annex I party to the UNFCCC because it is a developing country. Nigeria's obligations are to produce national communications, four in-depth reviews/summaries, a demonstrable progress report produced by the National Adaptation Programme of Action (NAPA), and a Global Climate Observing System (GCOS) Report. Following these obligations, Nigeria submitted its first national communication in 2003, its second national communication in February 2014, and recently submitted its third national communication in April 2020.[73] Also, Nigeria submitted its Intended National Determined Contributions (INDC) in November 2015 and updated NDC in 2021, pledging a 20% unconditional 47% conditional reduction of emission by 2030.[74]

Nigeria has continuously renewed its commitment to the UNFCCC. At the COP25 held in Madrid, Spain, from 2 to 13 December 2019, Nigeria listed eight areas to reduce carbon emissions. It stated its commitment to 45% conditional and 2% unconditional emission reduction of carbon, off-grid solar up to 13 Giga Watts, reforestation, climate-smart agriculture, promoting the use of buses instead of the massive use of cars, improving the electricity grid, attaining 30% energy efficiency by 2030, and ending gas flaring by 2030.[75] Nigeria is also committed to the Kyoto Protocol and the Paris Agreement. As a UNFCCC non-annex I party, Nigeria must not comply strictly with the protocol's provision. However, implementing the UNFCCC, the Kyoto Protocol, and the Paris Agreement, as well as participating in the Clean Development Mechanism (CDM), can accrue several benefits to Nigeria, such as—the acquisition of environmentally benign technologies, access to additional funds for climate change projects, capacity building of national institutions, and several sustainable development initiatives.[76]

Nigeria's policy position is that gas flaring is unacceptable. The FGN has initiated several actions to reaffirm its commitment to ending the practice of gas flaring in our oil fields. It is also a signatory to the Global Gas Flaring Partnership (GGFR)

---

[72] Non-Annex 1 Parties, Kyoto Protocol.

[73] UNFCCC (2021).

[74] UNDP (2025).

[75] O Adanikin (2016).

[76] United Nations Framework Convention on Climate Change (UNFCCC) (2020).

principles for global flare-out by 2030 while committing to a national flare-out target by 2020.

However, the core challenge in implementing Nigeria's obligations under these international agreements has been the lack of necessary frameworks and human and financial capacity to address the vulnerability of various country sectors to climate change. The country's policy framework for aligning human development and climate change management remains undeveloped. Nevertheless, the government has recognized the need for national policies, agencies, strategies, and plans to address climate change.[77]

5 Ministry and Agencies Involved in Climate Change in Nigeria

Several agencies in Nigeria have been established to tackle climate change, with the Federal Ministry of Environment as the supervisory body at the federal level. In the sub-nationals, there are state ministries of environment and agencies. However, this paper is examining the agencies established by federal laws.

5.1 *Federal Ministry of Environment*

The Federal Ministry of Environment (FMEnv), though created by executive decision and not an act of parliament, is the primary public body responsible for setting Nigeria's environmental agenda, including the coordination of national climate change responses. Since 1999, when it was established upon the return of Nigeria to democratic rule, the ministry has been leading the framing and implementation of laws and policies for environmental sustainability, conservation, and integrating environmental considerations into economic development. The importance of its roles and powers is seen majorly in the statutory powers some subsidiary laws confer on the Minister of Environment to appoint and approve. Also, the global challenge of climate change has further increased the strategic relevance of the ministry, as it is mandated to lead national actions on all plans, policies, and laws framed for the purpose of responding to the climate change challenges

Talking about its role in managing the national climate change policies, the ministry supervises government institutions that provide expert and policy advice to the government on directions and policies that make contributions to the global effort at responding nationally to climate change. Its work encompasses biodiversity conservation, pollution control, waste management, afforestation, and disaster risk management, all of which intersect with climate adaptation and mitigation strategies. Apart from agencies and parastatals such as the National Council on Climate Change

[77] Federal Republic of Nigeria (n 35).

which is not an enforcement institution, there are agencies such as the National Environmental Standards Regulation and Enforcement Agency (NESREA), which are under the oversight supervision of the Ministry of Environment. Another such agency is the National Oil Spill Detection and Response Agency (NOSDRA), which is mandated to prevent and manage oil spill incidents. Collectively, these institutions provide the technical and enforcement backbone of Nigeria's environmental governance.

5.2 National Environmental Standards and Regulations Enforcement Agency (NESREA)[78]

Established on 30 July 2007 by the National Environmental Standards and Regulations Enforcement Agency Act, NESREA is empowered to regulate and enforce environmental regulations, laws, policies, guidelines, and standards in Nigeria. Section 2 of the NESREA Act provides that: "the Agency shall, subject to the provisions of this Act, have responsibility for the protection and development of the environment, biodiversity conservation, and sustainable development of Nigeria's natural resources in general and environmental technology, including co-ordination and liaison with relevant stakeholders within and outside Nigeria on matters of enforcement of environmental standards, regulations, rules, laws, policies, and guidelines." Sections 7 and 8 of the Act provide for the functions of the Agency. The Agency is empowered to ensure compliance with national and international environmental laws through monitoring and regulatory measures.[79] However, it is essential to note that these regulatory functions have a monumental exception; none extends to the oil and gas sector.[80]

5.3 National Council on Climate Change (NCCC)

The National Council on Climate Change (NCCC), established by the Climate Change Act of 2021, represents Nigeria's most ambitious institutional reform for addressing climate change.[81] Being the only public institution that is specifically for addressing climate change, it leads national responses to climate change primarily by coordinating all climate-related policies, programs, and actions formulated and being implemented by the government.[82] The Climate Change Act 2021 provides

[78] Federal Republic of Nigeria (July 2007).

[79] NESREA Act Sections 7 and 8.

[80] NESREA Act Sections 7 (g), (h), (j), (k) and (l).

[81] Section 3, Climate Change Act, 2021.

[82] Section 1, Climate Change Act, 2021.

the statutory framework for the NCCC's creation, powers, and functions. To demonstrate the high importance of the Council and the seriousness of the government in addressing the challenge of climate change, the Council is chaired by the President of the Federal Republic of Nigeria, with the Vice President as Vice Chair.[83] Other members are ministers responsible for key ministries, namely, environment, finance, petroleum, power, solid minerals, women's affairs, transportation, water resources, national planning, agriculture, water resources, and justice, along with representatives from the private sector, civil society, the youths, and persons with disabilities. This multi-stakeholder structure is intended to ensure cross-sectoral coordination and inclusive participation in decision-making. The Act also provides for the appointment of a Director-General as the Council's Chief Executive Officer and secretary to the Council. The office is tasked with overseeing day-to-day operations, supervising technical staff, and implementing the Council's strategic decisions.

5.4 Nigerian Meteorological Agency (NiMet)

The Nigerian Meteorological Agency (NiMet) is the primary institution with the mandate to oversee every meteorological activity and service in Nigeria.[84] By providing essential meteorological information and data for different purposes and to relevant agencies such as the National Civil Aviation Agency, NiMet remains a cardinal institution in the management structure for climate change in Nigeria. Established by the NiMet Act, 2022,[85] the agency is mandated to provide weather, climate, and hydrometeorological information for safety, socio-economic development, and environmental management.[86] Although it is not a regulatory or enforcement body like the National Environmental Standards and Regulations Enforcement Agency (NESREA) or a policymaking institution like the National Council on Climate Change (NCCC), its scientific and technical functions are critical for climate change adaptation, disaster risk reduction, and evidence-based policymaking. NiMet's statutory functions cover the observation, collection, processing, and archiving of weather and climate data across Nigeria. Furthermore, It provides real-time and long-term meteorological services to various sectors, including agriculture, aviation, water resources, health, energy, and disaster management. Also, NiMet is tasked with the sole responsibility of conducting climate research and producing seasonal and annual forecasts that inform both policy and practice of government, businesses, and the population.

[83] Section 5, Climate Change Act, 2021.

[84] Section 8(2), NiMet Establishment Act 2022.

[85] Section 1, NiMet Establishment Act 2022.

[86] Section 7, NiMet Establishment Act 2022.

5.5 *Nigerian Upstream Petroleum Regulatory Commission (NUPRC) and Midstream and Downstream Petroleum Regulatory Authority (NMDPRA)*

The Petroleum Industry Act (PIA) of 2021 overhauled the regulatory and governing structure of the oil and gas industry in Nigeria by establishing the Nigerian Upstream Petroleum Regulatory Commission (NUPRC) and the Nigerian Midstream and Downstream Petroleum Regulatory Authority (NMDPRA) and repealed several laws, including the Department of Petroleum Resources. The objective of this transformational law is to promote transparency, good governance, and accountability in the management of Nigeria's petroleum resources[87] while addressing the environmental and climate implications of hydrocarbon development.

5.5.1 Nigerian Upstream Petroleum Regulatory Commission (NUPRC)

The NUPRC, created under the Petroleum Industry Act, 2021,[88] has the power to regulate the technical and commercial operations of the upstream petroleum section of the petroleum industry, including exploration, drilling, production, and decommissioning. Because of the contribution of upstream operations to greenhouse gas emission, particularly through gas flaring, the PIA mandates the NUPRC to ensure operators in the upstream section of the industry comply the provisions of the Act aimed at eliminating gas flaring, except in a few cases allowed under the Act.[89] Also, the commission oversees the Nigerian Gas Flare Commercialization Programme (NGFCP), relaunched in 2022 to align with the PIA. Through the auctioning of flare sites to investors, NUPRC aims to eliminate routine gas flaring and monetize associated gas, reducing greenhouse gas (GHG) emissions. In September 2023, it awarded 42 flare sites to private developers.[90] Also, by emission control regulatory mechanism for the upstream sector, the Commission contributes to Nigeria's compliance with the carbon budgets mandated under the Climate Change Act, 2021.

5.5.2 Nigerian Midstream and Downstream Petroleum Regulatory Authority (NMDPRA)

The NMDPRA is the second regulatory body established under the PIA.[91] It regulates midstream and downstream petroleum operations of the oil and gas industry, covering

---

[87] Section 1, Petroleum Industry Act (PIA) of 2021.

[88] Section 4, PIA, 2021.

[89] Sections, 104 and 108, PIA, 2021.

[90] Addeh, "History as NUPRC Picks 42 Winners for Its Gas Flare Commercialisation Bid Round." Available online at https://www.thisdaylive.com/2023/09/14/history-as-nuprc-picks-42-winners-for-its-gas-flare-commercialisation-bid-round/.

[91] Section 29 (1), PIA, 2021.

processing, storage, transportation, distribution, and marketing of petroleum products. The objectives of the law creating it include ensuring the efficient and sustainable development of infrastructure for healthy, safe operations.[92] Also as part of its regulatory functions, the Authority ensures that license is not granted to an operator unless its operational plan provides for the elimination of routine gas flaring.[93] Furthermore, to ensure the quality of petroleum products imported into the country conforms with the ECOWAS prescribed standard, the Authority enforces fuel quality standards, such as the Afri-5 Specification (50 ppm Sulphur), to reduce air pollution and greenhouse gas emissions from petroleum products.[94]

6 Nigerian Policies, Regulations and Laws on Climate Change

The climate change related laws in Nigeria are embedded in various laws for specific sectors or industries. However, there policies that are formulated to address environmental challenges which include climate change. There are also climate change related regulations that are subsidiary laws made pursuant to the provisions in an Act. The major policies, regulations and laws are briefly examined in this chapter.

6.1 National Environment Policy (NEP) 1999[95]

To adequately address environmental problems and challenges, the Nigerian government promulgated the National Environmental Policy in 1989. To accommodate new and emerging environmental concerns, this policy was revised in 1999. The goal of the revised NEP is to achieve sustainable development based on the proper management of the environment, to secure a quality environment adequate for all Nigerians' health and well-being, to ensure that natural resources are used for the benefit of present and future generations, to promote environmental education and public awareness, to restore and enhance the ecosystems.[96] The NEP proposes initiating and implementing complementary policies, strategies, and management approaches to ensure environmental concerns are adequately integrated into the economic decision-making process. The policy further stated that major developmental projects must incorporate ecological remediation costs, embrace environmentally friendly technologies, and ensure environmental impact assessment before embarking on any significant development project. The policy further adumbrates the importance of sustainable

[92] Section 30 (c), PIA 2021.

[93] Section 111 (d), PIA 2021.

[94] Section 317 (11), PIA 2021.

[95] National Environment Policy (2017).

[96] Oladipo (2010).

development principles such as the precautionary principle, pollution prevention principle, polluter pays principle, the user pays principle, the intra-generational and intergenerational equity principle, and the subsidiary principle.

6.2 National Water Policy (NWP) 2004[97]

This policy was enacted to optimize Nigeria's water resources for present and future generations. In light of this, it seeks to improve the nation's water resources management, including managing hydrological risks and vulnerabilities. The policy emphasized assessing water resources to enhance real-time forecasting of hydrological phenomena such as precipitation, floods, and droughts. A significant adaptation measure is required to reduce societal vulnerability to the impacts of climate change.[98] Section 5(3)(3) of the policy explicitly stated the guiding principles, some of which are—water policy must comply with the Constitution, all water is a national asset and resource common to all, and managing the nation's water resources to ensure optimum, long term, environmentally sustainable social and economic benefit. This section emphasized that individuals do not own water but have the right to equitable access to it. The policy also emphasizes the importance of awareness and participation of all users at all levels and the "polluter pays" principle, among others.

6.3 National Erosion and Flood Control Policy (NEFCP) 2005[99]

This policy was enacted to ensure coordinated and systematic measures to manage and control the hazards and risks of erosion and floods, reducing their impacts on people and the environment.[100] Key strategies for the implementation of the policy are to:

- Evolve a mechanism for forecasting, monitoring, and control of erosion and floods;
- review the land use laws and regulations;
- promote and strengthen training at all levels in erosion and flood prevention, management and control;
- create public awareness to encourage participation;
- protect the marginal lands by limiting utilization;
- subject resources users and developers to guidelines; and

---

97 Federal Republic of Nigeria (n 93).

98 Federal Republic of Nigeria (n 35).

99 As above.

100 Wonah (2017).

- provide early warning systems to avert the escalation of flood and erosion hazards.

An Action plan was developed to aid the implementation of this policy. The Action Plan encourages the active participation of all stakeholders in creating efficient institutions and laws that will ensure the prevention and management of erosion and floods and facilitate sustainable funding mechanisms for effective erosion and flood management.[101]

6.4 *National Forest Policy (NFP) 2006*[102]

This policy is geared towards ensuring sustainable forest management, promoting the participatory process in forest development, facilitating the private sector's involvement in forestry development, and adopting an integrated approach. The government is currently embarking on several afforestation programs, for example, the "Green Wall Initiative," in which a "green wall" of trees will be planted across the Sudano-Sahelian zone of the country (40 million trees annually in the next 10 years).[103] The National Forest Policy is consistent with the national objectives and the principles guiding sustainable development. It focuses on managing the nation's forestry sector so that the forests continue to deliver goods and services perpetually and meet this generation's needs without compromising future generations' rights.[104] The policy recognized past problems and stressed the integration of community, conservation, and commercial forest management for the benefit of all. Section 2 set out the guiding principles for the National Forestry Policy as follows:

- To revitalize the contribution of forests to economic development and growth.
- To encourage community and civil society in forestry development.
- Enhance partnerships with the private sector, Non-Governmental Organizations (NGOs), and Community-Based Organizations (CBOs).
- Implement the international forest policy initiatives, Intergovernmental Panel in Forests (IPF), and Intergovernmental Forum on Forests (IFF) proposals for sustainable forest management.

---

[101] Oladipo (n 95).

[102] Federal Republic of Nigeria 'National Forest Policy' (June 2006).

[103] OF Ati, E Agubamah & IB Abaje '(2018).

[104] (n 101) Section 2(1).

6.5 National Policy on Drought and Desertification (NPDD) 2007[105]

The Federal Ministry of Environment formulated the NPDD to implement the National Action Programme (NAP) that was submitted to the United Nations Convention to Combat Desertification (UNCCD).[106] This policy identifies how climate change can exacerbate drought and desertification, particularly in the northern part of the country. To curb this, the policy stresses the necessity of adequately equipping relevant institutions, agencies, stakeholders, and citizens to collect, analyze, and use data effectively to alleviate and combat drought and desertification.[107]

The policy specifically stated specific strategies that aid climate change adaptation preparedness, which are:

- strengthening of agencies, institutions, and facilities for collection and analyses of meteorological and hydrological information as well as dissemination of such information;
- ensuring the effectiveness of existing national early warning facilities;
- creating and facilitating programmes for formal and informal education to improve information on climate and environmental issues, and
- enhancing carbon dioxide sequestration by promoting appropriate land use.

6.6 Nigeria Vision 2020, 2010[108]

The National vision was introduced to strengthen environmental governance, promote education on the environment, and optimize economic advantages from sustainable ecological management. Section 1 stated that "the NV20: 2020 was developed by Nigerians for the Nigerian people and involved a process of thorough engagement with all stakeholders across all levels of government and society. The Vision is, therefore, a rallying point for all Nigerians, regardless of ethnicity, political leaning, economic status, or religion, behind a common cause of placing the country on a sustainable development path and transformation into a modern society better able to play a greater role in the comity of nations".[109]

The reduction of the impact of climate change on development processes and the environment is one of the vision targets. The vision acknowledges climate change as a critical challenge in Nigeria that can cause irrecoverable damage to infrastructure, food production, and water supplies, precipitating natural resource conflicts. There were also climate mitigation targets formulated under the vision, which are—to

---

105 Oladipo (n 95).

106 Oladipo (n 95).

107 O Fagbohun & FE Nlerum '(2011).

108 Federal Republic of Nigeria 'Nigeria Vision 20: 2020—Abridged Version' (December 2010).

109 As above.

increase by 2013 the energy mix of hydro-power to 25%, wind energy capacity to 10 MW, solar energy capacity to 10 MW, and biomass power generation capacity to 1000 MW. Forests cover from 6 to 10% and reduce the losses and impacts of floods and drought to 10% by 2013. It is noteworthy to state that as of 2020, these targets still need to be met.

6.7 National Environmental (Soil Erosion and Flood Control) Regulations (NESEFCR) 2011[110]

This policy was adopted in 2011 by NESREA according to Section 34 of the NESREA Act to establish technically feasible and economically reasonable standards and procedures to achieve the appropriate level of management and conservation practices to abate soil erosion, dilatation, and segmentation of the waters of Nigeria, which are caused by soil erosion and floods aggravated by non-agricultural earth-damaging activities.[111] These regulations ensure the sustainable protection and enhancement of the ecological integrity of flood plains, vulnerable lands, and water from the significant adverse effects of environmental degradation resulting from persistent soil erosion, flooding, and sediment deposition.[112] These regulations aim to protect human life and the environment, minimize losses caused by floods and erosion, and mitigate their impact on vulnerable areas. They achieve this by regulating land-disturbing activities and controlling accelerated soil erosion, flooding, and sediment deposition in water bodies and watercourses to prevent pollution of these water resources.[113]

6.8 National Environmental (Desertification Control and Drought Mitigation) Regulations, 2011[114]

This regulation was adopted in 2011 by NESREA under Section 34 of the NESREA Act.[115] Parts I and II of the regulation deal with provisions and regulations regarding desertification control, and parts III and IV deal with provisions and regulations regarding drought mitigation. Section 2 of the rules sets out the objectives regarding

[110] Federal Republic of Nigeria 'National Environmental Soil Erosion and Flood Control Regulations' (April 2011).

[111] (n 110) Section 3(1).

[112] (n 110) Section 3(2).

[113] (n 110) Section 2.

[114] Federal Republic of Nigeria 'National Environmental Desertification Control and Drought Mitigation Regulations' (April 2011).

[115] Federal Republic of Nigeria 'National Environmental Desertification Control and Drought Mitigation Regulations' (April 2011).

desertification control. These regulations are to provide practical and pragmatic regulatory frameworks for the sustainable use of areas already affected by desertification, the protection of vulnerable lands, sensitize the public on the causes and dangers associated with desertification, encourage the sustainable use of fuelwood, promote the use of alternative sources of energy, encourage reforestation, reseeding, afforestation and conservation of areas under desertification or vulnerable to duplicate for the rehabilitation as well as sustain areas under forest and tree cover through conservation, protection, rehabilitation of natural negations, tree planting and control of forest exploitation, to reverse desertification trend.

These regulations also explicitly state how it plans to tackle desertification, which are

- attaining 25% forest cover for the enhancement of ecological integrity and the reduction of the impacts of climate change;
- ensuring sustainable agriculture, improved animal husbandry, and management of water resources in desertification-prone areas to achieve sustainable livelihoods;
- poverty reduction and wealth creation through the introduction of modern and affordable production technologies to resource-poor farming communities;
- Promote cooperation with relevant international and non-governmental organizations through partnerships, knowledge sharing, and the domestication of conventions such as the UNCCD, UNCBD, and UNFCCC, to which Nigeria is a signatory.[116]

Section 12 of the regulations set out the objectives regarding drought mitigation. It states that the regulation is to ensure appropriate proactive measures are taken to alleviate the impacts of drought at its occurrence, to ensure proper criteria for declaring drought emergencies are established, to ensure adequate definition of the duties and responsibilities of all agencies concerning drought management, identify mitigation actions which can be employed to address vulnerabilities and reduce drought impacts and finally to identify the principal activities, groups or areas at risk during drought and develop mitigating actions to reduce the effects of drought.

6.9 National Adaptation Strategy and Plan of Action on Climate Change for Nigeria (NASPA-CCN), 2011[117]

This strategy focuses on climate change adaptation in Nigeria. Section 5(3) of the NASPA-CCN contains the strategy's vision, goals, and objectives. This Strategy envisions a Nigeria where climate change adaptation is an integrated development component. The focus of the plan is to ensure the reduction of the vulnerability and to enhance the resilience and adaptive capacity of all economic sectors and all people (mainly women, children, and resource-poor men) to the adverse impacts of climate

[116] As above.

[117] BNRCC (n 40).

change while also capturing the opportunities that arise as a result of climate change. The NASPA-CCN aims to take necessary actions that will help reduce the vulnerability to climate change impacts, increase the resilience and sustainable well-being of all, and minimize the risks by improving adaptive capacity, leveraging new opportunities, and facilitating collaboration outside Nigeria. This strategy aims to reduce the impacts of climate change through adaptation measures that can be undertaken by public and private institutions, communities, and individuals and increase awareness and preparedness for climate change impacts. Furthermore, it focuses on mobilizing communities for climate change adaptation actions, reducing the effects of climate change, and integrating climate change adaptation into all sectors of the country.[118]

6.10 National Environmental (Control of Vehicular Emissions from Petrol and Diesel Engines) Regulations, 2011[119]

This regulation was adopted in 2011 by NESREA under Section 34 of the NESREA Act. These regulations are to restore, preserve, and improve air quality. The rules are to ensure the protection of the air from pollutants, citizens' right of access to clean air, reduce and prevent air pollution through the improvement of the quality of automobiles that operate on the roadway, as well as improve the health of Nigerians, especially in the urban areas with a high incidence of air pollution due to increased number of automobiles that ply the roads.

6.11 National Environmental (Control of Bush/Forest Fire and Open Burning) Regulations, 2011[120]

Having identified how indiscriminate bush and forest burning is contributing to the pollution of the environment and destruction of ecosystem, the regulation is made pursuant to Section 34 of the NESREA Act 2007 which empowers the minister of environment to make such provision as necessary for achieving the objectives of the Act. The regulation prevents and minimizes the destruction of ecosystems through fire outbreaks and the burning of any material that may affect the ecosystem's health

[118] As above.

[119] Federal Republic of Nigeria 'National Environmental Control of Vehicular Emissions from Petrol and Diesel Engines Regulations' (April, 2011).

[120] Federal Republic of Nigeria 'National Environmental Control of Bush/Forest Fire and Open Burning Regulations' (April 2011).

by emitting hazardous air pollutants.[121] It prohibits the burning of bushes and forests without permit issued by the NESREA.[122]

6.12 National Climate Change Policy Response and Strategy (NCCP-RS) 2013[123]

The Federal Executive Council adopted this policy to ensure an effective national response to climate change's multi-faceted impacts. The NCCP-RS focuses on ensuring Nigeria becomes climate change resilient and ready for sustainable socio-economic development. The strategy aims to strengthen national initiatives to adapt and mitigate the impact of climate change. Also, mitigation measures should be implemented to promote low carbon emissions, increase the nation's capacity to adapt to climate change, and support and encourage climate change-related science, technology, research, and development. Furthermore, it focuses on increasing public awareness and the private sector's role in tackling climate change, strengthening institutions and mechanisms within the country to establish a suitable and functional framework for climate change governance.

Several policies have been developed due to this strategy; some of them are:

- The National Energy Policy offers the framework for sustainable energy development, with the private sector participating to provide clean, affordable, adequate, and reliable energy.
- The Sustainable Energy for All Action Agenda (SE4ALL-AA) ensures universal access to modern energy services, doubles energy efficiency, and doubles renewable energy use by 2030.[124]
- Nigerian Renewable Energy and Energy Efficiency Policy (NEP): This policy seeks to achieve a renewable electricity target of 16% by 2030, as opposed to the current 1.3%.[125]
- The Renewable Energy Master Plan (REMP) aims to articulate a roadmap for national renewable energy development.
- The Nigeria Feed-in Tariff for Renewable Energy Sourced Electricity is an optimal economic instrument for hydro schemes not exceeding 30 MW, all biomass

---

[121] As above.

[122] Section 4 (1), National Environmental Control of Bush/Forest Fire and Open Burning Regulations' (April 2011).

[123] Federal Republic of Nigeria 'National Climate Change Policy Response and Strategy' (October 2013).

[124] Federal Republic of Nigeria 'Sustainable Energy for All Action Agenda' (July 2016).

[125] Federal Republic of Nigeria 'National Renewable Energy and Energy Efficiency Policy' (April 2015).

cogeneration power plants, and solar and wind-based power plants, regardless of size.[126]

- The UN-REDD Programme aims to consolidate countries' efforts to reduce deforestation and forest degradation emissions, foster conservation and sustainable forest management, and enhance forest carbon stocks.
- Solid Waste Programme Interventions of the Federal Ministry of Environment—This is to implement an integrated solid waste management programme.
- The Natural Gas Flare Out Policy—This policy aims to phase out gas flaring in the country and utilize its economic advantage.

6.13 National Environmental (Energy Sector) Regulations, 2014[127]

This regulation was adopted in 2014 by NESREA under Section 34 of the NESREA Act. These regulations prevent or minimize pollution and encourage energy efficiency in all operations and ancillary activities of the energy sector to achieve sustainable economic development in Nigeria. Section 2(2) states that the regulations encourage electrical power efficiency in all electrical appliances by adopting green technologies based on the ice cycle approach, anchored on the 5Rs (reduce, reuse, repair, recover, and recycle) cradle to cradle. These regulations cover power generation, transmission and distribution from combustion processes fuelled by gaseous, liquid and soil fossil fuels designed to deliver electrical or mechanical power, steam, heat, or any combination of these, regardless of the fuel type, renewable and nuclear sources. The regulation stated that the energy sector should deploy cost-effective, energy-efficient processes that align with the Kyoto Protocol and encourage green investment in renewable energy and energy efficiency. Also, to reduce GHG emissions, address the problem of environmental degradation resulting from pollution, deforestation, and vegetation loss, and ensure that the energy sources are ecologically sustainable.

6.14 National Drought Plan (NDP) 2018[128]

The NDP provides an integrated document that addresses the challenges posed by water shortages in agriculture, navigation, energy production, drinking water supply, recreation, nature, and urban areas faced during droughts. The NDP aims to provide an effective and systematic means of assessing drought conditions and developing mitigating actions and programmes to reduce drought's potential vulnerability

[126] International Energy Agency 'Nigeria Feed-in Tariff for Renewable Energy Sourced Electricity' (23 August 2017).

[127] Federal Republic of Nigeria 'National Environmental Energy Sector Regulations' (December 2014).

[128] Federal Republic of Nigeria 'National Drought Plan' (November 2018).

and impacts. To achieve these aims, the NDP plans to—implement comprehensive drought monitoring and early warning systems, carry out detailed vulnerability assessments for sectors, populations, and regions vulnerable to drought, implement mitigation measures to limit the adverse impacts of drought and provide appropriate response measures when drought occurs.

6.15 *Flare Gas Regulations (Prevention of Waste and Pollution), 2018*[129]

This regulation was enacted to prohibit the flaring of gas in Nigeria. In light of this, this regulation seeks to reduce the environmental and social impact of gas flaring, ensure environmental protection, prevent the waste of natural resources, and create social and economic benefits from gas flare capture.[130] This regulation prohibits the flaring and venting of gas without having a certificate issued by the Minister.[131] It also provides the legal framework for the Nigerian Gas Flare Commercialisation Programme (NGFCP). The regulation introduces a new penalty for gas flaring by adopting the "polluter pays" principle as a carbon tax. The regulation changed the gas flare penalties from N10 per thousand standard cubic feet to USD 2.0 per 28.317 standard cubic metres (one thousand standard cubic feet) of gas in the case of any company producing 10,000 barrels of oil or more and USD 0.50 per thousand standard cubic square feet of gas, in the case of any producing less than 10,000 barrels of oil per day.[132] The failure of any oil company to make accurate flare data, provide access to flares or flare sites and sign a Connection Agreement carries additional fees of USD 2.50 per 28.317 standard cubic metres (one thousand standard cubic feet) of gas.[133]

6.16 *Nigerian Gas Flare Commercialisation Programme, 2018*[134]

This programme permits the Nigerian government to sell flare gas via competitive and transparent bidding. This programme aims to stop gas flaring through competent investors invited to participate in a transparent bidding process. These investors are expected to develop technically and commercially sustainable gas utilization

[129] Federal Republic of Nigeria 'Flare Gas Regulations (Prevention of Waste and Pollution)' (July, 2018).

[130] (n 128) Section 1.

[131] (n 128) Section 12.

[132] (n 128) Section 13.

[133] (n 128) Section 21.

[134] Federal Republic of Nigeria 'Nigerian Gas Flare Commercialisation Programme' (July 2018).

projects. The government strategically established this programme to implement its plans and policies for eliminating gas flaring by accruing commercial benefits for Nigeria. In light of the recently enacted Petroleum Industry Act (PIA), 2021, the Nigerian Gas Flare Commercialization Programme (NGFCP) was relaunched in 2022 to align it with the provisions of the PIA to reflect the current industry realities. The regulatory authority, the Nigerian Upstream Petroleum Regulatory Commission (NUPRC), announced the successful bidders for the NGFCP in October 2023. Out of 49 sites auctioned, 42 companies/entities were awarded flare sites. These include both stand-alone single flare site developments and cluster developments.

6.17 Nigeria's National Adaptation Plan Framework, 2020[135]

The National Adaptation Plan (NAP) Framework is an international initiative established under the Cancun Adaptation Framework (CAF) monitored by NAP Global Network. This framework provides a guideline for Nigeria to help it formulate, regulate, and enforce all the various plans, strategies, policies and laws that will enable it to address its medium and long-term adaptation needs. This Framework's objectives are to:

- Ensure proper clarification of the country's approach to its NAP process, such as articulating its vision and objectives on climate change adaptation, the roles and responsibilities among relevant stakeholders, and any priority actions to be undertaken. It also serves as the reference point for integrating adaptation planning efforts from various sectors and decision-makers (e.g., national, state, and local governments).
- Align the NAP process with existing laws, policies, strategies, regulations, and research.
- Focus on specific themes relevant and unique to Nigeria's context.

The two broad objectives of the NAP process are to:

- Reduce vulnerability to the impacts of climate change by building adaptive capacity and resilience.
- Facilitate the coherent integration of climate change adaptation into relevant new and existing policies, programmes, and activities, particularly development planning processes and strategies, within all relevant sectors and at different levels, as appropriate.

[135] Federal Republic of Nigeria 'Nigeria's National Adaptation Plan Framework' *Federal Ministry of Environment* (June 2020).

6.18 The Climate Change Act, 2021

The Act was enacted in 2021 in response to the call for a legal framework aimed explicitly at establishing and empowering a specialized agency for climate change and related affairs. By the Act, the National Council on Climate Change (NCCC) was established to set policy and regulatory pathways to achieving national emission targets. Also, the NCCC is responsible for coordinating the implementation of climate change policies and plans across government agencies to ensure strict adherence to Nigeria's international climate change commitments, such as the Paris Agreement.[136] One of the Act's key provisions requires the Federal Government to develop a 5-year carbon budget outlining sector-specific emission reduction targets.[137] It also mandates integrating climate action into national development plans, compelling public and private entities to adopt sustainable strategies.[138] Additionally, the law introduces a climate change fund to support mitigation and adaptation projects, particularly for vulnerable communities.[139]

6.19 The Petroleum Industry Act

The Petroleum Industry Act (PIA), signed into law in 2021, aims to reform Nigeria's petroleum sector by enhancing efficiency, transparency, and revenue generation while promoting local participation. For Nigeria's climate change and energy transition agenda, the PIA is particularly significant because it restructured the governance of petroleum operations in ways that directly intersect with environmental sustainability and climate policy: the Nigerian Upstream Petroleum Regulatory Commission (NUPRC)[140] for upstream operations and the Nigerian Midstream and Downstream Petroleum Regulatory Authority (NMDPRA)[141] for midstream and downstream activities. Additionally, it replaces the Nigerian National Petroleum Corporation (NNPC) with NNPC Limited, a commercially driven entity with enhanced operational autonomy.[142] These agencies replaced the Department of Petroleum Resources (DPR) and other defunct regulators, streamlining oversight across the value chain. Their mandates extend beyond commercial and technical regulation to include enforcement of environmental standards and the reduction of harmful practices such as routine gas flaring. The Act integrates environmental considerations into licensing and operational requirements, requiring petroleum companies to submit

[136] Section 2(1) CCA, 2021.

[137] Section 20 (1) CCA, 2021.

[138] Section 19 (1) CCA, 2021.

[139] Section 13 (1) CCA, 2021.

[140] Section 4 (1) PIA, 2021.

[141] Section 29 (1) PIA, 2021.

[142] Section 53 (1) PIA, 2021.

decommissioning and site remediation plans, thereby strengthening accountability for ecological damage.

One of the PIA's most important contributions to Nigeria's climate governance is its treatment of natural gas. The Act elevates gas to a central position in Nigeria's energy strategy, framing it as a "transition fuel" that can drive industrialization, expand electricity access, and reduce reliance on biomass for cooking. By supporting programs such as the Nigerian Gas Flare Commercialisation Programme (NGFCP), the PIA provides the legal backing for capturing and utilizing gas that would otherwise be wasted through flaring. This not only reduces greenhouse gas emissions but also creates opportunities for power generation and cleaner domestic energy use.

The PIA also introduced fiscal and governance provisions intended to promote sustainability. For instance, it established the Host Communities Development Trust, requiring petroleum companies to contribute a percentage of their operating expenditure to local development projects. While framed primarily as a conflict-prevention and community empowerment tool, this provision has indirect climate benefits, as it channels resources into environmental restoration and resilience-building in host communities disproportionately affected by oil exploration and climate impact. Furthermore, it includes environmental protection provisions, requiring oil companies to implement decommissioning and remediation plans to mitigate pollution and ecological damage.[143]

6.20 Energy Transition Plan 2022

The Nigeria Energy Transition Plan (ETP), launched in August 2022, represents the country's first detailed and long-term decarbonization strategy. It provides a comprehensive framework for achieving a balance between economic development, universal energy access, and climate change mitigation. The plan sets a clear national target of reaching net-zero emissions by 2060, aligning Nigeria with global climate ambition while recognizing the country's status as Africa's largest oil producer and a major gas economy. The omission of the ETP from the original manuscript leaves a critical gap, as it has since become the reference point for all of Nigeria's climate and energy-related policymaking.

At the core of the ETP are six pillars. First, it commits Nigeria to a net-zero target by 2060, a goal that marks a significant step forward from earlier climate pledges. Second, the plan seeks to expand the country's renewable energy capacity to 30 gigawatts by 2030, enabling a rapid scale-up of solar, wind, and hydropower projects. Third, it prioritizes universal electricity access, aiming to connect 100 million more Nigerians to modern energy services by the end of the decade. Fourth, the plan emphasizes the transition away from traditional biomass to clean cooking solutions, thereby addressing both public health and environmental concerns. Fifth, the ETP outlines a financing requirement of approximately $410 billion over the next four

[143] Section 102 (1) PIA, 2021.

decades, with at least $10 billion needed annually through 2030. Finally, and most controversially, the plan identifies Nigeria's vast natural gas reserves as a "transition fuel", framing gas as a bridge resource to industrialization and energy security before a complete shift to renewables.

The Energy Transition Plan has profound implications for Nigeria's climate governance and institutional framework. Agencies such as the Nigerian Upstream Petroleum Regulatory Commission (NUPRC) and the Nigerian Midstream and Downstream Petroleum Regulatory Authority (NMDPRA) are tasked with aligning petroleum operations with the plan's goals, particularly in reducing gas flaring and promoting gas-to-power infrastructure. The National Council on Climate Change (NCCC), as the apex coordinating body, is expected to ensure that the ETP's targets are mainstreamed into national development planning and linked to the carbon budgets mandated under the Climate Change Act of 2021. Other agencies, including NiMet and NESREA, provide technical and regulatory support to ensure that the plan's ambitions are grounded in scientific data and compliance mechanisms.

6.21 Presidential CNG Initiative (PCNGI) 2023

The Presidential Compressed Natural Gas Initiative (PCNGI), launched in 2023, represents a flagship component of Nigeria's efforts to transition its energy and transport sectors toward cleaner and more affordable alternatives. It is a direct outgrowth of the National Gas Expansion Programme (NGEP) and aligns closely with the Nigeria Energy Transition Plan (ETP, 2022), which identifies natural gas as a critical "transition fuel" in the country's pathway to net-zero by 2060. The PCNGI specifically targets the transportation sector, a major contributor to greenhouse gas emissions and urban air pollution, by encouraging the widespread adoption of compressed natural gas (CNG) in place of petrol and diesel.

The initiative is designed to reduce Nigeria's carbon footprint while simultaneously addressing economic and social challenges linked to fuel pricing. Following the removal of petrol subsidies in 2023, CNG was positioned as a more affordable and cleaner alternative. CNG produces approximately 25% less carbon dioxide (CO_2) per unit of energy compared to petrol and diesel, while also emitting significantly lower levels of nitrogen oxides, sulfur oxides, and particulate matter. This dual benefit of cost reduction and emissions mitigation situates the PCNGI within Nigeria's broader climate change mitigation framework, while also responding to pressing domestic demands for energy affordability and security.

Implementation of the PCNGI involves the development of nationwide infrastructure for CNG refueling stations, vehicle conversion centers, and maintenance facilities. The government has set ambitious targets, including the conversion of over one million vehicles to CNG by 2027, with an initial focus on public transport buses and commercial fleets. By prioritizing high-volume vehicles such as buses and taxis, the initiative aims to maximize both economic and environmental impact. The Nigerian Midstream and Downstream Petroleum Regulatory Authority (NMDPRA) plays a key role in licensing infrastructure for CNG distribution, while partnerships with private sector investors have been encouraged to accelerate deployment.

7 Challenges and Successes of Climate Change Targeted Laws in Nigeria

Nigeria's response to climate change threats in the context of policy development, legal framework, and implementation remains a significant challenge. According to the National Economic Empowerment and Development Strategy (NEEDS) study, the current approach needs to address climate change's effects in the future. A more comprehensive and coordinated approach to mitigating the impact of climate change in Nigeria is required.[144] Another challenge is the limited human and institutional capacity for climate change uncertainty and effects. Very few experts and professionals in the country are experts on this subject. Institutional and professional competencies are yet to be fully built to aid the development and implementation of appropriate climate change adaptation and mitigation actions. Furthermore, there needs to be more funding for climate change projects, and institutions in charge of mitigation and adaptation must be better funded.[145]

Despite the above challenges, Nigeria's response and contribution to the global commitment against climate change have not been all doom and gloom. The country has demonstrated a significant commitment to initiating and implementing adaptation and mitigation strategies within the past 10 years. After the initial failed attempt at establishing the climate change commission, the Nigerian Climate Change Act of 2021 established the National Council on Climate Change. The council has led the design and driving of critical policies to tackle climate change nationwide.

In addition, the Nigerian Gas Flare Commercialization Program has helped develop an emission control scheme to reduce gas flaring by attracting investments and creating a transparent market mechanism for allocating flared gas to third-party investors. Beyond addressing environmental concerns, the program promotes economic development by harnessing previously wasted resources for energy production.

Furthermore, in 2022 Nigeria ranked 162 out of 180 countries in the Environmental Performance Index. But in 2024, Nigeria was ranked 140 out of 180 countries.[146] That is a significant progress and also a proof that more can be achieved.

---

[144] Federal Republic of Nigeria (n 38).

[145] Oladipo (n 95).

[146] Environmental Performance Index, "2024 Environmental Performance Index—Nigeria."

8 Conclusion

Nigeria's climate change narrative has evolved significantly over the past two decades, moving from fragmented responses and outdated policy frameworks toward a more structured and coordinated governance system. The enactment of the Climate Change Act (2021), the Petroleum Industry Act (2021), and the launch of the Energy Transition Plan (2022) together mark a turning point in the country's approach. These instruments have elevated climate governance to the highest levels of policymaking, created specialized regulatory institutions, and mapped out a pathway toward net-zero emissions by 2060. At the same time, ongoing initiatives such as the Presidential Compressed Natural Gas Initiative (PCNGI) reflect the government's attempt to link climate goals with affordability, energy security, and socio-economic development.[147]

Also, the establishment of the National Council on Climate Change (NCCC) has centralized responsibility for climate action, providing coordination across ministries, agencies, and sectors. Complementary institutions such as the Federal Ministry of Environment, NESREA, NiMet, NUPRC, and NMDPRA play specialized roles in enforcement, data generation, regulatory oversight, and technical support. This institutional framework, though still developing, represents an important attempt to harmonize Nigeria's fragmented governance structures and align them with both domestic priorities and international commitments under the Paris Agreement and the UNFCCC.

Despite these reforms, Nigeria continues to face formidable challenges. Climate impacts such as flooding, drought, and desertification are increasing in frequency and severity, threatening food security, public health, and national stability. The devastating floods of 2022 underscored the urgency of building resilience and highlighted the gaps in preparedness, financing, and institutional capacity. Moreover, financing remains the Achilles' heel of Nigeria's climate agenda. The Energy Transition Plan alone requires over USD 400 billion,[148] an amount far beyond the government's current fiscal capacity, making international climate finance and private sector investment indispensable.

There are also deeper contradictions that complicate Nigeria's transition. While the Energy Transition Plan emphasizes renewable energy expansion, both the PIA and the CNG Initiative foreground natural gas as a "transition fuel." This approach is pragmatic given Nigeria's vast reserves and pressing energy poverty, but it also risks reinforcing long-term fossil fuel dependence. The tension between exploiting hydrocarbons for development and reducing emissions for climate stability reflects broader debates about justice, equity, and responsibility in the global energy transition. Nigeria's challenge is therefore not just technical or financial, but also the political will to reconcile immediate development needs with long-term sustainability.

---

[147] Egboboh, "Nigeria's CNG Initiative Attract over $500 m Investment in 2024—PCNGI."

[148] Sustainable Energy for All | SEforALL, "Nigeria Energy Transition and Investment Plan."

Moving forward, Nigeria's ability to confront climate change will depend on more than legislative frameworks or institutional reforms. It will require effective implementation, stronger enforcement, and greater inclusivity. State and local governments must be empowered to act, communities must be engaged in decision-making, and climate policies must be linked to poverty reduction, food security, and social justice. International partnerships, especially around climate finance and technology transfer, will remain vital. At the same time, Nigeria must harness its domestic capacities in renewable energy, youth innovation, and indigenous knowledge to craft solutions that are both context-specific and resilient.

Nigeria's climate change response stands at a crossroads. The country now has the laws, institutions, and policy blueprints needed to address the crisis. What remains is the political will, resource mobilization, and coordinated action to translate these frameworks into tangible results. If Nigeria succeeds, it will not only protect its people and ecosystems but also demonstrate how a developing country with deep hydrocarbon dependence can chart a just and inclusive path toward sustainability. If it fails, the costs will be borne not only in economic terms but also in human lives and social stability. The stakes could not be higher, and the time for decisive action is now. So far, the transport is leading actions that are helping reduce greenhouse gas emissions. Major truck haulage like Dangote company have invested so much in CNG technology as fuel for their trucks.[149]

References

Aaron Sayne (2011) Climate change adaptation and conflict in Nigeria. United States Institute of Peace, June 2011. https://www.usip.org/sites/default/files/Climate_Change_Nigeria.pdf

Adanikin O (2019) Gas flaring: Nigeria is not committed to the Paris agreement it signed in 2016, here is why. ICIR, 28 December 2019. https://www.icirnigeria.org/gas-flaring-nigeria-is-not-committed-to-the-paris-agreement-it-signed-in-2016-here-is-why/

Ahmad Salkida (2012) Africa's vanishing lake chad. UN, April 2012. https://www.un.org/africarenewal/magazine/april-2012/africa%E2%80%99s-vanishing-lake-chad#:~:text=In%20his%20book%20An%20Inconvenient,1%2C500%20square%20km%20in%202001

Akande A, Costa AC, Mateu J, Henriques R (2017) Geospatial analysis of extreme weather events in Nigeria (1985–2015) using self-organizing maps. Adv Meteorol. https://doi.org/10.1155/2017/8576150

Akinyemi O, Ogundipe A, Alege P (2012) Energy supply and climate change in Nigeria. Munich personal RePEc archive, 2 May 2012. https://mpra.ub.uni-muenchen.de/55820/1/MPRA_paper_55820.pdf

Amanchukwu RN, Amadi-Ali TG, Ololube NP (2015) Climate change education in Nigeria: the role of curriculum review. Education 5(3):71–79. http://article.sapub.org/10.5923.j.edu.20150503.01.html

Amangabara GT (2014) Understanding effective gully control measures in Imo State, Nigeria. Can Open Soil Erosion J 1(1):1

[149] Olawin, "CNG."

Anabaraonye B, Okafor CJ, Ikuelogbon OJ (2019) Educating farmers and fishermen in rural areas in Nigeria on climate change mitigation and adaptation for global sustainability. Int J Sci Eng Res 10(4):1391. https://www.ijser.org/researchpaper/Educating-Farmers-and-Fishermen-in-Rural-Areas-in-Nigeria-on-Climate-Change-Mitigation-and-Adaptation-for-Global-Sustainability.pdf

Ati OF, Agubamah E, Abaje IB (2018) Global climate change policies and politics: Nigeria's response. FUDMA J Polit Int Aff 106:1

Azaiki S (2007) Oil gas and life in Nigeria. Y-Books, p 280

Bell S, McGillivray D, Pedersen O, Lees E, Stokes E (2008) Environmental law. OUP, Oxford

BNRCC (2011) National adaptation strategy and plan of action on climate change for Nigeria (NASPA-CCN) (BNRCC, November 2011). http://csdevnet.org/wpcontent/uploads/NATIONAL-ADAPTATION-STRATEGY-AND-PLAN-OF-ACTION.pdf

Buma WG, Lee S-I, Seo JY (2016) Hydrological evaluation of lake chad basin using space borne and hydrological model observations. Water 8(5):205. https://www.mdpi.com/2073-4441/8/5/205

Central Bank of Nigeria, Crude oil price (US$/Barrel), production (mbd) and export (mbd). https://www.cbn.gov.ng/rates/crudeoil.asp

Climate Scorecard (2017) Nigeria checkup. Climate Scorecard, 4 July 2017. https://www.climatescorecard.org/2017/07/nigeria-checkup/

Co2.earth, Are we stabilizing yet? https://www.co2.earth/

Copenhagen Accord (2010) Report of the conference of the parties on its fifteenth session. Decision 2/CP.15, U.N. Doc. FCCC/CP/2009/11/Add.1

Ebele NE, Emodi NV (2016) Climate change and its impact in Nigerian economy. J Sci Res Rep 10(6):1–13. http://www.journaljsrr.com/index.php/JSRR/article/view/21917/40737

Edmonds J, Joos F, Nakicenovic N, Richels RG, Sarmiento JL (2004) Scenarios, targets, gaps, and costs. In: Field CB, Raupach MR (eds) The global carbon cycle: integrating humans, climate and the natural world. Island Press, London, pp 77–102

Egboka BCE, Nwankwor GI, Orajaka IP (1990) Implications of Palaeo- and neotectonics in gully erosion-prone areas of southeastern Nigeria. Nat Hazards 3(3):219

Egboka BCE, Orji AE, Nwankwoala HO (2019) Gully erosion and landslides in south-eastern Nigeria: causes, consequences and control measures. Global J Eng Sci GJES 2(4):1–11. https://irispublishers.com/gjes/fulltext/gully-erosion-and-landslides-in-southeastern-nigeria-causes-consequences-and-control-measures.ID.000541.php

EIA (2020) Country analysis executive summary: Nigeria. EIA, 25 June 2020. https://www.eia.gov/international/content/analysis/countries_long/Nigeria/NigeriaCAXS_2020.pdf

Elisha I, Sawa BA, Lawrence EU (2017) Evidence of climate change and adaptation strategies among grain farmers in Sokoto State, Nigeria. IOSR J Environ Sci Toxicol Food Technol 11(3):1–7. https://www.iosrjournals.org/iosr-jestft/papers/vol11-issue%203/Version-2/A1103020107.pdf

Enete I (2014) Impacts of climate change on agricultural production in Enugu State, Nigeria. J Earth Sci Clim Change 5(9):234. https://www.journaljsrr.com/index.php/JSRR/article/view/21917/40737

European Environment Agency, Climate change mitigation. Available at: https://www.eea.europa.eu/themes/climate

Ezeibe KK (2011) The legislative and institutional framework of environmental protection in the oil and gas sector in Nigeria—a review. NAUJILJ 2:39. http://www.oalib.com/paper/1345793#.WyNufDMzZ-U

Fadare T (2019) Buhari declines assent to digital rights climate change bills. Order paper, 20 March 2019. https://www.orderpaper.ng/buhari-declines-assent-to-digital-rights-climate-change-bills/

Fagbohun O, Nlerum FE (2011) Implementing an effective scheme for climate change in Nigeria: the role of law. NIALS J Environ Law 1:267. http://elri-ng.org/Implementing%20An%20Effective%20Regulatory%20Scheme.pdf

Federal Government of Nigeria (2022) Nigeria energy transition plan

Federal Republic of Nigeria, Associated gas re-injection act. FRN, September 1979. http://extwprlegs1.fao.org/docs/pdf/nig150976.pdf

Federal Republic of Nigeria (2004) National water policy. FRN, July 2004. http://extwprlegs1.fao.org/docs/pdf/nig158231.pdf
Federal Republic of Nigeria (2006) National forest policy. FRN, June 2006. http://www.fao.org/forestry/15148-0c4acebeb8e7e45af360ec63fcc4c1678.pdf
Federal Republic of Nigeria (2010a) National environmental economic and development study (NEEDS) for climate change in Nigeria. FRN, September 2010. https://unfccc.int/files/adaptation/application/pdf/nigerianeeds.pdf
Federal Republic of Nigeria (2010b) Third national communication (TNC) of the federal republic of Nigeria under the United Nations framework convention on climate change (UNFCCC). UNFCCC, March 2010, 21. https://www4.unfccc.int/sites/SubmissionsStaging/NationalReports/Documents/187563_Nigeria-NC3-1-TNC%20NIGERIA%20-%2018-04-2020%20-%20FINAL.pdf
Federal Republic of Nigeria (2010c) Nigeria vision 20: 2020—abridged version. FRN, December 2010. https://www.nigerianstat.gov.ng/pdfuploads/Abridged_Version_of_Nigeria%20Vision%202020.pdf
Federal Republic of Nigeria (2011a) National environmental control of bush/forest fire and open burning regulations. FRN, April 2011. https://www.nesrea.gov.ng/wp-content/uploads/2020/02/CONTROL_OF_BUSH_FOREST_FIRE_AND_OPEN_BURNING.pdf
Federal Republic of Nigeria (2011b) National environmental control of vehicular emissions from petrol and diesel engines regulations. FRN, April 2011. https://www.nesrea.gov.ng/wp-content/uploads/2020/02/Control_of_Vehicular_Emissions_from_Petrol_and_Diesel_Engines_Regulation%202011%20.pdf
Federal Republic of Nigeria (2013) National climate change policy response and strategy. FRN, October 2013. https://climatechange.gov.ng/wp-content/uploads/2020/09/national-climate-change-policy-1-1.pdf
Federal Republic of Nigeria (2014a) Nigeria's second national communication under the UNFCCC. Federal Republic of Nigeria, February 2014. https://unfccc.int/sites/default/files/resource/nganc2.pdf
Federal Republic of Nigeria (2014b) National environmental energy sector regulations. FRN, December 2014. https://www.nesrea.gov.ng/wp-content/uploads/2020/02/Energy_Sector.pdf
Federal Republic of Nigeria (2015) National renewable energy and energy efficiency policy. FRN, April 2015. https://www.se4all-africa.org/fileadmin/uploads/se4all/Documents/Country_PANEE/Nigeria_National_Energy_Efficiency_Action_Plans.pdf
Federal Republic of Nigeria (2016) Sustainable energy for all action agenda. FRN, July 2016. https://www.seforall.org/sites/default/files/NIGERIA_SE4ALL_ACTION_AGENDA_FINAL.pdf
Federal Republic of Nigeria (2018a) Nigerian gas flare commercialisation programme. FRN, July 2018. https://ngfcp.dpr.gov.ng/media/1120/flare-gas-prevention-of-waste-and-pollution-regulations-2018-gazette-cleaner-copy-1.pdf
Federal Republic of Nigeria (2018b) National drought plan. FRN, November 2018. https://knowledge.unccd.int/sites/default/files/country_profile_documents/1%2520FINAL_NDP_Nigeria.pdf
Federal Republic of Nigeria (2018) Flare gas regulations (prevention of waste and pollution). FRN, July 2018. https://ngfcp.dpr.gov.ng/media/1120/flare-gas-prevention of waste-and-pollution-regulations-2018-gazette-cleaner-copy-1.pdf
Federal Republic of Nigeria (2020a) Third national communication (TNC) of the federal republic of Nigeria under the UNFCCC. Federal Republic of Nigeria, March 2020, at 7–9. https://www4.unfccc.int/sites/SubmissionsStaging/NationalReports/Documents/187563_Nigeria-NC3-1-TNC%20NIGERIA%20-%2018-04-2020%20-%20FINAL.pdf
Federal Republic of Nigeria (2020b) Nigeria's national adaptation plan framework. Federal Ministry of Environment, June 2s020. https://napglobalnetwork.org/wp-content/uploads/2020/06/napgn-en-2020-Nigeria-National-Adaptation-Plan-Framework.pdf
Federal Republic of Nigeria National Assembly 'Bill Tracker'. National Assembly of Nigeria. https://www.nassnig.org/documents/bill/10740

Federal Republic of Nigeria, Nigeria's intended nationally determined contribution. https://www4.unfccc.int/sites/ndcstaging/PublishedDocuments/Nigeria%20First/Approved%20Nigeria%27s%20INDC_271115.pdf

Floodlist News (2019) Nigeria—almost 20,000 displaced by Benue and Niger river floods. Floodlist News, 9 October 2019. http://floodlist.com/africa/nigeria-floods-september-2019cross-river-kogi-niger-taraba

Floodlist News (2020a) Nigeria—3 dead or missing after flash floods destroy a bridge in Kwara State. Floodlist News 16 June 2020. http://floodlist.com/africa/nigeria-floods-kwara-june-2020

Floodlist News (2020b) Nigeria—7 dead after heavy rainfall in Niger state. Floodlist News, 25 July 2020. http://floodlist.com/africa/nigeria-floods-niger-state-july-2020

Floodlist News (2020c) Nigeria—damaging flash floods hit Lagos. Floodlist News, 19 June 2020. http://floodlist.com/africa/nigeria-floods-lagos-june-2020#:~:text=Almost%2090mm%20of%20rain%20fell,as%20a%20result%20of%20flooding

Floodlist News (2020d) Nigeria—dozens of homes destroyed after floods in Akwa Ibom. Floodlist News, 14 June 2020. http://floodlist.com/africa/nigeria-floods-akwa-ibom-june-2020

Floodlist News (2020e) West Africa—rains and floods damage IDP camps in Nigeria and Burkina Faso. Floodlist news, 11 June 2020. http://floodlist.com/africa/floods-nigeria-burkina-faso-june-2020

GFP (2020) Oil production by country. Global Fire Power. http://globalfirepower.com/oil-production-by-country.asp

Giddens A (2015) The politics of climate change. Policy Polit 43(2):155

Haider H (2019) Climate change in Nigeria: impacts and responses. K4D 10 October 2019. https://assets.publishing.service.gov.uk/media/5dcd7a1aed915d0719bf4542/675_Climate_Change_in_Nigeria.pdf

Hassan A, Kouhy R (2013) Gas flaring in Nigeria: analysis of changes in its consequent carbon emission and reporting. Account Forum 37(2):124–134

International Crisis Group (2017) Herders against farmers: Nigeria's expanding deadly conflict. Africa Report N°252, 19 September 2017. https://d2071andvip0wj.cloudfront.net/252-nigerias-spreading-herder-farmer-conflict.pdf

International Energy Agency (2017) Nigeria feed-in tariff for renewable energy sourced electricity. Int Energy Agency, 23 August 2017. https://www.iea.org/policies/5974-nigeria-feed-in-tariff-for-renewable-energy-sourced-electricity

Johnson IA, Okunola BT (2017) Pastoralism as a new phase of terrorism in Nigeria. Global J Human Soc Sci 17(4):51

Johnston R, Deeming C (2015) A reflection on "Giddens's paradox". Policy and politics blog, 5 May 2015. https://policyandpoliticsblog.com/2015/05/05/a-reflection-on-giddenss-paradox/

Kindzierski WB (2000) Importance of human environmental exposure to hazardous air pollutants from gas flares. Environ Rev 8:41–62

Ko J-Y, Day J, Barras J, Morton R, Johnston J, Steyer G, Kemp GP, Clairain E, Theriot R (2013) A review of ecological impacts of oil and gas development on coastal ecosystems in the Mississippi Delta. Ocean Coast Manag 47(11):597

Kyoto Protocol (2005) Decision 4/CMP.1 guidance relating to the clean development mechanism FCCC/KP/CMP/2005/8/Add.1

Le Page M (2016) Highest ever annual rise in carbon dioxide levels recorded. Newscientist, 8 March 2016. https://www.newscientist.com/article/2079995-highest-ever-annual-rise-in-carbon-dioxide-levels-recorded/

Leahy S (2007) Environment: 21st century may belong to the shrub. IPS inter press service, 31 August 2007

McMichael AJ, Campbell-Lendrum DH, Corvalan CF, Ebi KL, Githeko AK, Scheraga JD, Woodward A (eds) Climate change and human health: risks and responses. WHO, London, p 322

NACGOND, The Nigerian State and gas flaring in the Niger Delta, pp 258–261. www.nacgond.org/downloads/nacgond-report-pdf-version.pdf

National Environment Policy. https://www.nesrea.gov.ng/wp-content/uploads/2017/09/National-Policy-on-Environment.pdf

Nemet GF, Jakob M, Steckel JC, Edenhofer O (2017) Addressing policy credibility problems for low-carbon investment. Global Environ Change 42:47. https://www.sciencedirect.com/science/article/abs/pii/S0959378016305829

Nigeria—Crude Oil Reserves. https://countryeconomy.com/energy-and-environmen/crude-oil/reserves/nigeria

Nigerian Red Cross Society (2019) Emergency plan of action (EPoA) Nigeria: floods. NRCS, 07 October 2019. https://reliefweb.int/sites/reliefweb.int/files/resources/MDRNG028do.pdf

Nigerian Upstream Petroleum Regulatory Commission (NUPRC) (2023) NGFCP bid round results

Nwankwo C, Nwankwoala HO (2018a) Gully erosion susceptibility mapping in Ikwuano local government area of Abia state using GIS techniques. Earth Sci 2(1):8–15. Also see Adekalu KO, Osunbitan J et al (2007) Grass mulching effect on infiltration, surface runoff and soil loss of three agricultural soils in Nigeria. Bioresour Technol 98(4):912

Nwankwo C, Nwankwoala HO (2018b) Analysis of a 36-year rainfall data (1980–2015) for erosivity potential in Ikwuano local government area of Abia State, Southeastern Nigeria. Nigerian J Cartography GIS 87 12(1):2

Nwilo PC, Olayinka DN, Uwadiegwu I, Adzandeh AE (2011) An assessment and mapping of gully erosion hazards in Abia state: a GIS approach. J Sustain Develop 4(5):21

Ochayi C (2019) New policy on erosion control, flood MGT under way—FG. Vanguard, 18 June 2019. https://www.vanguardngr.com/2019/06/new-policy-on-erosion-control-flood mgt under-way-fg/

Oladipo E (2010) Towards enhancing the adaptive capacity of Nigeria: a review of the country's state of preparedness for climate change adaptation. Heinrich Böll Foundation Nigeria. https://ng.boell.org/sites/default/files/uploads/2013/10/nigeria_prof_oladipo_final_cga_study.pdf

Onu NN, Opara AI (2012) Analysis and characterization of Njaba river gully erosion, Southeastern Nigeria: deductions from surface geophysical data. Aust J Basic Appl Sci 6(4):122

Paris Agreement in UNFCCC (2016) COP report no. 21, Addendum, at 21, U.N. Doc. FCCC/CP/2015/10/Add, 1 (hereinafter Paris Agreement)

Patz J, Campbell-Lendrum D, Holloway T, Foley JA (2005) Impact of regional climate change on human health. Nature 438(17):310

Premium Times (2012) Hundreds displaced in Adamawa as Fulani herdsmen attack two communities. Premium Times, 13 May 2012. https://www.premiumtimesng.com/news/5088 hundreds_displaced_as_fulani_herdsmen_attack_communities.html

Priscilla Offiong (2019) Nigeria president issues climate change action plan: expansion of green bonds program; stakeholder capacity building efforts. Climate Scorecard, 17 December 2019. https://www.climatescorecard.org/2019/12/president-issues-climate-change-action-plan-expansion-of-green-bonds-program-stakeholder-capacity-building-efforts/

Rozowska A, Murphy-McGreevey C (2020) Global gas flaring jumps to levels last seen in 2009. World Bank Group, 21 July 2020. https://www.worldbank.org/en/news/press-release/2020/07/21/global gas-flaring-jumps-to-levels-last-seen-in-2009

Sahara Reporters (2020) Residents cry for help as floods hit Lagos, disrupt businesses. Sahara Reporters, 19 June 2020. http://saharareporters.com/2020/06/19/residents-cry-help-floods-hit-lagos-disrupt-businesses

Shell in Nigeria Briefing Notes April 2020. https://www.shell.com.ng/media/nigeria-reports-and-publications-briefingnotes/_jcr_content/par/pageHeader.stream/1600103569657/ea58c56d3d22ac11f793533d976a03bba4dea568/nbn-2020.pdf

The Federal Republic of Nigeria (2007) National environmental standards and regulations enforcement agency. FRN, July 2007. http://extwprlegs1.fao.org/docs/pdf/nig120569.pdf

The Federal Republic of Nigeria (2011a) National environmental soil erosion and flood control regulations. FRN, April 2011. https://www.nesrea.gov.ng/wp-content/uploads/2020/02/Soil_Erosion_and_Flood_Control_Regulations%202009.pdf

The Federal Republic of Nigeria (2011b) National environmental desertification control and drought mitigation regulations. FRN, April 2011. https://www.nesrea.gov.ng/wp-content/uploads/2020/02/Desertification_Control_and_Drought_Mitigation%20Regulation%202011.pdf
The Preamble of the Associated Gas Re-Injection Act
Thomas CD, Cameron A, Green RE, Bakkenes M, Beaumont LJ, Collingham YC, Erasmus BFN, de Siqueira MF, Grainger A, Hannah L, Hughes L, Huntley B, van Jaarsveld AS, Midgley GF, Miles L, Ortega-Huerta MA, Peterson AT, Phillips OL, Williams SE (2004) Extinction risk from climate change. Nature 427:145
UN OCHA (2022) Nigeria: flooding situation report, Oct 2022. https://www.unocha.org/publications/report/nigeria/nigeria-floods-response-flash-update-1-last-updated-19-october-2022
UNDP Global Climate Promise "Nigeria". https://climatepromise.undp.org/what-we-do/where-we-work/nigeria#:~:text=NDC%20Status,gender%20concerns%20and%20youth%20engagement
UNFCCC (2021) Nigeria's updated nationally determined contribution. https://unfccc.int/documents/497790
UNFCCC, The clean development mechanism. https://unfccc.int/process-and-meetings/the-kyoto-protocol/mechanisms-under-the-kyoto-protocol/the-clean-development-mechanism
UNFCCC, National communication submissions from non-annex I parties. https://unfccc.int/non-annex-I-NCs
United Nations Environment Programme on Climate Change is available at https://www.unep.org/unep-and-climate-emergency
United Nations Framework Convention on Climate Change (9 May 1992) S. Treaty Doc No. 102-38 (1992), 1771 U.N.T.S. 107
Varrella S (2020) Contribution of oil sector to GDP in Nigeria 2018–2020. Statista, 31 August 2020. https://www.statista.com/statistics/1165865/contribution-of-oil-sector-to-gdp-in-nigeria/#:~:text=Nigeria's%20oil%20sector%20contributes%20to,the%20same%20period%20of%202019
WHO (2021) Climate change and health fact sheet. https://www.who.int/publications/m/item/fast-facts-on-climate-change-and-health
Wonah EI (2017) The state, environmental policy and sustainable development in Nigeria. Global J Arts Hum Soc Sci 5(3):25
World Bank (2023) Global gas flaring tracker report 2023. World Bank. https://www.worldbank.org/en/programs/gasflaringreduction/global-flaring-data#:~:text=Global%20Flaring%20Data%20Resources&text=PDF-,The%20World%20Bank's%20Global%20Gas%20Flaring%20Tracker%2C%20a%20leading%20global,of%20associated%20gas%20was%20flared
World Bank (2025) Global gas flaring tracker report. https://www.worldbank.org/en/programs/gasflaringreduction/publication/2024-global-gas-flaring-tracker-report
Yale Center for Environmental Law and Policy (2022) Environmental performance index. Yale University, New Haven

Work Practices in Nigeria: Environmental Health and Development Implications

Akeem Ayofe Akinwale

Abstract The environmental protection policies are not effective in Nigeria. This is evident in Nigeria's low global environmental performance index scores. Therefore, this study examined the environmental health implications of work practices in Nigeria, focusing on the following research questions: (1) what are the environmental health hazards in the dominant work practices in Nigeria? and (2) how should the Nigerian government strengthen its environmental health policies? These questions were addressed through analysis of secondary data and 52 in-depth interviews. Findings revealed that the environmental health hazards in the dominant work practices in Nigeria comprised: (1) greenhouse gas (GHG) emissions and environmental degradation from the production of crude oil and natural gas, (2) water pollution from dumping of industrial wastes, and (3) air pollution from bush burning, transportation, and manufacturing industries. Work practices have contributed to environmental health problems differently, while Nigeria's policies and institutional arrangements to protect environmental health have underperformed. Therefore, the Nigerian government should establish and enforce emission standards and provide comprehensive and inclusive environmental health education for all stakeholders across workplaces and industries in Nigeria.

1 Introduction

Human consciousness of the interface between work practices and environmental health must address the shortcomings in adapting and mitigating the climate change crisis. This observation is consistent with human–environment interactions, which studies have examined from different perspectives, such as environmental determinism, environmental possibilism and environmental probabilism.[1]

[1] Fekadu (2014, 139) and Önal (2018, 92).

A. A. Akinwale (✉)
Department of Employment Relations and Human Resource Management, University of Lagos, Lagos, Lagos State, Nigeria
e-mail: aakinwale@unilag.edu.ng

D. Oghenerobor Agelebe (ed.), *Climate Ethics in Africa*, Global Perspectives on Law and the Environment 1, https://doi.org/10.1007/978-3-032-08757-7_8

Environmental determinism focuses on how the environment has affected human activities, while environmental possibilism deals with how human activities affect the environment. Environmental probabilism depicts an analysis of mutual interactions between human activities and the environment. A study of the environmental health implications of work practices is tenable in this context.

Work practices could enhance or damage the environment, which Barry[2] described as nature, including the nonhuman world and human surroundings, although he also noted that environment is a relational concept or idea. People in the Western world often describe the environment in terms of the physical nonhuman environment, such as the countryside, forests, animals, rivers, etc. In contrast, in other cultures, the environment may include those mentioned above and non-physical phenomena such as the spirits, the ghosts of ancestors, and cultural heritage, as the case may be.

This study is based on a non-Western description of the environment, focusing on the following research questions: (1) What are the environmental health hazards in the dominant work practices in Nigeria? and (2) How should the Nigerian government strengthen its environmental health policies? These questions were addressed through analysis of secondary data and 52 in-depth interviews.

2 A Global Interest in Environmental Health

Environmental health is the total well-being of the physical and non-physical environment, including the various aspects of public health.[3] Some environmental health problems include climate change, global warming, and biodiversity loss.[4] The aforementioned environmental health problems are associated with greenhouse gas (GHG) emissions from burning fossil fuels in industries. Consequently, the average worldwide surface temperature has exceeded 0.74 °C, causing widespread melting of snow and rising sea levels, adversely impacting ecosystems.[5]

The need to promote environmental health has attracted global attention by establishing the United Nations Environment Programme (UNEP) in 1972. The UNEP has been calling for international cooperation between nations and peoples to solve or cope with global environmental problems, the most pressing of which is climate change. The Conference of Parties (COP), an organ of the United Nations, is responsible for monitoring and reviewing the implementation of the United Nations Convention on Climate Change. The first COP (COP 1) was held in Berlin, Germany, in March 1995, while the latest COP (COP 30) was held in Brazil in November 2025.

---

[2] J Barry (2007).

[3] Raimi et al. (2021, 2(3) 122).

[4] Barry (n2) & Nriagu (2008).

[5] Raimi et al. (n2).

Available records show that developed and developing countries suffer environmental health problems. Still, developing countries are more vulnerable to environmental health problems due to the prevalence of poverty, social exclusion, underdevelopment, and vulnerabilities. This situation shows the need to address some questions in the global environmental health politics:

> Who is supposed to take responsibility for climate change? Who will set the standards for controlling the emissions of ever more greenhouse gases (GHGs)? How can the necessary actions be implemented to protect and maintain the environment?[6]

Consistent with the abovementioned questions and the increasing awareness and visibility of environmental health problems, some countries have intensified efforts towards reducing GHG emissions and taking control of the expanding threat of global climate change. Unfortunately, several factors such as poverty, poor environmental education, lack of awareness, underdevelopment in developing countries and development of industrial capitalism in developed countries have contributed to the spread of global environmental health problems worldwide.

3 The Interplay of Environment and Work Practices

Historically, environmental determinism has been a dominant perspective in explaining human–environment interactions in the context of rudimentary technology. Ibn Khaldun, a scholar from Tunisia in North Africa, investigated the effects of weather conditions on human character and the foundation of states by establishing a relationship between civilizations, climate, and physical environmental conditions.[7]

Similarly, Ratzel confirmed the superiority of the physical environment over human beings. At the same time, Barry[8] opined that "humans, in common with every other living species, depend on their environment to survive and flourish." Moreover, some German and American geographers, such as Humboldt, Ritter, Davis, Semple, and Huntington, argued that the environment controlled human activities on Earth.[9] The following illustration of environmental determinism is instructive:

> Determinism in geography argues that to be successful in their struggle for life, man has to adapt to the rules of nature. In this context, determinists claim that the topic of geography is the exploration of the methods used by man to react to the physical environment.[10]

In contrast to the main argument in environmental determinism, some human beings have exhibited different socio-economic characteristics in regions with similar environmental conditions. Physical environmental conditions show that the physical environment's influence on human activities is not absolute. This trend resonates with

[6] Uddin (2017, 106).

[7] Önal (2018, 8(1) 92).

[8] Barry (n3) 17.

[9] Friedman (2003, 110) and Şahin and Belge (2016).

[10] Önal (n3) 76.

the belief that humans can use their knowledge of technology to shape the dictates of the physical environment.[11]

In light of the foregoing, some researchers have examined the validity of a different perspective known as environmental possibility. This perspective primarily focuses on the power of human agency and how this power is enacted in human–environment interactions, primarily through infrastructure development and other innovative ways of controlling the physical environment. The impact of industrial revolutions on the environment illustrates the main argument for environmental possibility.

> The Industrial Revolution refers to the changes in European economic life regarding concepts, theories, ideas and actual practice from the sixteenth to the nineteenth centuries. These changes laid the basis for the emergence and development of modern industrial society and a capitalist economy. Britain is often seen as the cradle of the Industrial Revolution.[12]

The Industrial Revolution prompted the rise of industrial capitalism, which launched the Modern Age and ripped into the available raw materials without regard for environmental consequences.[13] This is consistent with the belief that people can overcome environmental constraints through applications of science and technology. Unfortunately, applying science and technology has promoted the exploitation of the natural environment. The following illustration shows the influence of the Industrial Revolution on the environment:

> Central to the Industrial Revolution was a mainly instrumental attitude towards the natural environment. The environment was seen as a collection of means for human ends, raw materials for the factories, machines, and new productive technologies being invented. Science was seen as unlocking nature's secrets, developing new insights into its inner workings, and, in conjunction with technology, providing more effective ways for humans to exploit it.[14]

In a nutshell, environmental determinists argue that 'man is entirely under the influence of nature.' In contrast, ecological possibilists believe that man is never entirely free from the influence of the environment. Still, there is room for man's effort, such as technology, attitude, habits, and human values, to influence the physical environment.[15] This analogy follows the socio-technical system theory, which focuses on workplace interaction between employees and technology.[16] The development of the socio-technical system theory was based on observations of working conditions, employees' performance and job satisfaction in the coal mining industry in England during the Second World War.

Consistent with the main ideas in environmental possibilism, the interface of work practices and environmental health can be observed in different ways, as shown in the following illustration by Taylor[17]:

---

[11] Uddin (n2) 106.

[12] Barry (n4) 44.

[13] As above.

[14] As above.

[15] Fekadu (n2) 132 at 139.

[16] Trist (1981).

[17] Taylor (2011, 5).

> The effects of human culture and technology on the planetary biosphere are becoming ubiquitous. Due to the emergence of large-scale industrialization in the past century, the recent rise in the growth rate of the human population and the expansion of economies that stimulate and depend on high consumption levels, human presence is now felt worldwide. It is not only where we have taken over land areas to grow our food crops and build our towns and cities that ecological changes have been brought about; modern civilization's physical, chemical, and biological concomitants can be found everywhere.

As mentioned, the illustration supports the link between ecological imperialism and global environmental problems.[18] Unfortunately, the regions with the highest likelihood of exposure to environmental health hazards are disproportionately located in Africa, South Asia, and Latin America.[19] The global environmental performance index assesses the level of environmental health across countries.

The 2022 Environmental Performance Index (EPI) ranks 180 countries on their national efforts to protect environmental health and ecosystem vitality. It focuses on 40 performance indicators that show institutional commitment to environmental health, ecosystem vitality, and climate change mitigation.[20] Identifying the countries with the highest and lowest scores in the 2022 EPI is necessary to assess the relevance of government policies in protecting environmental health and ecosystem vitality.

The countries with the highest scores in the 2022 EPI, including Denmark (77.9%), the United Kingdom (77.7%) and Finland (76.5%), among others, have demonstrated adequate institutional commitment to environmental health, ecosystem vitality and climate change mitigation, while the countries with the lowest scores in the 2022 EPI, including India (18.9%), Ghana (27.7%), Indonesia (28.2%), Nigeria (28.3%), and China (28.4%), among others, have demonstrated lack of adequate institutional commitment to environmental health, ecosystem vitality and climate change mitigation.

A recent study on the environmental health situation in Nigeria showed that various industrial and other anthropogenic activities have immensely polluted the ground water in Nigeria, thereby revealing the health implications associated with the intake of contaminated water.[21] An illustration of this situation is depicted as follows in Fig. 1.

The abovementioned situation is similar to an earlier study by Ukah et al. (2019)[22] showing that the ecological risk assessment and health hazard index analysis of contaminated groundwater in the Ajao industrial area of Lagos State of Nigeria revealed the following outcomes; (i) a total of 85.7% and 14.3% of the sampled water indicated low and moderate ecological risks, respectively (ii) children are more exposed to non-carcinogenic chronic health risks due to ingestion of contaminated

[18] Frame (2022, 534).

[19] Obeng-Odoom (2022, 472).

[20] Wolf et al (2022) 2022 environmental performance index Yale centre for Environmental Law and Policy.

[21] MA Balogun et al. 'Environmental health impacts and controlling measures of anthropogenic activities on groundwater quality in Southwestern Nigeria' (2022) 194(5) *Environmental Monitoring and Assessment* at 384.

[22] Ukas (2019, 303).

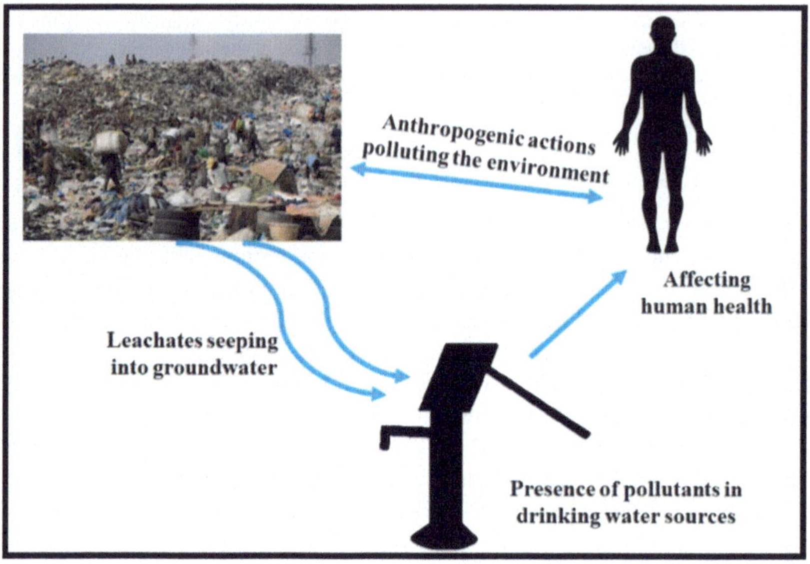

Fig. 1 Implications of polluted groundwater in Nigeria. *Source* Balogun et al. (2022)

groundwater than the adult population, (iii) the probability of cancer risk (PCR) revealed that 19.1% of the sampled water indicated a high risk of exposure to cancer for both adult and children, (iv) the presence of the heavy metals in water was mainly attributed to anthropogenic activities rather than natural processes.

In his study on livelihoods and environmental challenges in coastal communities of Nigeria, Akinwale[23] discovered that people's awareness of the need for environmental protection was generally low across rural communities of Nigeria, as only a few discussants perceived their economic activities as dangerous to the aquatic environment. An illustration of this situation is presented in the next paragraph for clarity:

> Significant proportions of men and women understood local sources of environmental challenges in the coastal environment. While men occasionally indulge in bush burning or use chemicals to catch fish, women fetch and burn firewood for fish processing or domestic cooking.[24]

Akinwale[25] also documented that several socio-economic activities, such as construction, farming, gas flaring, oil exploration, and transportation, have affected the physical environment in Nigeria. He noted that the above activities constitute most Nigerians' primary revenue sources.

---

[23] Akinwale (2011, 5673).

[24] Akinwale 5669.

[25] As above.

Furthermore, Akinwale[26] observed a disconnection between adverse consequences of socio-economic activities on the environment and the need to protect the environment, thereby showing that despite environmental protection agencies in Nigeria, ecological challenges remain high in Nigeria. He adduced the situation to several years of neglect and poor socio-economic conditions in Nigeria, where people largely contravene ecological laws with impunity in their struggle for survival.

4 Theoretical Framework

Theory systematically explains a phenomenon to enhance scientific knowledge based on logic and observations. Babbie[27] noted that logic is a system of principles for determining the validity of an explanation, thereby showing that theories seek to provide logical explanations. Therefore, the theoretical framework of this study comprises two theories: the open systems theory and the socio-technical systems theory. These theories are suitable for understanding the linkages between work practices and environmental health in Nigeria and elsewhere.

Craig and Solomon[28] developed an open systems theory to explain the flow of materials, energy and information across the boundaries of interactions between a working system and its environment. The main concerns or outputs of any industrial relations system are the allocation of rewards to employees and the physical and other conditions under which work is performed. The outputs and the processes they arrive at provide a basis for developing open systems theory. The system's outputs, such as collective agreement and job satisfaction, may affect the subsequent processes in the system, and this may also have implications for the environment and environmental sub-systems, which are primary external inputs into the industrial relations system.

In their open systems theory, Craig and Solomon[29] classified the environmental subsystems, which they described as external inputs, into the ecological environment, the economic climate, the political climate, the legal environment, and the socio-cultural environment. The ecological environment comprises the physical environment, natural resources, and climate. Thus, the open systems theory shows that the ecological environment can affect the industrial relations system, including work practices, and vice versa.

The open systems theory focuses on the behaviour of the industrial relations actors and how they translate both the constraints imposed by the environmental sub-systems and their goals, values and power into the outputs of an industrial relations system. A close observation of the open systems theory demonstrates that workforce interaction with the environmental sub-systems over the years has resulted in

---

[26] As above.

[27] Babbie (2021).

[28] Craig and Solomon (1996).

[29] As above.

changes in technologies and the ecological sub-systems. In adapting to the environment, the workforce has transformed their habitats through technologies for improved living standards with amenities, including networks of roads, electricity, water supply, establishment of industries, and communication facilities, among others.

However, despite the suitability of the open systems theory for an understanding of the environmental health implications of work practices in Nigeria, the theory did not explain how the industrial relations actors can contribute to efforts to protect environmental health, ecosystem vitality and climate change mitigation. Craig and Solomon[30] noted that "it is only through the action of concerned groups—such as "environmentalists" who are concerned that our environment (forests, fish, and so on) be preserved—that the ecological system stands a chance of being protected from overly zealous pursuit. Therefore, the socio-technical systems theory is included in the theoretical framework.

Trist shows that[31] the socio-technical systems theory focuses on workplace interaction between employees and technology. It also deals with the interaction between complex infrastructures and human behaviour in society. The main principles in the socio-technical systems theory are as follows: The interaction of social and technical factors creates the conditions for successful or unsuccessful organizational performance. In contrast, optimizing only the social or technical factors tends to generate adverse effects on the performance of a system. This clearly shows that the socio-technical systems theory focuses on the importance of joint optimization and how the interactions between social and technical factors can enhance organizational productivity and well-being.

The socio-technical systems theory's main concepts include responsible autonomy, adaptability, and tasks. Responsible autonomy focuses on effective leadership, regulation, and supervision of different teams or groups' activities to ensure a system's effectiveness. Adaptability focuses on understanding the dynamics of a system's internal structure. Tasks refer to the operations and responsibilities of different system elements.

In a nutshell, the socio-technical systems theory comprises an analysis of work design and the networks of relationships in the workplace, including the following issues: (i) the work system, (ii) the organization, and (iii) the macro-social phenomena. Trist[32] examined the relationship between these phenomena in a historical context. His historical analysis of the socio-technical systems is presented thus:

> As a society's historical process unfolds, individuals change their values and expectations concerning work roles, changing organizational design parameters. Conversely, changes in technology bring about changes in values, cognitive structures, lifestyles, habitats, and communications, which profoundly alter society and its chances of survival.[33]

---

[30] Craig and Solomon (n2) 15.

[31] Trist (n2).

[32] As above.

[33] Trist (n3) 11.

The above-mentioned analysis of the socio-technical systems theory shows the need for a thorough understanding of the work system and networks of relationships in workplaces in efforts to protect environmental health, ecosystem vitality and climate change mitigation in Nigeria and elsewhere. For instance, focusing only on a given work system without giving adequate attention to the environmental health implications of the work system is tantamount to painting an incomplete picture of social reality.

Also, it is essential to note that the main argument in the socio-technical systems theory is consistent with the use of innovative technologies to protect environmental health, ecosystem vitality, and climate change mitigation. Such innovative technologies may include using renewable energy instead of overdependence on fossil fuels, green practices across organizations and industries, and other environmentally friendly technologies. Renewable energy comprises solar, wind, hydro, tidal, geothermal, and biomass. Each aspect of renewable energy requires suitable knowledge and application of relevant technology.

5 Methods of Data Collection and Analysis

This study was based on content analysis of secondary and primary data. The secondary data were derived from a systematic literature review on environmental health and work practices. In contrast, the primary data were collected through in-depth interviews with some managers in the manufacturing, transportation, hospitality, and oil and gas industries in Lagos State. A total of 52 managers, including 13 managers from manufacturing companies, 13 managers from transport companies, 13 managers from hotels, and 13 managers from oil and gas companies, were purposively selected and interviewed based on different criteria such as availability, work experience, and knowledge of environmental impact assessment of business operations in Nigeria. The in-depth interviews were transcribed and analyzed in line with the research questions.

6 Findings

The findings comprise the socio-demographic characteristics of the respondents, followed by the respondent's views on the environmental health hazards in the dominant work practices in Nigeria and the assessment of environmental health policies in Nigeria. Of the 52 respondents who participated in the interviews, 40 were men, while 12 were women. This shows that both men and women participated in the interviews.

The respondents' highest educational qualifications ranged from a Senior Secondary School Certificate to Bachelor's and Master's Degrees. A total of 32 respondents had Master's Degrees, 14 had Bachelor's Degrees, and six had Senior Secondary School Certificates.

The finding on the educational qualifications of the respondents shows that all the respondents had acquired formal education, and they could clearly express their knowledge of the environmental health implications of work practices in Nigeria.

All the respondents were senior staff in their respective organizations. Their ages ranged from 30 to 60. A total of 35 respondents were in the 40–50 age bracket, 10 were in the 30–39 age bracket, and seven were in the 51–60 age bracket. This finding shows that all the respondents were middle-aged adults.

The work experience of the respondents ranged from 10 to 25 years. A total of 28 respondents had been working in their respective organizations for up to 20 years, while 14 respondents had been working for their respective organizations for up to 25 years, and 10 respondents had worked for their respective organizations for at least 10 years. This finding shows that three out of every four respondents were highly experienced in their respective organizations. The respondents' work experience provided an opportunity to obtain valuable information on the environmental health implications of work practices in Nigeria.

The respondents' income ranged from N120,000 to N2,500,000 per month. A total of 28 respondents earned between N120,000 and N250,000 per month, 15 earned between N250,000 and N1000,000 per month, and nine earned between N1,000,000 and N2,500,000 per month. The respondents with higher monthly incomes identified themselves as managers in oil and gas companies. This finding shows that oil and gas companies provide high levels of remuneration for their managerial employees.

Regarding their views on the dominant work practices in Nigeria, three out of every five respondents noted that crude oil and natural gas production, distribution, and marketing dominated Nigeria's manufacturing and oil and gas industries. One-third of the respondents mentioned the dominant work practices in the extractive sector, focusing on agriculture, fishing, and lumbering. Also, all the respondents recognized other dominant work practices in Nigeria, focusing on transportation, telecommunication, banking, hospitality, and informal sector economic activities.

Concerning their views on the environmental health hazards in the dominant work practices in Nigeria, the respondents presented different instances of ecological health hazards across occupations and work practices as follows:

- Burning of fossil fuels occurred across industries in Nigeria
- Gas flaring resulted in the emission of greenhouse gases
- Dumping of industrial wastes from manufacturing activities resulted in water pollution, land pollution, air pollution, and environmental degradation
- Unhealthy agricultural practices resulted in air pollution and soil erosion.
- Excessive cutting of timbers resulted in deforestation and biodiversity loss
- Predominance of old vehicles resulted in air pollution

All the respondents expressed awareness of some improvements in Nigeria's environmental health policies. However, three out of every five respondents observed that the existing framework for environmental protection in the social and environmental impact assessment of business operations across industries in Nigeria is below the global best practices in environmental health, ecosystem vitality, and climate change mitigation.

Some narratives concerning the awareness of environmental impact assessment of business operations in Nigeria include the following:

> Impact assessment is a prerequisite and essential to the success of any development project. Often, there are concerns over different environmental issues that stakeholders feel are worthy of serious attention from the multinational companies in the oil and gas industry. We engage with stakeholders in the communities to contribute towards environmental health. We deploy our initiatives to promote ecological health.[34]

> Impact assessment of manufacturing activities is essential, and we factor the assessment into our operations. Environmental permits are obtained from both State and Federal Environmental agencies. These evaluations help determine the desired impact our activities are expected to have in the communities.[35]

> I am unaware whether hotels or other organizations in the hospitality industry carry out environmental impact assessments of their business operations in the communities. Concerning the evaluation of the implications of the hospitality business on ecological health, we are doing our best to keep the environment clean. Still, we cannot do without burning fuels such as diesel. This is because electricity is not always available from the national grid, and we must make our clients enjoy their stay in the hotel.[36]

7 Discussion of Findings:

7.1 Environmental Health Hazards in the Dominant Work Practices in Nigeria

The findings on the environmental health hazards in the dominant work practices in Nigeria are consistent with the accumulated knowledge in the literature. For instance, some researchers have shown that extracting natural resources such as crude oil and natural gas is Nigeria's primary revenue source for governments. Crude oil exports account for 13% of the GDP, while 95% of all foreign exchange earnings come from the oil sector in Nigeria.[37]

However, since the discovery and production of crude oil in Nigeria, there have been several cases of greenhouse gas (GHG) emissions and environmental degradation, with a rising spate of carbon emissions from producing and consuming fossil fuels such as coal, gas, and petroleum. More than 60% of the business organizations in Nigeria constantly use diesel generators for electrification, thereby reinforcing the use of fossil fuels, which contributes to environmental health impairment.

[34] IDI Oil and Gas Company (2022).

[35] IDI Manufacturing Company (2022).

[36] IDI Hospitality Industry (2022).

[37] Elum et al. (2016).

Also, less than one-third of the Nigerian states—mainly Lagos State, Sokoto State, Delta State, Oyo State, Bauchi State and Enugu State—have deployed renewable energy technologies as alternative sources of electricity.[38] Moreover, a recent study by Uddin[39] showed that human activities such as industrialization, consumerism, burning of fossil fuels, tropical deforestation, and ever-increasing use of automobiles are strongly connected with environmental degradation as these contribute to the production of greenhouse gases (GHGs), which in turn accelerates climate change.

7.2 Efforts to Strengthen Environmental Health Policies in Nigeria

The findings on the efforts to strengthen the environmental health policies in Nigeria are consistent with the prevailing situation in the growth and development of ecological health in Nigeria. For instance, Section 20 of the 1999 Constitution of Nigeria empowers the state to protect and improve the environment and safeguard the water, air, land, forest, and wildlife. As shown in the literature such as Abubakar[40]; Agency Report[41] Ajibola et al.[42] Nwankwo,[43] and Raimi et al.[44] other institutional arrangements that the Nigerian Government established to protect environmental health in Nigeria include the following:

- The Federal Environmental Protection Agency (FEPA) Act of 1988
- The Environmental Impact Assessment Act of 1992
- The National Environmental Standards and Regulations Enforcement Agency (NESREA) Act of 2007
- The National Council on Climate Change and Energy Transition Plan of 2022

The most recent policy and institutional arrangement to protect environmental health in Nigeria is the Energy Transition Plan (ETP) and the National Council on Climate Change (NCCC), which the Nigerian government established to demonstrate its commitment towards achieving net-zero emissions by 2060 following the Paris Agreement.[45] The ETP focuses on five critical sectors: power, cooking, oil and gas, transport, and industry. An estimate of the financial costs for effective implementation of the ETP shows that the Nigerian government would need 410 billion dollars, including 10 billion dollars per annum, to deliver the Transition Plan by 2060. The World Bank and Sun Africa, a renewable energy organization, pledged

[38] Osunmuyiwa et al. (2018) 20(2) *156*.

[39] Uddin (n3) 106 at 127.

[40] Abubakar (2022).

[41] Agency Report Energy transition (2022).

[42] Ajibola et al. (2020) 36.

[43] Nwankwo (2022) 147.

[44] Raimi et al. (2020) 84.

[45] Agency Report (n2).

1.5 billion dollars each, totalling an initial 3 billion dollars in investment to support the implementation of the Nigerian Energy Transition Plan.[46]

However, the policies and institutional arrangements to protect environmental health have underperformed in Nigeria, hence the growing cases of ecological problems such as oil spillage and flaring of gas, which constitute threats to the ecosystems in the Niger Delta region of Nigeria. A close observation of the policies and institutional arrangements to protect environmental health in Nigeria shows that the implementation of environmental law in Nigeria is generally loose, uncoordinated, and fraught with conflict of roles, standards, and responsibilities among the different organs and agencies of government.[47]

8 Conclusion

This study examined the linkages between work practices and environmental health issues in Nigeria, focusing on the environmental health implications of the dominant work practices in Nigeria and how to strengthen the ecological health policies in Nigeria. The study has shown the dominant work practices in Nigeria, including production and marketing of oil and gas, manufacturing activities, agriculture, lumbering, transportation, telecommunication, banking, hospitality, and informal economic activities.

The dominant work practices have contributed to environmental health problems, including greenhouse gas (GHG) emissions, environmental degradation, water pollution, air pollution, land pollution, deforestation, and biodiversity loss. These environmental health problems have affected the health and safety of many people, including children, adults, and the elderly, thereby showing the need to establish relevant policies to adequately protect environmental health, ecosystem vitality, and climate change mitigation in Nigeria.

The Nigerian government has been active in the renewal of interests in formulating policies to protect environmental health, ecosystem vitality, and climate change mitigation in Nigeria. The most recent efforts of the Nigerian government comprise the Energy Transition Plan and the National Council on Climate Change, which demonstrate the Nigerian government's commitment towards achieving net-zero emissions by 2060 in compliance with the Paris Agreement on Climate Change Mitigation. However, the existing policies and institutional arrangements to protect environmental health and ecosystem vitality have performed below expectations in Nigeria, thereby showing the need for private–public partnerships and stakeholder engagement to improve the situation.

In support of efforts to protect environmental health and ecosystem vitality, there has been a clamour for trades and economies that can support the conservation and

[46] Agency Report (n3).

[47] Agelebe et al. (2019).

restoration of the environment, including the protection of biodiversity and ecosystems. This situation has prompted the pursuit of a new agenda known in some organizations as green growth policies. Green growth policies comprise innovative strategies and approaches to mitigate environmental crises, including investment in clean industries, pollution control, and trade in green goods.

Therefore, the following recommendations are put forward for consideration by different stakeholders in efforts to promote environmental health, ecosystem vitality and climate change mitigation in Nigeria and elsewhere because the impact of environmental health problems in Nigeria can spread to other parts of the world if the current situation in Nigeria is not addressed urgently:

- All business organizations in Nigeria should make their processes and products environmentally friendly to protect the country's environmental health.
- The Nigerian government should require multinational companies to make their processes and products environment-friendly to sustainably protect environmental health, ecosystem vitality, and climate change mitigation.
- The Nigerian government should establish and enforce emission standards to strengthen its environmental health protection policies.
- Environmental protection stakeholders, especially government officials and industry captains, should promote environmental sustainability by developing human capital and using appropriate technologies to protect ecological health in Nigeria.
- A comprehensive and inclusive environmental health education should be provided for the general public in Nigeria.
- Finally, all stakeholders in Nigeria's educational institutions should be involved in promoting environmental health, ecosystem vitality, and climate change mitigation.

References

Abubakar Z (2022) President Buhari appoints DG national council on climate change. Voice of Nigeria, London

Agelebe D, Prityi M, Ignacio J, Nielsen G (2019) Environmental impact assessment systems. Routledge, Berlin

Agency Report Energy Transition: Nigeria Proposes Debt-for-Climate Swap Deal (2022)

Ajibola AF, Raimi MO, Steve-Awogbami OC, Adeniji AO, Adekunle AP (2020) Policy responses to addressing the issues of environmental health impacts of the charcoal factory in Nigeria: necessity today; essentiality tomorrow. Commun Soc Med 3(3):1–36

Akinwale AA (2011) Livelihoods and environmental challenges in coastal communities of Nigeria'. Afr J Food Agric Nutr Dev 11(7):5661–5673

Babbie E (2021) The practice of social research. Cengage Learning Inc., New York

Balogun MA, Anumah AO, Adegoke KA, Maxakato NW (2022) Environmental health impacts and controlling measures of anthropogenic activities on groundwater quality in Southwestern Nigeria. Environ Monit Assess 194(5):384

Barry J (2007) Environment and social theory. Routledge, Berlin

Craig AWJ, Solomon NA (1996) The system of industrial relations in Canada. Prentice Hall Canada Inc., London

Elum ZA, Mopipi K, Henri-Ukoha (2016) A oil exploration and its socioeconomic effects on the Niger Delta region of Nigeria. Environ Sci Pollut Res 23(13):80–89

Fekadu K (2014) The paradox in environmental determinism and possibilism: a literature review. J Geography Region Plan 7(7):132–139

Frame ML (2022) Ecological imperialism: a world-systems approach. Am J Econ Sociol 81(3):503–534

Friedman SW (2003) Historical geography and historical sociology: our honest toil and the spatial turn. In: Delanty G, Isin EF (eds) Handbook of historical sociology. Sage Publications, Berlin, p 110

Nriagu JO (2008) The encyclopedia of environmental health. Elsevier Press, London

Nwankwo CM (2022) Diversification of mono-economies: how legislation manages the environmental impact of foreign investments in Nigeria. In: Junker KW, Farah PD (eds) Globalisation, environmental law, and sustainable development in the global south challenges for implementation. Routledge, London, pp 122–147

Obeng-Odoom F (2022) Mainstream economics and conventional environmental policies. Am J Econ Sociol 81(3):443–472

Önal H (2018) Reflections of environmental determinism in the questions prepared by geography teacher candidates. Rev Int Geograph Educ Online 8(1):74–92

Osunmuyiwa O, Biermann F, Kalfagianni A (2018) Applying the multi level perspective on socio-technical transitions to rentier states: the case of renewable energy transitions in Nigeria. J Environ Plan Policy Manag 20(2):143–156

Raimi MO, Okoyen E, Omidiji AO, Ebuete AW (2020) Governing the environmental impact of dredging: Consequences for marine biodiversity in the Niger Delta region of Nigeria. Insights Min Sci Technol 2(3):76–84

Raimi MO, Odubo TV, Omidiji AO (2021) Creating the healthiest nation: Climate change and environmental health impacts in Nigeria: a narrative review. Sustain Environ 6(1):61–122

Şahin C, Belge R (2016) The geographical determinism. In: Khaldun İ (ed) Globethics

Taylor PW (2011) Respect for nature: a theory of environmental ethics. Princeton University Press, Princeton

Trist E (1981) The evolution of socio-technical systems: a conceptual framework and an action research programme. Occasional Paper No. 2, Ontario Quality of Working Life Centre, Ontario Ministry of Labour.

Uddin MD (2017) Climate change and global environmental politics: north-south divide. Environ Policy Law 47(3–4):106–127

Ukah BU, Egbueri JC, Unigwe CO, Ubido OEX (2019) Extent of heavy metals pollution and health risk assessment of groundwater in a densely populated industrial area, Lagos, Nigeria. Int J Energy Water Resour 3(3):291–303

Wolf MJ, Emerson JW, Esty DC, de-Sherbinin A, Wendling ZA, Northonsmith Q, Brigham S, Zaharieva S, Li L, Jacob M, Bradford E (2022) Environmental performance index. New Haven, CT: Yale Centre for Environmental Law and Policy

The Efficacy of Climate Change Adaptation Strategies for Enhancing Trade and Economic Development in Kenya

Aloyce Peter Ndege

Abstract Trade has an impact on the environment. Trade undeniably exacerbates climate change. Climate change impacts such as drought, floods, severe weather, and sea-level rise are likely to result in food shortages, increases in insecurity, infrastructure damage, and the degradation of natural resources upon which livelihoods are based. The negative impacts of climate change hit poor people and poor countries disproportionately. In other words, climate change threatens development objectives and is affected by development choices. Furthermore, the impacts of climate change are likely to become progressively more significant in the years and decades beyond the 2030 target date for achieving the Sustainable Development Goals. In principle, various development activities could help reduce vulnerability to many climate change impacts. In some cases, however, 'development as usual' may inadvertently increase vulnerability mainly because of maladaptation. The risk of maladaptation points to the need for countries to systematically assess climate risks and vulnerabilities and to include potential adaptation measures in development policies, plans and projects. Yet, many development policies, plans and projects currently fail to consider climate variability, let alone climate change. While efforts to integrate climate change adaptation into development will be led by developing country partners, donor agencies have a critical role in supporting such efforts and integrating such considerations within their organizations. To this end, partners and donors alike need adequate guidance. This paper examines policies and structural areas that could help Kenya make strides in adapting to climate change by building resilience and improving coping mechanisms through sound policies and strategies.

A. P. Ndege (✉)
Faculty of Law, University of Nairobi, Nairobi, Kenya
e-mail: aloyce.ndege@students.uonbi.ac.ke

D. Oghenerobor Agelebe (ed.), *Climate Ethics in Africa*, Global Perspectives on Law and the Environment 1, https://doi.org/10.1007/978-3-032-08757-7_9

1 Introduction

Like all human and economic activities, trade impacts the environment. Trade involves the production, movement, exchange and consumption of goods and services within and across borders, and that is the foundation and a prerequisite of modern society. Through this process, we get the energy that powers our lives by getting us to work each day, delivering the food we need, delivering the appliances we have come to depend on, and providing the medical supplies that help make us safer and healthier. Yet, each process stage entails a fresh contribution to greenhouse gas emissions. Whereas it is undeniably true that trade exacerbates climate change, it is equally valid that trade is disrupted by climate change. Extreme weather events often devastate transport and logistics infrastructure essential for trade. These events eat into capital stock, weaken export capacity, damage agriculture, and disrupt food security—all with adverse consequences for long-term development outcomes. Trade, in short, is a critical node to mobilize if the world is to achieve green, resilient, and sustainable development in the coming years. As this paper demonstrates, it is a central element of the solution to climate change—because it can enhance mitigation and adaptation efforts.

1. Trade can help shift production to areas with cleaner production techniques. As the world transitions to a low-carbon economy, export comparative advantages will change, compelling countries to adapt and seize new opportunities.
2. Trade promotes the spread of critical environmental goods and services that can help reduce emissions and improve environmental management—today, global trade in environmental goods is estimated at more than US$1 trillion annually and is rising.[1]
3. Imports are critical to immediate recovery from a natural disaster when essential items such as food and medicines are in short supply. In a world increasingly shaped by climate change, trade will be crucial to addressing food insecurity, supporting adaptation, and enabling recovery from natural disasters.

In contrast, in the last decades, some developing countries have emerged as fast-growing emitters of greenhouse gases, poorer countries have continued to remain the most minor contributors to emissions. This is despite the fact that the poorest countries have continued to suffer the most from climate change. Natural disasters disproportionately affect the most vulnerable people and the smallest firms, most of which are in poor developing countries. All developing countries, therefore, have an essential stake in adapting to climate change—especially in agriculture, which is central to food security, employment, and trade and is also the most sensitive to climate change. Tourism, the mainstay of the economy of many developing states in Africa and small islands, is also particularly sensitive to climate change. Diversifying away from the most vulnerable sectors to climate shocks will allow these economies to become more resilient over the long term.

---

[1] Brenton and Chemutai (2021).

2 Impacts of Climate Change on Trade and Development

Smit and Wandel affirm that climate change is a permanent shift in the standard patterns of climate change, possibly due to natural or human-initiated processes. This shift can have negative or positive impacts.[2] The positive and negative effects differ from one adaptation option to another and from region to region. Still, it has become increasingly clear that the anthropogenic element in climate change supersedes all the differences and costs involved. For example, whereas agriculture offers solutions through adaptation and mitigation processes, it increases adaptation costs by emitting two greenhouse gases (GHG): Methane and Nitrous oxide. Methane emanates from the livestock digestion process and is stored in animal manure, whereas nitrous oxide is discharged from organic and mineral fertilizers.[3]

The new normal of extreme weather events and a changing climate is putting pressure on communities, organizations and governments around the globe. In sub-Saharan Africa, the impact of climate change mainly affects trade, generally, through the following ways:

2.1 *Infrastructure*

Infrastructure is vital to the global community, and climate change highlights vulnerabilities that disrupt these critical systems. Infrastructure is the set of fundamental facilities and systems supporting households and firms' sustainable functionality. Infrastructure comprises public and private physical structures such as roads, railways, bridges, tunnels, water supply, sewers, electrical grids, and telecommunications (including Internet connectivity and broadband access), vital for trade and economic development. Infrastructure is defined as '*the physical components of interrelated systems providing commodities and services essential to enable, sustain, or enhance societal living conditions and maintain the surrounding environment.*'[4] The transportation sector, for example, is the backbone of the economy. It is responsible for the movement of people and the delivery of goods and services from the production points to the consumers.

2.2 *Extreme Weather Events*

Our changing climate exacerbates extreme weather events such as hurricanes, wildfires, droughts, and floods. These often lead to disasters that displace communities

---

[2] Smit et al. (2006, 292).

[3] Sere et al. (1996, 127).

[4] Fulmer (2009, 32).

and disrupt businesses and supplies. They also destroy infrastructure, cutting off producers and consumers.

2.3 National Security Implications

Climate change has many national security implications, mainly relating to public safety caused by disasters and, to an indirect extent, food insecurity. Disasters accelerate instability, cause displacements of people, thereby increasing the need for humanitarian and disaster relief, and affect infrastructure and defence facilities. The catastrophes, in turn, lead to food insecurity mainly due to low food production and insufficient supply caused by inadequate or destroyed means of transport and communication. Trade and economic development cannot thrive in such situations.

3 Adaptation to Impacts of Climate Change

There are two broad categories of responses to climate change: mitigation and adaptation. Both mitigation and adaptation aim at reducing the risks of climate change. Mitigation seeks to avoid or at least limit climate change itself by reducing the emissions of GHGs, for instance, by promoting energy efficiency, the use of renewable energy such as solar and wind power, and avoiding deforestation. Adaptation consists of deliberate actions to reduce adverse consequences and harness beneficial opportunities. More specifically, mitigation consists of activities that aim to reduce GHG emissions, directly or indirectly, by avoiding or capturing GHGs before they are emitted to the atmosphere or sequestering those.

Already in the atmosphere by enhancing 'sinks' such as forests. Such activities may entail, for example, changes to behaviour patterns or technology development and diffusion. On the other hand, adaptation is defined as adjustments in human and natural systems in response to actual or expected climate stimuli or their effects that moderate harm or exploit beneficial opportunities. IPCC defines adaptation as an adjustment in natural or human systems responding to actual or expected climatic stimuli or their effects, which moderates harm or exploits beneficial opportunities that can be carried out in response to (ex-post) or anticipation of (*ex-ante*) changes in climatic conditions.[5] It entails a process by which measures and behaviours to prevent, moderate, cope with and take advantage of the consequences of climate events are planned, enhanced, developed and implemented.[6]

Both mitigation and adaptation to climate change are essential and complementary. The extent and scope of regional climate change impacts will depend on the degree of mitigation. Reductions in GHG emissions will delay and reduce damages

[5] IPCC (2007).

[6] As above.

caused by climate change, thus reducing the amount of adaptation that will be necessary. However, while climate change mitigation is crucial to limit long-term impacts, climate change is already happening. It is bound to continue simply because of the GHGs that have already been emitted. Furthermore, GHG mitigation relies on difficult policy choices and further technological development, so emissions are bound to keep rising before they may eventually taper off. For those reasons, climate change will inevitably continue and even accelerate, at least for several decades. Adaptation to these already 'locked-in' impacts is therefore critical. This paper focuses on adaptation and how to integrate it within development processes.

3.1 Adaptation Within International Negotiations on Climate Change

Adaptation is a relatively recent concern in international negotiations on climate change. Adaptation is mentioned in the United Nations Framework Convention on Climate Change (UNFCCC), negotiated in 1992, and the Kyoto Protocol, negotiated in 1997. The implementation of adaptation has come into a much sharper focus since the Seventh Conference of the Parties (CoP-7) in Marrakech in 2001, which established three funds dealing with adaptation:

1. **The Least Developed Countries Fund** addresses the particularly low adaptive capacity of the least developed countries (LDCs). The Marrakech Accords established the fund to help such countries prepare their National Adaptation Programs of Action (NAPAs), establishing and prioritizing adaptation needs. The fund also supports institutional capacity building and other activities.
2. **The Special Climate Change Fund** finances many mitigation and adaptation activities in all developing countries. The activities can be specific to sectors—energy, transport, industry, agriculture, forestry, and waste management—or aimed directly at adaptation, technology transfer, and economic diversification.
3. **The Adaptation Fund**—the only Marrakech fund linked to the Protocol rather than the Convention—provides funding only to Protocol parties. Like the other two funds, its resources come from voluntary contributions. Still, it also benefits from a 2% share of the proceeds of certified emissions reductions from projects under the Protocol's Clean Development Mechanism (CDM).

These funds are part of a more complex architecture of international funding sources for adaptation, including the Strategic Priority on Adaptation of the Global Environmental Facility and the World Bank Group Climate Investment Funds and bilateral initiatives. Successive rounds of negotiations have sought to develop a more comprehensive approach to addressing adaptation. This includes a 5-year (2005–10) Programme—the Nairobi Work Programme on Impacts, Vulnerability and Adaptation to Climate Change—that aimed at fostering a wide range of initiatives to help governments improve their understanding of the impacts, vulnerability and adaptation to climate change, and take informed decisions on practical adaptation actions. The programme aimed to achieve the following outcomes:

1. Improve capacity at international, regional, national, sectoral and local levels to identify further and understand impacts, vulnerability, and adaptation responses to effectively select and implement practical, efficient and high-priority adaptation actions;
2. Enhance and improve the level and amount of information and advice to the UNFCCC Conference of the Parties and its subsidiary bodies on the scientific, technical and socio-economic aspects of impacts, vulnerability and adaptation;
3. Enhance the degree of dissemination and utilization of knowledge from practical adaptation activities;
4. Enhance cooperation among parties, relevant organizations, businesses, civil society and decision-makers to advance their ability to manage climate change risks and
5. Enhance the integration of adaptation into sustainable development plans.

The Nairobi Work Programme serves as a platform for disseminating scientific and technical knowledge and aims to support and catalyze the implementation of adaptation.

3.2 What Does Adaptation Entail?

Adaptation reduces the impacts of climate stresses on human and natural systems. It consists of a multitude of behavioural, structural and technological adjustments. Adaptation measures can vary in their timing (anticipatory vs. reactive; *ex-ante vs. ex-post*), scope (short-term vs. long-term; localized vs. regional), purposefulness (autonomous vs. planned; passive vs. active) and adapting agent (private vs. public; societies vs. natural systems). Adaptation measures include changing crop varieties and altering farming practices, developing heat- and drought-resistant crops, diversifying livelihoods, building flood defences and land-use planning.

The need to adapt to changing environmental and climatic conditions is not a new one. Societies throughout history have had to adapt to variations or changes in their climate through various strategies and by using knowledge accumulated through experience of past climatic events. In addition, societies have also had to cope with and respond to extreme weather events, such as droughts and floods. For example, the Kano plains region of Kisumu County in Kenya has historically frequently faced extreme climatic variability and events such as droughts alternating with floods. Therefore, communities in the Kano plains have had to regularly adapt to unreliable rainfall and drought conditions, alternating with excessive rainfall and floods, for example, by diversifying their livelihoods and adopting new crop varieties. IPCC defines adaptation as an adjustment in natural or human systems responding to actual or expected climatic stimuli or their effects, which moderates harm or exploits beneficial opportunities that can be carried out in response to (*ex-post*) or anticipation of (*ex-ante*) changes in climatic conditions.[7] It entails a process by which measures and behaviours to prevent, moderate, cope with and take advantage of the consequences of climate events are planned, enhanced, developed and implemented.[8]

A wide range of adaptation measures can be applied to respond to observed and anticipated climate change. Such measures include altering farming practices and crop varieties, building new water reservoirs, enhancing water use efficiency, changing building codes, investing in air-conditioning, and constructing sea walls. To better understand this diversity, it is possible to classify adaptation measures into the following generic options or categories:

1. **Bear losses**. All adaptation measures may be compared with the baseline response of 'doing nothing' except bearing or accepting the losses. In theory, bearing loss occurs when those affected cannot respond in any other way, for example, in impoverished communities or where the costs of adaptation measures are considered high about the risk or the expected damages.
2. **Share losses:** This adaptation response involves sharing the losses among a wider community. Such actions occur in traditional and the most complex, high-tech societies. In traditional societies, many mechanisms exist to share losses among a wider community, such as extended families and village-level or similar small-scale communities. At the other end of the spectrum, large-scale societies share losses through public relief, rehabilitation, and reconstruction paid for from public funds. Sharing losses can also be achieved through insurance.
3. **Modify the threat**: For some risks, it is possible to exercise a degree of control over the environmental danger itself. When this is a 'natural' event such as a flood or a drought, possible measures include flood control works (dams, dikes, levees). For climate change, the primary modification possibility is to slow the rate of climate change by reducing GHGs and eventually by stabilizing GHG concentrations in the atmosphere (i.e. mitigation).
4. **Prevent effects**: A frequently used set of adaptation measures involves steps to prevent the impact of climate change and variability. For example, in agribusiness, such measures include changes in crop production and management practices, such as increased irrigation water, additional fertilizer use, and pest and disease control.
5. **Change use:** Where the threat of climate change makes the continuation of an economic activity impossible or hazardous, consideration can be given to changing the use. For example, a farmer may substitute a more drought-tolerant crop or switch to varieties with lower moisture. Similarly, cropland may be returned to pasture or forest, or other uses, such as recreation, wildlife refuges, or national parks, may be found.
6. **Change location:** A more extreme response is to change the location of economic activities. There is considerable speculation, for example, about relocating major

[7] IPCC (2007).

[8] (n 10) 23.

crops and farming regions away from areas of increased aridity and heat to areas that are currently cooler and may become more attractive for some crops.

7. **Research**: Research into new technologies and adaptation methods can also advance the adaptation process.
8. **Encourage behavioural change:** This may be through education, information and regulation. It is believed that disseminating knowledge through education and public information campaigns may lead to behavioural change. Though such activities have received little recognition and have received low priority in the past, this paper believes that they will likely assume increased importance as the need to involve more communities, sectors and regions in adaptation is becoming apparent.

3.3 Adaptation Versus Regular Development

Although adapting to climate change represents a new challenge, overlaps do exist. Between 'business-as-usual' development strategies and adaptation. In some cases, activities undertaken to achieve development objectives can automatically lead to adaptation benefits. For instance, decisions taken as part of development activities can have considerable bearing on societies' vulnerability to the potential impacts of climate change. In principle, an array of development activities geared towards poverty reduction, improved nutrition, education, infrastructure and health would synergize with climate change adaptation. This is the fundamental principle of adaptive capacity in that better-developed societies possess more adaptive capacity than less-developed societies and, therefore, have lower vulnerability to climate change. Furthermore, in situations where vulnerability is primarily contextual, adaptation might require emphasis on baseline or business-as-usual economic development activities—alleviating poverty and improving nutrition, health care, livelihoods, etc.—as these activities will also boost the capacity for coping with climate change. However, while there are overlaps between adaptation and development, activities with a clear focus on adaptation and climate change will also be required. For example, in cases where the biophysical impact of climate change significantly exacerbates vulnerability, adaptation will likely need more explicit consideration of climate risk in development activity.

However, dealing with climate change adaptation is necessary for the present, as the impacts of a changing climate often demand an immediate response. For this reason, it mustn't be separated from other development priorities but mainstreamed into development planning, programs, and projects.[9] However, it is critical to prefer climate-smart strategies, given that responses aimed at adapting to climate change may negatively affect other vital areas of development, such as food security. Methods taken to increase food security may exacerbate emissions, as discussed earlier.[10] Whereas mitigation can be more cost-effective using an integrated approach that

[9] Wekesa and Karani (2009).

[10] Climate Change Agriculture and Food Security (2009).

combines measures to reduce energy use, decarbonize energy supply, reduce net emissions and enhance carbon sinks in land-based sectors, adaptation answers to the immediate survival needs synonymous with developing regions such as Kenya.[11] Studies done by the Centre for European Policy Studies on adaptation to climate change underscored the usefulness of adaptation policy and how to implement it.[12] The results showed that adaptation reduces the economic impacts of climate change. In some cases, adaptation was seen to lessen the impacts or turn climate change into a positive opportunity.

4 Kenya's Adaptation Strategies

Kenya, a middle-class developing country, is one of the most vulnerable countries to climate variability and climate change due to its heavy reliance on rain-fed agriculture and other primary commodities and natural resources and its limited capacity to deal with potential adverse impacts. Kenya's challenges include the underdevelopment of water resources, weak health services, a high population growth rate, low economic development, inadequate road infrastructure, especially in drought-prone areas, weak institutional structure, and lack of climate risk awareness.[13] Kenya has a complex geo-ecological setting, and virtually all her ecosystems are fragile. Key environmental and climate change issues especially relevant to the rural areas of Kenya include climate variability and change, land degradation and desertification, and water scarcity and stress. Climate variability and climate change impacts on rural populations and livelihoods, particularly in ASALs, which comprise roughly 75 percent of Kenya's landmass, are a priority concern. Extreme weather events are becoming more frequent, including intense droughts and floods in some areas, leading to lower crop yields or total crop failure, smaller livestock herds or more loss of livestock, diminished livelihoods of smallholder farmers and pastoralists, and heightened food insecurity. The nexus of climate change and trade is particularly relevant for Kenya for the following reasons:

Climate change poses significant social, political, and economic risks to Kenya since the country is an agricultural economy, with most of its population employed in agriculture. Notably, climate change will alter comparative advantages and market dynamics in the agriculture sector to the detriment of low- and middle-income countries and major food-importing countries— of which Kenya is both[14]; increase production costs and worsen the country's terms of trade and balance of payments; increase unemployment; and increase the risk of armed conflict and fragility, especially in the pastoralist areas.

[11] IPCC (2014, p 151).

[12] Aheim et al. (2008).

[13] World Bank Group (2021).

[14] United Nations Conference on Trade and Development (2021).

Kenya is one of the fastest-growing economies in Africa and the world, with a nascent industrial sector undergoing rapid advancement. In line with the theories on the climate-trade nexus, the time is appropriate for aligning the 'greening' of the economy while it is in take-off. Hence, Kenya is opportune to green its nascent industrial policy based on lessons learned from industrialized countries and best international practices. Greening the industrial sector is pertinent, as market access is tied increasingly to environmental necessities, mainly in high-income countries such as the European Union (EU), the biggest export market for Kenya. Expanding the scope of the EU Carbon Border Adjustment Mechanism (CBAM) policy, designed to reduce the risk of carbon leakage to agricultural goods, could potentially affect Kenya's exports. CBAM targets imports of carbon-intensive products, in full compliance with international trade rules, to prevent offsetting the EU's greenhouse gas emissions reduction efforts through imports of products manufactured in non-EU countries, where climate change policies are less ambitious than in the European Union.

An analysis of Kenya's challenges and opportunities could spur Kenya's interest in addressing climate change and trade. Low—and middle-income countries have hardly participated in the negotiations of environmental goods and services, and conducting a cost–benefit analysis is critical, as is examining the country's interests and how best to secure and promote them.

4.1 The Impact of Climate Change on Trade in Kenya

4.1.1 Effects on Productivity

The physical impacts of climate change will include adverse conditions for crop cultivation and livestock rearing. These impacts can have substantial effects not only on the production of primary agricultural and agroforestry commodities but also on the broader economy. In national economies with strong downstream linkages like Kenya's, reductions in primary agricultural production can trickle down and affect productivity in downstream sectors, such as food processing and leather manufacturing, by reducing the availability and increasing the prices of primary input. The World Bank estimates that Kenya will lose more than 6 percent of each year's agricultural output if the current decline in average annual rainfall levels of primary farm zones continues to mid-century.[15] Rising temperatures and shifting rainfall patterns may increase soil erosion, make it harder to grow many crops, and shorten the growing season. The livestock sector, a key economic sector for Kenya, is likely affected by extreme weather conditions. Heat stress has many detrimental effects on livestock, but it can include milk production and reproduction reductions, particularly for dairy cows. Extreme events, such as heat waves, may particularly affect

[15] World Bank Group Kenya.

beef and dairy cattle.[16] All regions worldwide lose from reduced crop productivity (although a few individual countries may gain), but some are more than compensated by terms of trade gains. Because total food demand is inflexible, an average crop productivity decrease will substantially increase food prices. Whereas net-food-exporting regions like Latin America, Northern America, and Oceania will benefit, the burden of adjustment to climate change will fall on net-food-importing regions. Despite its comparative advantage in agricultural production, Kenya is a significant food importer and is increasingly dependent on aid.[17] Its balance of payments and the welfare of its people are also likely to be negatively affected. Kenya is expected to lose its comparative advantage in producing some agricultural products, although its comparative advantage in producing others may be enhanced. Climate change will reshuffle global exporters and Importers of agricultural products. Competitiveness effects are also likely to occur.[18]

4.1.2 Effects on Communities

Remote and marginalized communities are likely to be the most affected by changes in future trade patterns due to climate change. They depend predominantly on pastoral and agropastoral livelihoods and are exposed to fragile, conflict-affected neighbouring states. These regions include West Pokot, Turkana, Marsabit, Isiolo, Wajir, Mandera, Garissa, Lamu, Tana River, Samburu, Elgeyo Marakwet, Baringo, Migori, Kajiado and Narok. Standard features in the regions include remoteness from the centre and proximity to fragile neighbouring states (Somalia, South Sudan, North-Western Tanzania, Eastern Uganda, and Ethiopia), predominance of pastoral and agropastoral livelihoods, and limited access to public services such as schools and clinics and infrastructure such as roads, resulting in low levels of literacy, formal education, and public health; and widespread poverty.[19]

4.1.3 Effects on Transport

As already discussed, transport is likely to be affected substantially by extreme weather events. While this sector contributes heavily to greenhouse gas emissions, it also plays a crucial role in supporting Kenya's overall development strategy and supports its export and Import networks. Given Kenya's topographical variations, characterized by extremely rugged terrain, the construction of transportation infrastructure tends to be physically demanding and costly. Continued repairs and maintenance are necessary to keep road networks functional. Such infrastructure could be problematic due to higher rainfall, flooding, and rising temperatures. Concerning

[16] As above.

[17] World Bank (2020).

[18] (n 20) at 15.

[19] 'Socialist Republic of Viet Nam (2019).

marine transport, sea-level rise is projected to lead to the loss of a sizable proportion of the coastlines due to a combination of inundation and erosion, with consequential risks for port infrastructure at the Mombasa port.

4.2 Key Policy Developments Addressing Climate Change

The government of Kenya has responded to various national and international challenges by enacting policies, legislations, and strategies to address them and meet international obligations. The government has given climate change prominence and developed policies and strategies to address its associated challenges. The United Nations Framework Convention on Climate Change (UNFCCC) requires that countries report to their National Communication (NC) their progress in reducing vulnerability to the impacts of climate change. Kenya submitted its first NC in 2002. Kenya has also completed its National Climate Change Response Strategy[20] (NCCRS) and National Climate Change Action Plan (NCCAP).[21] Despite these compliance measures, there are concerns that adaptation preparedness remains inadequate in most African countries, even those that have prepared NAPAs and NCs.[22] NAPAs are not easy to implement because they only list the country's priority adaptation needs through programs and/or projects. They do not consider the overarching policy framework, such as mainstreaming adaptation into national plans. As a result, adaptation programs and projects are often implemented in African countries as standalone activities, which render them less effective.[23] In Kenya, the existing guidelines on climate change adaptation were first contained in a draft environment policy, culminating in the Environmental Management and Coordination Act (EMCA) of 1999 and the NCCAP. However, the EMCA and the current Environmental Policy do not have strong provisions for climate change adaptation and mitigation, which can be attributed to the time they were made. For instance, the policy states that a climate change strategy should be developed. Although completing the NCCRS is a positive step toward addressing climate change, Kenya needs strong policies that address mitigation and adaptation, guiding the integration and mainstreaming into all national sectors and institutions. Despite this shortcoming, however, concerns about the possible impacts of climate change generated strong support for policies, leading to the setting up of the first climate change unit in the then Office of the Prime Minister in 2008. The office provided high-level political support to climate change activities, leveraged financial support, harmonized ongoing and future activities on climate change and integrated them into different government departments. The office, which is now under the Ministry of Environment and Forestry, is still faced with the challenges of sufficient funding and qualified staff.

[20] Government of Kenya Ministry of Environment and Mineral Resources (2010).

[21] GoK *Long-term national low-carbon climate-resilient development pathway* (2013).

[22] Pandey (2002, 6).

[23] GoK Revised REDD (2010).

Other environment-related policies are similarly weak. More particularly, while the energy, forest, and ASAL (arid and semi-arid lands) policies are heavily influenced by environmental management, climate change issues go beyond ecological management. The impacts of these policies on society and ecosystems' vulnerability to climate change and their potential to contribute to global carbon sequestration are reviewed in detail below, along with an analysis of the interactions between these policies.

4.2.1 National Climate Change Response Strategy (NCCRS) and National Climate Change Action Plan (NCCAP)

The strategy is the framework that guides the integration of climate concerns into development priorities, government planning and budgeting in Kenya.[24] The processes of formulating the NCCRS and its implementation action plan were participatory and consultative, and all the key sectors of the economy were addressed; climate change was viewed as a challenge that cuts across all industries and segments of society in Kenya, hence the need to have inputs from diverse stakeholders. The stakeholders comprised development partners, representatives from the private and public sectors and a parliamentary committee dealing with climate change.

The NCCRS highlights various measures for adaptation and mitigation of the impacts of climate change in all sectors of the economy.[25] For example, in agriculture, the strategy proposes applying multiple innovative technologies such as irrigation, early maturing and high-yielding crop varieties, drought and pest-resistant crop varieties, and disease-resistant livestock. The NCCRS also advocates diversification of livelihoods, adaptation of agricultural technologies from analogue environments, and enhancing early warning systems with drought monitoring and seasonal forecasts for better food security. The action plan prioritizes protecting and rehabilitating water towers and increasing forest cover through farm forestry and afforestation in the environment and water sectors. The NCCRS includes indicative budgets and plans for line ministries. The NCCRS is meant to guide the government in all activities and interventions aimed at addressing issues related to climate change; it consolidates all national efforts and focuses on climate change adaptation and mitigation. To operationalize the NCCRS, the government, in March 2013, finalized the development of the National Climate Change Action Plan (NCCAP). This was developed through a consultative process involving government and private sector actors. And civil society. The NCCAP aims to operationalize the NCCRS by providing the analysis and enabling mechanisms for successful implementation. It also supports efforts toward the implementation of the Kenya Constitution 2010 and the attainment of Vision 2030. It encourages people-centred development, ensuring climate change actions help the country achieve its long-term development goals. In particular, the NCCAP sets out a vision for a low-carbon climate-resilient development

[24] GoK Ministry of Environment and Mineral Resources (2010).

[25] As above.

pathway; summarises the analysis of mitigation and adaptation options and recommended actions; recommends an enabling policy and regulatory framework; and sets out the next steps for knowledge management and capacity development, technology requirements, a financial mechanism and a National Performance and Benefit Measurement System (NPBM).[26]

4.2.2 National Environment Policy, 2013

The National Environment Policy (NEP) 2013 treats climate change and disaster management as emerging environmental issues. The government seeks to adopt mitigation and adaptation approaches to deal with climate change in this policy document. This policy recognizes that many of Kenya's natural disasters, such as floods, droughts, landslides, and fires, are climate-related and that their negative impacts cut across all key sectors of the economy.

The policy anticipates that as the climate changes further, the frequency and intensity of these extreme weather events will increase. The NEP recommends the following measures for dealing with disasters related to climate change, most of which can be said to embrace adaptation strategies:

1. Develop and implement a national climate change strategy.
2. Identify and raise awareness of opportunities for adaptation measures by promoting appropriate technology transfer and capacity building.
3. Develop and implement the Kyoto Protocol's Clean Development Mechanism programs and projects, encouraging significant investment and technology transfer for sustainable development.
4. Develop an integrated, improved early warning and response system for climate change and disaster risks, with a clear strategy for dissemination of information at the grassroots level.
5. Build and strengthen research capacity on climate change and related environmental issues. The policy statement contains the building blocks of a legal framework for addressing environmental concerns beyond the current EMCA 1999.

4.2.3 National Policy for Disaster Management in Kenya, 2009

The policy aims to institutionalize disaster management and mainstream disaster risk reduction in the country's development initiatives. The policy seeks disaster management and minimizes risks, such as loss of life, economic loss, and property. Climate change is one of the issues highlighted in the policy. The policy also recognizes that more than 70% of natural disasters in Kenya are related to climate events (flooding, famines, landslides, etc.). The policy, therefore, calls for preparedness to

[26] GoK 'Long-term National Low-carbon Climate Resilient Development Pathway (2013).

reduce climate change's causes and negative impacts. It proposes that the government harmonizes all the policies and laws, promotes environmental safeguards and encourages sustainable management of natural resources. The policy proposes mainstreaming disaster risk reduction activities in national and county plans and policies with appropriate budgetary allocation. However, the policy does not give specific strategies to mitigate climate change in fragile ecosystems except for urging collaboration among relevant agencies. Thus, neither adaptation nor mitigation is proposed as a strategy in the policy.

4.2.4 Constitution of Kenya 2010

In 2010, Kenya promulgated a new Constitution, which provides for a participatory system of government. Article 42 of the Constitution deals with issues related to the environment and claims the right to a clean and healthy environment for all citizens. This includes the right to have an environment protected for the benefit of the present and future generations through legislative and other measures. In addition, Articles 69 and 70 of the Constitution deal with environmental issues such as climate change mitigation and adaptation. For example, Article 69 emphasizes the sustainable use, management and conservation of the environment and natural resources to ensure equitable sharing of the accruing benefits. It also encourages the people of Kenya to achieve and maintain a tree cover of at least 10% of the land area. The article encourages the public to participate in environmental management, protection and conservation. Article 70, conversely, enforces the right to a clean environment and recognizes actions that responsible people may take to ensure a clean and healthy environment. It states that the government may compensate citizens deprived of their right to a clean and healthy environment. The Kenyan Constitution thus has elaborate provisions for using adaptation strategies, amongst others, to combat climate change.

4.2.5 Climate Change Act, 2016

This Act is applied to develop, manage, implement, and regulate mechanisms to enhance climate change resilience and low-carbon development for the sustainable development of Kenya. The Act places duties on the national and county governments to mainstream climate change responses into development planning, decision making and implementation and to respond in various other ways to climate change. The Act sets out principles for climate change planning and the implementation of measures. The National Climate Change Council is established as an unincorporated body. The Cabinet Secretary, for the time being responsible for environment and climate change affairs, shall be the secretary to the Council. The Council shall provide an overarching national climate change coordination mechanism. It shall, among other things, approve and oversee the implementation of the National Climate Change Action Plan. The Act also establishes a Climate Change Directorate and a Climate Change Fund. The Directorate shall be the government's lead agency on national climate change plans and actions to deliver operational coordination and report to the Cabinet Secretary. A Director of Climate Change shall head the Directorate. The Cabinet Secretary shall formulate a National Climate Change Action Plan. The Council may, on the recommendation of the Cabinet Secretary and in consultation with relevant Cabinet Secretaries and county government, impose duties relating to climate change on any public entity at all levels of government and on private entities. The Council shall make regulations governing the nature and procedure for reporting on performance by private entities, including the authority to monitor and evaluate compliance. The Council, on the recommendation of the Cabinet Secretary, shall each year identify priority strategies and actions for disaster risk reduction related to climate change. The Kenya Private Sector Alliance (KEPSA) should be part of the council's membership. The entrepreneurs working around climate change must ensure that they are in associations and are part of KEPSA to ensure they can be nominated to the council, especially with the specialization in the area. The membership of entrepreneurs in the Climate Change Council will allow them to voice the needs of the industry players. The Act also provides public participation and access to information and intergenerational and gender equity in all aspects of climate change responses. On behalf of the Council, the National Environmental Management Monitoring Authority shall monitor, investigate, and report on compliance with the assigned climate change duties and regulate, enforce, and monitor compliance with levels of greenhouse gas emissions set by the Council under this Act.

4.2.6 Kenya Vision 2030

Kenya's Vision 2030 was launched in 2007 as the long-term national development blueprint for the country to create a globally competitive and prosperous nation with a high quality of life by 2030 by transforming Kenya into a newly industrialized, middle-income country providing a high quality of life to all its citizens in a clean and secure environment. It was implemented as a replacement for the Economic Recovery Strategy for Wealth and Employment Creation for the 2003–2007 is grounded on three key pillars: political, economic, and social. The blueprint's goals and strategies are effectively implemented through prescribed medium-term plans.

Notably, an environmental pillar was not included in the preparation of Vision 2030. It was merely considered a sub-division of the social pillar. The relegation of environmental matters to the social pillar, and not a stand-alone thematic pillar, limits the ability to ensure ecological/climate change mainstreaming within the development framework as a whole. It is challenging to balance the conflicting interests of maintaining environmental integrity and facilitating economic growth. The inability to achieve an effective balance in the foundations of socio-economic transformation promotes the emergence of policy incoherence and discourse within the national development framework goals. For example, energy is a prerequisite to socio-economic development. However, the use of energy sources in Kenya is both directly and indirectly associated with long-term adverse environmental impacts, which contribute to global warming, such as forest degradation (due to cutting forests for biomass products) and the emission of Greenhouse gases (due to the use of fossil fuels), among others.

4.2.7 Kenya Climate Smart Agriculture Implementation Framework 2018–2027 (KCSAIF)

The agriculture sector is a major contributor to Kenya's economy in terms of food and nutrition security, employment creation, and the supply of raw materials and foreign exchange earnings from the agro-based industry. About 98% of Kenya's agricultural activities are rain-fed and highly susceptible to climate change and variability.[27] The susceptibility of the agricultural sector to climate change is likely to compromise the attainment of the sector's contribution to food and nutrition security, national economy and the global Sustainable Development Goals (SDGs).

KCSAIF provides a guide to various innovative and transformative initiatives and best practices that will strive to address challenges brought about by climate change. It is envisioned to ensure increased agricultural productivity and sustainably build resilience of the national agricultural systems. It also aims to provide various options for implementing the Kenya Climate Smart Agriculture Strategy 2017–2026 (KCSAS). The summary actions outlined in the framework provide broad guidelines from which the Counties and other stakeholders in the sector are expected to develop appropriate activities towards building the resilience of the agricultural systems and value chain actors to cope with the effects of the current and projected climate change patterns based on specific climate change vulnerabilities.

4.2.8 Summary Overview and Policy Gaps in Climate Change Adaptation Strategies

This analysis has shown that existing policies and legislation are relatively weak and inadequate to deal with climate change issues. Very few sections address climate change adaptation, and they are not exclusively devoted to climate change adaptation. Thus, an exclusive and comprehensive climate change policy and legislative framework is needed to create or set out the mandate for a leading institution to spearhead the nation's efforts in climate change adaptation. In addition, the policy should have a clear and comprehensive implementation framework to ensure that funds are channelled into projects that address the most vulnerable social groups and regions.

Furthermore, the policies are scattered, disjointed and contradictory, contributing to their inadequacy in addressing climate change adaptation issues. For example, while the policy for forests emphasizes the need to increase tree cover, the livestock policy encourages livestock keepers to clear trees on rangelands to encourage

[27] GoK *Kenya Climate Smart Agriculture Implementation Framework* (2018).

grass growth for improved pasture. Various actors working in sectors related to the environment have noted the need to harmonize these policies.[28]

The various policies and institutional frameworks have led to weak coordination in basic approaches to addressing the nation's challenges regarding climate change, and institutional rivalries have exacerbated this due to overlapping mandates among the ministries. The policy incoherence identified at the national level is that while Kenya is remarkably committed to tackling challenges posed by climate change, coordination and management of strategic activities are fragmented between different actors. This poses a threat of interfering with institutional frameworks and possibilities of weak enforcement of policies and legislation at all levels of governance, both national and county. If coordination is not synchronized at all levels, there will be conflicts and paralysis at local levels, especially with devolution.

4.3 Kenya Climate Change Adaptation Strategies

The risks climate change poses to trade and economic development are well recognized in Kenya's National Climate Change Action Plan 2018–2022. Climate impacts are predicted to be particularly pronounced in rural areas where the economy and rural livelihoods depend highly on climate-sensitive activities. Impacts are already being felt, with increasingly frequent and severe droughts and floods, more erratic rainfall and higher average temperatures affecting food production and water availability. These climate changes will likely significantly impact business and trade in rural Kenya, primarily dry, poor, and underdeveloped. Over 70% of Kenya is dry. For example, projections show that by the 2030s, 30% of the rural areas and 7 out of the 21 ASAL counties will have more than 50% of their cattle range above 30 °C, the temperature at which cattle production will be physiologically affected by reduced feed intake and reproduction.[29] Associated water stress, limited forage availability and the spread of pests and diseases will directly affect the quantity and quality of livestock production. As well as increasing the risks of food insecurity and malnutrition in most of the country, the economic costs due to reduced cattle productivity and the potential cattle losses are likely to be significant. It is estimated that 1.7 million cattle will be affected, equal to a possible loss of 34 and 68 billion Kenyan shillings (cattle prices range between Kshs 20,000 and Kshs 40,000 per head).[30] The IPCC Special Report on Global Warming at 1.5 °C states that globally, a 7–10% decline in livestock is expected at about 2 °C of warming, with associated economic losses of between $9.7 and $12.6 billion.[31]

[28] Ongugo PO, Langat D, Oeba VO, Kimondo JM, Owuor B, Njuguna J, Okwaro G and Russell AJM. (2014) 155.

[29] Said et al. (2022).

[30] Said (n 36).

[31] Boone et al. (2018, 1393).

Climate change will exacerbate other drivers of change, such as land fragmentation, migration and population growth, and increase environmental pressure, such as water demand and soil quality. The privatization and fragmentation of the rangelands are creating small land parcels, inhibiting mobility and the ability to access seasonal or drought pastures, thus exacerbating vulnerability to climate change. Land tenure change may provide investment options to build climate resilience for those who can secure land. Still, it also marginalizes women and other vulnerable groups and can reduce their ability to absorb climate shocks.

Harnessing the opportunities in a changing climate sector where production is rooted in the rural areas is vulnerable and exposed to climate risk. Still, these systems also have inherent adaptive capacity that can form the basis of climate-resilient economic development.[32] The demand for agricultural and livestock products such as crops, fruits, meat, and milk grows as urban centres expand and wealth levels increase.

For example, there are options for investing in a climate-resilient agricultural and livestock sector that can create employment and revenue and realize growth in the future while increasing food security and the share of manufacturing in the industry. This includes the production of hay or other sources of feed for fattening cattle that can fulfil demand in expanding high-end markets; breeding programs to promote climate-resilient and commercially attractive traits; investments in cold storage in the meat chain that allow the development of a 'cold chain,' beyond the high-end markets; increased value addition to beef, especially low-quality meat (for the low-end markets); and vertically-integrated private or community livestock enterprises that link livestock directly to markets (for the high-end markets).[33]

Access to finance, pricing information and early warning systems is limited in most parts of Kenya, especially the drier rural areas. Although some actors in the chain may have access to informal credit arrangements, the limits on credit availability and a lack of capital were identified as barriers to implementing adaptation options.[34] More supporting services that are better designed for agricultural production systems can help producers and traders engage with and invest in business opportunities.

However, not all adaptation strategies by private actors are sustainable or sufficient to deal with current or future climate risks, and governments have a role in providing the right enabling conditions to support private adaptation. This includes policy intervention that supports better access to finance and markets and enhanced availability of climate information and early warning systems.

Temporary or permanent migration is a vital resilience strategy against the impacts of climate change, especially in the ASALs, through diversification into other economic livelihoods and the transfer of remittances. Livestock mobility, enabled through social networks that customary institutions drive, is an essential source

[32] Carabine and Simonet (2018).

[33] Bedelian et al. (2016).

[34] As above.

of resilience to drought and other climate shocks in the ASALs.[35] High rates of migration into the ASALs can be attributed to emerging opportunities.

Land tenure changes in the rural areas, mostly where pastoralism is practised and ASALs, from communal to privatized land, also open up new market and value addition opportunities for enhanced investment and the expansion of agriculture, trade and infrastructure in the areas.[36] Rural-to-urban migration brings significant remittances that contribute to poverty reduction and investment in the ASALs.

Small and Medium Enterprises (SMEs) play an important role in Kenya's economy, employing around 80% of the workforce and contributing 20% to the GDP[37] Yet, they are highly vulnerable to climate change and need government support to adapt to its impacts. In Kenya, women are significant in running SMEs that support household livelihoods and build resilience.[38] Female-led SMEs are more likely to adopt sustainable adaptation strategies in response to climate risk.[39] Social networks, such as women's groups and table banking initiatives, are essential adaptation tools for women. They create opportunities for sharing knowledge, identifying new opportunities to diversify enterprises, and pool resources to facilitate the introduction of new climate-resilience technologies and build market linkages.[40]

5 Recommendations

As sites of considerable projected climatic and demographic change, the Kenya government and development partners must support climate change adaptation and investments to achieve climate-resilient economic development. This paper makes the following recommendations:

1. Kenya could consider the following trade policy measures to mitigate and adapt to the impact of climate change.

   (a) *Deepen its participation in the WTO*: WTO membership is essential for many reasons: it enables Kenya to diversify its markets and exports, which is imperative for coping with climate shocks on trade (exports exceptionally), and would provide a forum for reaping the benefits of special and differential treatment to which Kenya is entitled as a low-income country. Differential treatment will be critical for securing more time for transitioning to green standards set by its main trade partners, such as the EU, as well as for accessing resources (monetary and technical) for the transition. The WTO provides a forum for eliminating trade measures and barriers,

[35] As above.

[36] Moiko et al. (2019).

[37] Bauer and Raikundalia (2015).

[38] Atela (2018).

[39] Crick et al. (2018, 168).

[40] World Bank Group (n 16).

such as agricultural subsidies and tariff escalation, in the country's main export markets—factors that threaten Kenya's economic diversification of production. WTO membership also gives Kenya a seat at the table where the global rules for environmental goods and services are being negotiated—giving Kenya a feasible avenue for securing its international interests in ecological goods and services. WTO membership drives trade liberalization in environmental goods and services through tariffs and commitments encompassed in the accession process. Notably, WTO liberalization allows the government to tie domestic reforms for building a green economy with WTO accession. This is critical for countering domestic resistance to the green transition since the reforms are essentially an international obligation under international law. The WTO (through its dispute settlement mechanism) offers a legal avenue for enforcing and fulfilling international environmental-climate treaties. WTO legal jurisprudence is critical for ensuring that progressive green jurisprudence becomes part of the domestic legal policy framework and would bolster efforts to reinforce the national legal framework for purposes of climate change adaptation and mitigation.

(b) *Enhance participation in sub-regional and regional trade organisations and Agreements to which Kenya is a party:* The multilateral trade system (underpinned by the WTO) has generally performed poorly in liberalizing environmental goods and services, as evidenced by the stalled progress of the Environmental Goods Agreement negotiations. The complications at the WTO are exacerbated by the current trade tensions among the most influential members of the organization. Regional economic organizations afford unique opportunities to diversify goods and services (specifically, changes in comparative advantages in climate-related trade). The most excellent prospects are broadening or refocusing the spectrum of export markets and redirecting exports to new markets where value-added production is in the most significant demand. For these reasons, enhanced sub-regional and regional trade offers substantial opportunities.

(c) *Closely monitor developments in the environment-climate requirements of its Main trading partners: Monitoring would allow Kenya to strategically anchor its trade policy or legal framework with its main trade partners, who are increasingly interested in environmental sustainability. For example, the exports of low*—and middle-income countries that implement a national carbon tax system will be exempt from carbon border fees imposed by other countries.[41]

2. The national and county governments need to change the narrative of rural and ASALs to capture their potential as the new frontiers of development in Kenya. This will help raise the profile of the regions and recognize them as areas that will be increasingly important for climate-resilient economic growth and investment.

[41] (n 51).

As part of this, government and development partners should emphasize the importance of ASALs, particularly the livestock sector, in achieving the Big Four Agenda and Vision 2030 targets.

3. There is an urgent need to consolidate all policies, legislation, and regulations to reflect the national strategy for climate change adaptation and mitigation and the national climate change action plan.
4. The national and county governments, supported by development partners, will need to support adaptation within the range of private actors already thriving in Kenya's rural areas and ASALs. The Big Four Agenda recognizes SMEs as key to value addition and manufacturing. The ability of private actors to respond to climate risks will depend on factors shaped by the implementation of supporting policies. These interventions must target and be accessible to the full range of private sector actors in rural and ASAL areas, including informal actors, producer associations, and women-led enterprises.
5. The national and county governments, supported by development partners and the private sector, must ensure that adequate investments and an appropriate enabling environment for public and private adaptation are developed to achieve national and county development goals and increase resilience to climate shocks. What is clear is that public and private sectors must partner together to deliver services and infrastructure for agricultural and livestock sector development, including:

   (i) Increased investments in agricultural and livestock production research will provide opportunities to improve the value chain in these sectors.
   (ii) Develop agricultural and livestock marketing infrastructure, such as abattoirs, weighing machines, and holding grounds adjacent to markets, to improve the efficiency of market operations and encourage better returns.
   (iii) Upgrading and investing in the hides and skins value chains to produce sufficient good quality leather products. This requires adequate quality and quantity of livestock to provide the necessary by-products of the hides and skins. There is currently minimal processing of hides and skins in Kenya. The leather industry was a priority sector under manufacturing in the Big Four Agenda that was championed by the previous administration of 2017–2022, and which the current president had pledged to continue with and was part of his United Democratic Alliance (UDA)-Kenya Kwanza Policy of 2022 general elections.[42] Upgrading the hides and skins value chain can contribute to economic growth by expanding leather product exports.[43]
   (iv) Financial services and credit systems should be developed to be better designed and tailored to agricultural and pastoral production systems and informal SMEs.

[42] Kwanza (2022, 13).

[43] Hansen et al. (2015, 9).

6. The Meteorological Department and the National Drought Management Authority (NDMA) will develop and disseminate timely and appropriate climate services, including early warning systems and climate information, to facilitate early action and decision-making about food and nutritional security.
7. Several urgent priorities for building adaptive capacity in Kenya should be fast-tracked, notably monitoring, forecasting, and information (as these underpin future prediction and analysis) and early warning systems, as well as information provision, monitoring (indicators), and supporting science-policy networks and sectoral focal points. These early priorities are part of a broad strategy to increase the knowledge base, including education and training and strengthening existing programs.
8. The national and county governments should develop and implement policies to facilitate the mobility of people and livestock. These policies and associated implementation strategies must incorporate and consider the inherent adaptive capacity of rural areas and ASALs—including flexibility, mobility, and informality—which will be important in managing increasing climate risk. This includes supporting sustainable land and pasture management through resource-sharing mechanisms, such as livestock corridors and strengthened social networks.
9. The national and county governments, supported by their development partners, should adopt and implement inclusive policies that support and enhance the adaptive capacity of women, youth and other vulnerable groups. Although the national government recognizes the need to support female-led enterprises in various policies, including climate change legislation, these policies are currently poorly implemented and mainstreamed locally. More support and priority are needed for the benefits of these policies to be realized in real terms.
10. The national government should investigate what the impacts of climate change will be for the Bottom-Up Economic Transformation Agenda.[44] This should include a full assessment of the potential impacts of climate change on livestock and crops in achieving food and nutritional security, as well as an understanding of the economic cost of not taking adaptation or mitigation action against the impacts of climate change.
11. The National Climate Change Action Plan 2018–2022 aligns well with the Bottom-Up Economic Transformation Agenda. The Bottom-Up Agenda should now be aligned with Kenya's Climate-Smart Agriculture Implementation Framework 2018–2027, which aims to promote the growth of climate-resilient carbon and sustainable agriculture that ensures food security and contributes to national development goals. To ensure the implementation of the Bottom-Up Economic Transformation Agenda at the county level, county governments should strive to align their County Integrated Development Plans (IDPs) and other development plans and strategies with the bottom-up priorities.

[44] (n 53).

12. Climate change risk screening. Future climate change risk screening must be built into development and planning at a sectoral and regional level. Information on climate, resources, adaptation strategies and options should be mainstreamed into all sectoral plans. The study recommends that a national knowledge management system be developed with easy access for all stakeholders.
13. Building Capacity. Access to substantial adaptation funds must be assured. However, mechanisms, institutions and governance systems for practical use must be developed to allow Kenya to access these funds. This requires early and concerted action to build capacity across stakeholders and with the affected communities themselves. This is an early priority.
14. A national adaptation facility should assess the potential for climate-resilient growth across all areas of the economy and mainstream adaptation in government departments and with Kenya's development plan. A multi-stakeholder trust fund would enable early and timely action and is an early priority, encouraging learning by doing and establishing the basis for scaling up to sectoral resilience

References

Aheim A et al (2008) Adaptation to climate change, why is it needed and how can it be implemented. ADAM Newsl 2008:1

Atela J (2018) Climate change adaptation among female-led micro, small and medium enterprises in semi-arid areas: a case study from Kenya. In: LW Filho (ed) Handbook of climate change resilience, pp 1–18

Bauer S, Raikundalia S (2015) Closing the gap Kenya: update on key challenges for the "missing middle" in Kenya. Intellecap Available at: http://intellecap.com/wp-content/themes/intellecap/pdf/26.10.15.pdf

Bedelian C et al (2016) Land tenure influences pastoralist grazing management and climate change adaptation strategies: gender and equity implications

Boone RB et al (2018) Climate change impacts on selected global rangeland ecosystem services. Global Change Biol 24(3):1382–1393

Brenton P, Chemutai V (2021) The trade and climate change nexus: the urgency and opportunities for developing countries

Carabine E, Simonet C (2018) Value chain analysis for resilience in drylands: identification of adaptation options in key sectors. Pathways to resilience in semi-arid economies (PRISE) working paper available at: http://prise.odi.org/research/value-chain-analysis-for-resilience-in-drylands-vc-arid-identifeication-of-adaptation-options-in-key-sectors-/

Climate Change (2014) Synthesis report. In: Pachauri RK, Meyer LA (eds) Contribution of working groups I, II and III to the fifth assessment report of the intergovernmental panel on climate change in IPCC 2014

Climate Change Agriculture and Food Security (2009) A strategy for change. Alliance of the CGIAR centres

Crick F et al (2018) How do African SMEs respond to climate risks? Evidence from Kenya and Senegal, p 108. *World Development*

Fulmer J (2009) What in the world is infrastructure? PEI infrastructure investor July/August 30 at 32

Government of Kenya (2010) Revised REDD readiness preparation proposal Kenya submitted to the forest carbon partnership facility Kenya forest service.
Government of Kenya (2018) Kenya climate smart agriculture implementation framework. Available https://repository.kippra.or.ke/handle/123456789/2065
Government of Kenya Long-Term National Low-Carbon Climate-Resilient Development Pathway: Climate Risk Assessment of Kenya's Flagship Projects. Rehabilitation and protection of indigenous forests in five water towers. National Climate Change Action Plan (2013)
Government of Kenya Ministry of Environment and Mineral Resources (2010) National climate change response strategy
Hansen ER et al (2015) Republic of Kenya Kenya leather industry: diagnosis, strategy and action plan. Ministry of Industrialization and Enterprise Development
Intergovernmental Panel on Climate Change (2001) Climate change 2001: impacts, adaptation and vulnerability
IPCC (2007) Climate change: mitigation. contribution of working group III to the fourth assessment report of the intergovernmental panel on climate change
IPCC (2014) Climate change: synthesis report. contribution of working groups I, II and III to the fifth assessment report of the intergovernmental panel on climate change
IPCC (2018) Summary for policymakers. In: Global warming of 1.5 °C. An IPCC special report on the impacts of global warming of 1.5 °C above pre-industrial levels and related global greenhouse gas emission pathways in the context of strengthening the international response to the threat of climate change, sustainable development, and efforts to eradicate poverty
Kwanza K (2022) The plan: the bottom up economic transformation agenda 2022–2027
Moiko S et al (2019) Land use transformation in Kajiado county, Kenya: implications of a changing climate. PRISE policy brief
Pandey N (2002) Gender economics of the Kyoto protocol. Conservation ecology at 6(1): r14 http://www.consecol.org/vol6/iss1/resp14/
Said MY et al. Projected climate change impacts on cattle in Kenya's arid and semi-arid lands based on RCP 2.6, 4.5 and 8.5
Sere C et al (1996) World livestock production systems: current status, issues and trends. Animal Product Health Pap 1996:1
Smit B et al (2006) Adaptation, adaptive capacity and vulnerability. Glob Environ Chang 16(3):1
Socialist Republic of Viet Nam: Country Strategic Opportunities Programme (COSOP) 2019–2025 (2019) International fund for agricultural development Washington
United Nations Conference on Trade and Development (2021) Trade climate readiness for developing countries. Trade and environment review
Wekesa M, Karani I (2009) A review of the status of emergency water trucking in Kenya's arid and semi-arid districts. Food and Agriculture Organization of the United Nations, Nairobi
World Bank Group (2020) Climate risk country profile: Ethiopia
World Bank Group (2021) Climate risk country profile: Kenya

Climate Change Policies in Southeast Asia and African Countries: Approaches to Trade and Development

Ibnu Sitompul and Dennis Oghenerobor Agelebe

Abstract Unless Southeast Asian countries drastically reduce glasshouse gas emissions, global warming in this region will result in rising seas, dangerous flooding, forest fires, and changing rain patterns, potentially leading to a more severe natural disaster that threatens trade and development. Climate change mitigation efforts by the region's intergovernmental organization, the Association of Southeast Asian Nations (ASEAN), have been ineffective, causing losses in various aspects of each member's economy and possibly decreasing Southeast Asia's overall GDP. Africa has been grappling with some of the most adverse effects of climate change despite being one of the least contributors to the cause of the climate change crises. The high development deficit weakens Africa's preparedness for responding to the catastrophic effects of climate change. As Africa continues to push more for economic growth in the face of the global climate challenge by increasing intra-trade through the Africa Continental Free Trade Agreement and other regional economic policies, there are concerns about how economic interest and climate change commitments of African countries may be at odds without a regional framework on how the commitments can be factored into economic and development plans. Although all ten ASEAN members have committed to combating climate change by ratifying the 2015 Paris Agreement and establishing various national climate change adaptation and mitigation actions, implementation remains challenging. This paper provides a general overview of the similarities and differences in climate change policies in both regions and a critique of the current framework, particularly in the Southeast Asian and African regions.

I. Sitompul (✉)
Faculty of Law, The Institute of Law, Emergencies and Disasters (LEAD), University of Canterbury, Christchurch, New Zealand
e-mail: ibnu.sitompul@canterbury.ac.nz

D. O. Agelebe
Environmental Law Center, Faculty of Law, University of Cologne, Cologne, Germany
e-mail: Dageleb1@uni-koeln.de

D. Oghenerobor Agelebe (ed.), *Climate Ethics in Africa*, Global Perspectives on Law and the Environment 1, https://doi.org/10.1007/978-3-032-08757-7_10

1 Introduction

Asia is more exposed to climate change than others without adaptation and mitigation. The Southeast Asian region, in particular, is one of the world's most vulnerable regions to climate change. This region is considered one of the most dynamic and fast-growing regions in the world, with long coastlines and a high concentration of population and economic activity along the coastal area. This includes the ten ASEAN member states: Indonesia, Malaysia, Singapore, Thailand, Myanmar, Vietnam, the Philippines, Brunei, Cambodia, and Laos.[1] They rely on agriculture to provide livelihoods for a large population segment and depend on natural resources and forestry to drive development. Unless Southeast Asian countries drastically reduce glasshouse gas emissions, global warming in this region will result in sea level rise, dangerous flooding, forest fires, and changing rain patterns, potentially leading to a more severe natural disaster that threatens various aspects of regional and national interest, including trade and development. These natural disasters caused economic losses and potentially resulted in a drastic incline in Southeast Asian countries' overall GDP.

To date, climate change mitigation efforts and policies by the Association of Southeast Asian Nations (ASEAN), the region's intergovernmental organization, have been ineffective. It has failed to bring together ASEAN member states to cooperate and to put together effective and efficient strategies to mitigate the impact of climate change. Despite various commitments from all ten ASEAN members in relation to combating climate change, including the ratification of the 2015 Paris Agreement and the establishment of various national climate change adaptation and mitigation programs, implementation remains complex and challenging. Arguably, these challenges are mostly related to trade and the development of countries in this region.

However, this paper acknowledges difficulties and challenges concerning implementing national climate change adaptation and mitigation policies and programmes in other regions, including Africa. This region has also been hit hard by the impacts of climate change. Over the years, Africa has faced more frequent climate-related disasters, more severe weather, unpredictable rainfall, and rising sea levels, resulting in human tragedy, social upheaval, and economic disruption.

[1] As of 26 October 2025, ASEAN comprises 11 Member States with the admission of Timor-Leste. The Declaration on the Admission of Timor-Leste into ASEAN was signed during the Opening Ceremony of the 47th ASEAN Summit and Related Summits in Kuala Lumpur on 26 October 2025, marking ASEAN Leaders' consent to admit Timor-Leste as the 11th Member State of ASEAN.

2 Climate Issues in Southeast Asia

Based on various research, reports, and analyses, Asia is more exposed to climate change than other parts of the world, particularly in the absence of adaptation and mitigation efforts. By 2050, it is predicted that parts of Asia may see extreme events, including increasing average temperatures, heat waves, severe hurricanes, drought, and other extraordinary natural events that are new or have not been seen for decades.[2] It is predicted that by then, between 600 million and 1 billion people in Asia will live in areas with a non-zero annual probability of lethal heat waves, compared to a global total of 700 to 1.2 billion people. Economically, between $2.8 trillion and $4.7 trillion of GDP in Asia annually will be at risk from an effective loss of outdoor working hours because of increased heat and humidity, which can cause various health issues.[3] This includes the Southeast Asian region, one of the planet's most vulnerable regions to climate change.[4] This archipelagic regional bloc is predicted to be hit by climate change impacts, such as rising sea levels, heat waves, droughts, severe flooding, and forest fires, which can result in haze pollution. In 2019, Southeast Asia recorded 9.6 million people internally displaced by cyclones, floods, and typhoons, representing almost 30% of all global displacement in that year. In addition, 19 out of 25 cities most exposed to a one-meter sea-level rise are in Asia; seven are located in the Philippines.

Unfortunately, although Southeast Asian nations are projected to suffer among the most brutal effects of climate change, most countries in this region do not have effective climate change prevention strategies. Most Southeast Asian countries do not have carbon reduction strategies that can effectively mitigate the severity of the climate risks. As a result, natural resource exploitation in Southeast Asian countries occurs without proper constraints and pressure. Most of these countries continue to rely heavily on nonrenewable energy sources. Indonesia, the largest economy in Southeast Asia, is still 60% dependent on coal to power the country despite plans to achieve net-zero emissions by 2060. In addition, deforestation continues on a large scale, and slash and burn land clearing practises continue to be used, resulting in forest fires and, ultimately, haze pollution every year, which, most of the time, transfers to neighbouring countries. Deforestation is a significant source of greenhouse gases. In Indonesia and Malaysia, which have the world's largest forestlands, trees are cut down to make way for farms to feed growing populations, as well as for the production of pulp and paper and palm oil, both of which are major export revenue generators for the country. About half of Indonesia's emissions are from deforestation and more than fossil fuel emissions. Other Southeast Asian countries, such as Malaysia, intend to increase their use of coal power. This shows that environmental issues, including climate change, are not a top priority for governments in these countries to address.

[2] Woetzel et al. (2020, 3).

[3] At 7.

[4] Asian Development Bank (2009) at 4–5; Ding and Beh (2022, 1).

It cannot be denied that climate change issues hurt the region's development and economy. Therefore, an effective climate change policy is indispensable to maintain trade and development for every country in Southeast Asia. On the contrary, according to the Asian Development Bank, Southeast Asia could suffer more significant losses than the rest without effective climate change policies. Climate change could shave 11% off the region's GDP by the end of the century, devastatingly impacting key sectors such as agriculture, tourism, education, fishery, human health, and labour productivity.

3 Regional Climate Change Related Policies in Southeast Asia

Knowing the severe impact of climate change that is expected to hit Southeast Asia, particularly in terms of the region's development and economy, Southeast Asian countries and the region's intergovernmental organization, ASEAN, have put in place a number of legal frameworks to mitigate the effects of climate change. The ten Southeast Asian countries are all signatories to the United Nations climate change regime, which includes the UN Framework Convention on Climate Change,[5] the Kyoto Protocol,[6] And the Paris Agreement.[7] These countries have ratified these international treaties and have incorporated the treaties' obligations into their national legislation. This means that all ASEAN member states are committed to preventing and mitigating climate change and its impacts and are legally bound to implement obligations under the UN climate change regime at the national level through various implementation strategies.

As a result, all ten ASEAN member states have submitted their Nationally Determined Contributions (NDCs) required by the Paris Agreement, which set targets for reducing greenhouse gas emissions within a specific time frame. Indonesia, as the largest economy in ASEAN but also the most significant contributor to GHG emission in the region, submitted their latest NDC in 2022, committed to a higher GHG reduction target of 31.89% (unconditional) and 43.2% (conditional) as compared to 29% and 41% respectively in the previous version of NDC. This latest NDC targeted various areas and sectors, including agriculture, coastal zones, ecosystems, water, forestry, energy, and settlement. Additionally, the NDC has committed to restoring 2 million hectares of peatlands and rehabilitating 12 million hectares of degraded lands by 2030. To reach its NDC commitment, Indonesia developed various strategies which will be implemented through several national programmes, including

[5] UNGA, *UNFCCC: resolution/adopted by the General Assembly*, 20 January 1994, A/RES/48/189.

[6] Kyoto Protocol to the UNFCCC Dec. 10, 1997, 2303 U.N.T.S. 162.,

[7] Paris Agreement to the United Nations Framework Convention on Climate Change, Dec. 12, 2015, T.I.A.S. No. 16-1104.

programmes directly related to two sectors of the country's most significant sources of economic income: energy and forestry.

In the energy sector, Indonesia has implemented several mitigation measures, including increasing the use of renewable energy, improving energy efficiency, focusing on low-carbon emitting fuels, using clean coal technology and gas power plants, and post-mining reclamation. Meanwhile, Indonesia intends to reduce deforestation and forest degradation by controlling illegal logging, increasing timber plantation production, increasing land rehabilitation, improving peat water management, and strengthening peatland restoration efforts. If these measures are correctly implemented, they will positively impact the environment at the national, regional and global levels. However, any strict restriction in this sector would arguably affect various sectors, including the country's trade and development.

Another ASEAN member state, Malaysia, submitted its latest NDC in 2021 and committed to an unconditional target to cut carbon intensity against GDP by 45% by 2030 compared to 2005. This latest NDC also covers seven greenhouse gases compared to Malaysia's first NDC, which only covered three. Additionally, under this NDC, the government of Malaysia expanded the adaptation component, focusing on protecting biodiversity and mainstreaming climate resilience into urban training. Looking at the NDCs of these two ASEAN member states, it is clear that ASEAN member states have made an effort to address the global climate change issue. Countries in this region have implemented numerous policies to monitor and ensure their commitments, strategies, and programmes to prevent and mitigate climate change and its impacts are adequately implemented to reach their objectives by the targeted deadline.

Moreover, the ASEAN organization plays a significant role in assisting ASEAN member states in their efforts to comply with the international climate change regime by implementing several regional initiatives. These regional initiatives are intended to impact national efforts, development, and the economy positively. These initiatives include the establishment of the 2002 ASEAN Agreement on Transboundary Haze Pollution (AATHP), which focuses on mitigating the region's ongoing forest fires and transboundary haze pollution issue, which is one of the most significant contributors to greenhouse gas emissions that can directly cause climate change; the ASEAN Socio-Cultural Community (ASCC) Blueprint, which includes concrete strategies and action plans for various sectors including the environment and climate change mitigation and adaptation; the ASEAN zero burning guideline; ASEAN Joint Statement on Climate Change; ASEAN State of Climate Change Report, etc.

Under the AATHP, ASEAN member states are required to undertake measures to prevent and control activities related to forest fires that can cause transboundary haze pollution issues in the region.[8] In this sense, activities related to forest fires can include deforestation, forest degradation, peatland drainage, illegal logging, and the slash-and-burn land-clearing method. By preventing and controlling these activities, ASEAN member states also directly or indirectly comply with their NDCs to cut CO_2

[8] AATHP Article 9.

emissions and ultimately comply with the Paris Agreement's overarching objective to hold the global average temperature increase.

Furthermore, ASEAN member states reemphasized the need for developed countries to enhance international cooperation and assistance, including financing and technology transfer, in the ASEAN Joint Statement on Climate Change. This is an important and relevant point because most ASEAN member states would have to sacrifice some aspects of their country's trade and development to implement their climate change policies or comply with regional climate change policies. As a result, cooperation and financial and technological support from developing countries is required to alleviate the economic burden that ASEAN countries will bear when implementing their commitment to combat climate change.

Of course, establishing a policy is one thing; successfully implementing it is quite another. Although a number of regional initiatives have been established, implementation remains a challenge for ASEAN member states, even with financial and technological support from developing countries. This implementation challenge is especially evident in countries that rely heavily on natural resources for much of their economic income. Environmental issues are still viewed as secondary to other problems, such as economics, poverty, education, and development. Therefore, to date, environmental destruction continues to occur in the majority of ASEAN member states to this day.

Take, for example, the annual forest fires and haze pollution issues in the region, mostly occurring in the Indonesian forests of Sumatra and Kalimantan. Most of these fires are caused by human activity, such as converting millions of hectares of rainforest to agriculture and plantations. Indonesia and Malaysia account for over 68% of the world's 175,000 square km of palm oil cultivation land. These forest fires caused a transboundary haze pollution issue in the region. From 2001 to 2015, Indonesia's average CO_2 emissions were 505 million tonnes per year, making it the world's fifth-largest emitter of greenhouse gases. In 2019, Indonesia ranked tenth with a higher emission of 615 million tonnes of CO_2. These catastrophes cause national, regional, and global environmental damage. Also, the recurring environmental issue of forest fires has an adverse impact on the economies of ASEAN member states.

Due to the haze pollution caused by Indonesian forest fires, neighbouring countries such as Singapore and Malaysia suffered millions of dollars in economic losses. For example, Singapore lost $1 billion per week in tourism revenue because of haze pollution covering the country for months. In addition, in Malaysia, the total damages from the tourism sector amount to 321 million dollars. In Indonesia, the economic cost of forest fires exceeded USD 16.1 billion in 2015, a significant increase from 5.8 billion in 2013 to 2014. This billion-dollar loss suffered by ASEAN member states has directly harmed ASEAN member states' trade and development. Additionally, this demonstrates the ineffectiveness of various regional and national environmental policies, including the 2002 ASEAN Agreement on Transboundary Haze Pollution, which has the potential to directly impact mitigation efforts on climate change issues in the region. Some factors likely contribute to this ineffectiveness, reflecting the overall problems with Southeast Asia's climate change policy implementation.

4 Implementation Issues of Climate Change Policy in Southeast Asia

It is well known that effective implementation of climate change policies is always a challenge for many countries worldwide, and Southeast Asian countries are no exception. However, the causes of ineffectiveness vary and depend on many factors, such as political dynamics, involvement of interest groups, economic needs, enforcement mechanisms, and monitoring and reporting. It involves various elements of government, including legislative, administrative, and judicial bodies, and it is heavily reliant on the effective coordination and cooperation of these elements within governments. In this sense, the effectiveness of ASEAN climate change policies depends on effective national implementation collaboration and coordination among member states. For this region, cooperation and coordination are guided by the ASEAN Way, which is the fundamental principle guiding the relationship between the 10 ASEAN member states. In addition, it is important to look at natural resources and economic reasons to understand the implementation issues of climate change policy in Southeast Asia.

5 The ASEAN Way Principle

Regarding cooperation and coordination, the relationship between ASEAN member states is guided by the ASEAN way principle, a value firmly held by member states that promotes social procedure over domestic political procedure. This means that ASEAN member states' cooperation, as well as ASEAN practises and declarations, are based on a shared culture. These principles are codified in the Treaty of Aminy and Cooperation in Southeast Asia (TAC), established in 1976 and legally binding on ASEAN member states, focusing on peaceful coexistence and friendly cooperation among Southeast Asian states.

As a result, as stated in Article 2 of the Treaty of Amity and Cooperation in Southeast Asia, any issues that arise between ASEAN member states, including environmental issues, will be resolved in a non-legal manner through consultation and consensus, emphasising state sovereignty, nonintervention, and upholding informal peaceful dispute settlement.[9] This method of resolving disputes and conflicts among ASEAN member states has been used for many years, long before ASEAN was formed, and it has many successful examples. However, that peaceful method of dispute resolution, or the nonintervention principle of the ASEAN Way, has become debatably irrelevant in recent years, as the scale of issues has grown far beyond what occurred many years ago, including environmental issues. Some ASEAN Way principles may impede the implementation of several obligations under various climate

[9] TAC Article 2.

change policies in ASEAN countries. It also softens any joint statement or declaration related to climate change in Southeast Asia, which has a negative impact on the state's development and disrupts trade among ASEAN member states.

Furthermore, the ASEAN way principle contributes to ASEAN states' reluctance to bring environmental conflict between them to the international forum because they commit to resolving issues among themselves regionally and in a non-formal manner, particularly environmental problems that can affect the effectiveness of climate change policy. For example, when Singapore raised transboundary haze pollution issues caused by forest fires in Indonesia at the UN General Assembly in 2006, calling for a more significant effort that included international expertise to address this problem, Indonesia was quick to state that haze was a domestic problem and described Singapore's move as an interference in Indonesia's domestic affairs and sovereignty, and that this problem should be addressed at the ASEAN regional level. Article 9 of TAC states that "... [ASEAN member states] shall maintain regular contacts and consultation on international and regional matters with a view to coordinating their views, actions and policies."[10] Additionally, Article 2(c) states that "In their relations with one another, [ASEAN member states] shall be guided by ... fundamental principles [including] Non-interference in the internal affairs of one another..."[11] Based on these, rather than viewing Singapore's move as a positive step towards improving the effectiveness of the haze agreement in combating climate change, Indonesia sees it as an attack on its domestic affairs and a violation of the ASEAN way principle. The haze causes economic harm to neighbouring countries and disrupts their economic development and development in general, including countries like Singapore, Malaysia, Brunei, and Thailand.

Of course, this is not always the case. There are other disputes among ASEAN member states in which the ASEAN Way is not followed, particularly regarding territorial disputes related to state sovereignty, such as Malaysia vs Indonesia in the case of sovereignty over Sipadan and Ligitan Island.[12] This case was dealt with in the International Court of Justice (ICJ). However, the ASEAN Way principle is always put forward when it comes to most issues, including issues related to breaches of environmental and climate change policy, that have a direct negative impact on the trade and development of ASEAN states. This is arguably because most ASEAN countries rely heavily on natural resources and fossil fuels to improve their economies and develop. As a result, they are constantly looking for ways to limit the involvement of other countries, and the ASEAN way is viewed as a tool that can provide freedom for ASEAN states to conduct their domestic affairs without interference from other members.

---

[10] TAC Article 9.

[11] TAC Article 2(c).

[12] ICJ, Case of Pulau Ligitan and Pulau Sipadan *(Indonesia/Malaysia)*, Judgement of 17 December 2022.

6 Natural Resources and Economic Reasons

Another factor impeding the effective implementation of climate change policy in Southeast Asia is the region's economic dependence on natural resources. ASEAN countries are major export and import partners in natural resources. Malaysia is one of Indonesia's largest importers of palm oil. The Philippines is Vietnam's second-largest importer of rice. Malaysia and Vietnam are also major importers of Indonesian paper and so on. Politically, this situation impacts the effectiveness with which climate change policy obligations are implemented. ASEAN countries are failing to follow through on their commitments to climate change policies and are getting away with it. This is a problematic situation because their noncompliance may have benefited other sectors of the economy, but it may also have harmed their trade sectors and development.

There are many similarities between climate change policy in Southeast Asia and Africa. Strategies are placed around mitigation, adaptation, resilience, cooperation, education, financial support, and awareness, ultimately intended to benefit economic development. Climate change policy in Africa also faces similar political challenges from groups that may have an interest in natural resources. In addition, corruption is still an issue that may impact implementation, particularly in relation to international financial and technical support.

Despite these similarities, one of the points in my research that arguably distinguishes the two regions is the implementation and the existence of the ASEAN way principle in Southeast Asia. According to my findings, African countries are more direct in settling disputes among themselves, including disputes over environmental issues, which may strengthen implementation and positively impact the effectiveness of climate change policy at the regional and national levels. African countries are not restricted to bringing environmental issues, including noncompliance issues with climate change policy, to international forums such as international courts and tribunals.

In addition, environmental issues are considered a top priority in a number of African countries, which is demonstrated by the fact that some countries have established an environmental court in their legal systems. This could serve as the foundation for an African regional environmental court. This is not the case in ASEAN, though I argue that the presence of a regional environmental court will be critical in ASEAN as they want to keep and resolve all environmental disputes at the regional level.

7 Climate Change in Africa: Challenges and Responses

There is undoubtedly no more significant threat to the well-being and development of Africans and Africa than the present and future dangers the continent faces due to climate change's adverse effects. The threats encompass human health, food, water security, and socio-economic development in the countries in Africa.[13] Despite being the lowest contributor to global greenhouse gas emissions, no continent is currently experiencing the devastating effects of climate change as Africa. According to a report presented at the United Nations COP26 by the World Meteorological Organization (WMO), near-surface air temperatures across a vast expanse of tropical Africa for the first 9 months of 2021 exceeded the 1981–2010 averages.[14] The continent's biophysical makeup and low socioeconomic resilience to natural disasters make it vulnerable to climate change. Unfortunately, the primary sectors driving Africa's economic growth are climate-sensitive. For example, between 2002 and 2018, service, energy, industry, and agriculture contributed substantially to the 3.37% economic growth recorded during the period.[15] These sectors are critical to the economic well-being of the people as they provide the highly needed employment required to keep income coming in for individuals and their homes.

In specific terms, the high volatility of African countries' economies can be seen as symptomatic of how disasters linked to climate change roll back development efforts and, in some cases, successes recorded with the aid of adaptative policies. Nigeria, being a resource-dependent country, has suffered from several effects of climate change in the form of increasing temperatures, droughts, and flooding. The implications of these effects have further manifested in dire consequences, such as violent conflicts between farmers and herders and a drastic drop in water level, which has threatened the food security of the country and its over 200 million population. Rising global temperature is threatening the lives and ecosystem of coastal Nigerian communities as the Atlantic Ocean has increased its ferocity. During a webinar organized by the Nigerian Economic Summit Group, the group discussed how Nigeria's over 830 km of coastline is vulnerable to the impact of climate change. These impacts, such as flooding, coastal erosion, and pollution, are "causing significant socio-economic disruptions in the Niger Delta, with ocean encroachment and rising sea levels threatening homes and farmlands."[16]

Countries like Djibouti in the Horn of Africa are experiencing extreme effects of climate change. Djibouti is a resource-scarce country that depends on importation for almost everything it consumes. It is prone to natural disasters such as floods, earthquakes, droughts, and volcanic eruptions. With a population of less than 1,000,000 people, the country is critical to the economic survival of several landlocked countries that depend on its ports for importation and exportation. The implication is that apart from the distinct effect of climate change on countries like Ethiopia, Rwanda, Eritrea, Mali and other countries in the Sahel that depend on the port in Djibouti, the occurrence in Djibouti of any natural disaster caused by climate change will adversely affect the referenced countries.

According to the WMO, in 2022, over 110 million people were directly impacted by weather, climate, and water-related hazards. The economic toll of addressing each

---

[13] Intergovernmental Panel on Climate Change (2021).

[14] State of the Global Climate 2023.

[15] Institute for Economics & Peace (2020).

[16] Reported by Leadership Newspaper, 6th August 2024,

setback is evident as Africa grapples with the pressing need to bridge the development deficit gap. This cost of over US$ 8 billion could have been invested in progressive and social projects aimed at improving the standard of living for the population. As agricultural challenges persist in Africa, the continent faces significant difficulties in achieving self-sufficiency in food production. This struggle has led to a substantial rise in import expenditures, with developed nations reaping the benefits. According to the United Nations Economic Commission for Africa, it is projected that Africa's spending on importing food will surge from the current US$ 35 billion to a staggering US$ 110 billion by 2025.[17]

8 Africa's Response to the Challenge of Climate Change

Given the size of the African continent and the number of sovereign states, Africa can be adjudged to be one of the regions that should have it very tough to conceptualise policies and strategies that can address the climate change crises. The African Union and the eight other sub-regional bodies have initiated various programs that are products of policies and documents on climate change. Also, at the national level, states have launched either self-conceptualized policies or as part of their obligation under an international treaty. For example, each state that has ratified the Paris Agreement, as part of its obligation under the treaty, is required to submit their Intended Nationally Determined Commitments (NDCs) every 5 years to the United Nations Framework Convention on Climate Change (UNFCCC) secretariat.[18]

At the regional level, the African Union has taken multifaceted action plans, policies, frameworks, and multilateral collaborative initiatives involving nations and development financing institutions (DFIs). The African Union Climate Change and Resilient Development Strategy and Action Plan are founded on, amongst other principles, an emphasis on a people-centred approach, which aligns with the goals of the African Union's Agenda 2063, which aims to build a peaceful and prosperous Africa by Africans.[19] Considering the diverse strategies, goals, and models for achieving being employed by different States in pursuing their NDCs under the Paris Agreement, the action plan serves as an all-encompassing guide crafted to assist both governmental and non-governmental entities in tackling the effects of climate change in Africa. It emphasizes, among other priorities, the importance of strengthened climate collaboration and the advancement of sustainable, climate-resilient growth. Also, the plan recognises that for the continent to attain a level of development that is inclusive, sustainable, and stable, our development approach must integrate climate-resilient strategies.[20]

---

[17] Hoste and Anderson (2022).

[18] This is accordance with Article 2 and 4.1 of the Paris Agreement of 2015.

[19] African Union Agenda 2063.

[20] African Union climate change and resilient development strategy and action plan (2022–2032).

The plan emphasises fostering enhanced collaboration on climate issues to facilitate sustainable, climate-resilient development in the long term. This includes outlining specific principles, priorities, and guidelines to aid governmental and non-governmental parties in tackling climate change in Africa. Additionally, the plan underscores the significance of unified and synchronized efforts to alleviate the impacts of climate change and facilitate the shift toward a low-emission future.

Another notable way the African Union has responded to climate change is by launching the African Union Green Recovery Action Plan (AUGRAP).[21] Amidst the widespread impact of the COVID-19 pandemic, Africa's growing economy experienced the effects of the global economic downturn. As the world began to emerge from the pandemic and ease lockdown restrictions, there was a collective recognition among leaders and experts of the opportunity to maintain the reduced emissions observed during the height of the global lockdown. Answering this call for a sustainable and environmentally conscious recovery, the African Union launched the AUGRAP. The comprehensive plan provides a clear roadmap for tackling climate change while supporting economic recovery and inclusive growth. It places strong emphasis on improving climate finance, transitioning to renewable energy, implementing nature-based solutions for biodiversity and conservation, bolstering resilient agriculture, and developing green and resilient cities.

Despite the above and several efforts put into managing the challenge of climate change in Africa, there are still factors limiting the continent's potential, even in contemporary times. The leading factors are financial and technological limitations.[22] Governments and non-governmental organizations in Africa have identified a lack of required funds for investment in climate adaptation and mitigation projects that are too large for domestic financing institutions to fund under high-risk credit ratings. At the Conference of the Parties 27, the continent said climate policy initiative projects are left unimplemented because of lack of finance. Africa does not have the economic capacity to raise the $2.8 trillion needed between 2020 and 2030 to execute critical climate remediation and resilient projects that could help get the continent on the path of sustainability for present and future generations.[23] According to the Climate Policy Initiative findings, Africa has an over $200 billion gap in climate financing. States' implementation reports of NDCs will reflect this gap more.

Every state has a peculiar climate challenge contrasting with the prevailing economic situation. Though the NDCs requirement in the Paris Agreement acknowledges the disparity in the technical and financial capacity amongst states, African States have attracted substantial climate finance in a few countries, thereby widening the gap in the implementation of adaptation and mitigation measures. For example, Africa received about 25% of the global budget for climate-related development financing between 2018 and 2020. However, three countries, Kenya, Ethiopia, and

[21] African Union Green Recovery Action Plan.

[22] Chapter 9, IPCC Report for Africa.

[23] Meattle et al., Landscape of Climate Finance in Africa, Climate Policy Initiative (2022); Tamasiga et al. (2023).

Morocco, account for 27.3% of all that Africa received in the period reviewed. Institutional, human resources, and political capacity can affect how countries access finance.

Another major challenge Africa faces is low technology capacity in implementing climate change response strategies. African countries lack the technical capacity, expertise, and infrastructure to develop tailor-made technologies that fit well into their response mechanism. Experts have identified innovative technology as key to climate action. The Technology Executive Committee (TEC) of the UN Climate Change body stated in 2021 that the current level of ambitions set out in the NDC of countries is too low if the global community meets the target set by the Paris Agreement. The TEC further stated that if ambition is to be raised, technology is a key factor that must be mainstreamed and developed progressively. The AU Climate Change Strategy highlighted the challenge of low promotion and support of in-country technology development necessary for building climate resilient and remediation infrastructure.

9 Climate Change, Trade and Development in Africa

The development of Africa depends so much on the continent's ability to adapt to and mitigate the effects of climate. Considering the population of over 1 billion and the increasing need for pressure on the countries that have natural resources to diversify their economy while non-resource-rich economies are seeking ways to commercialize their agricultural sector further, the vulnerability of Africa is more pronounced. Many African countries cannot feed themselves and resort to importing food to feed their growing population, even when Africa has abundant arable land for growing agricultural products. Drought and flood are some of the climate factors responsible for declining agricultural productivity. The United Nations Development Programme (UNDP) reports that climate change remains the most urgent of the factors responsible for low productivity and poor development.[24] Even though the export of agricultural products is still very low compared to the capacity of Africa, the agriculture sector remains the pulse line for the economy of many African countries. Experts have projected that Africa would suffer a downturn in its Gross Domestic Product (GDP) due to the negative impact of climate change. According to a report by the African Development Bank (AfDB), the vulnerability of Africa will increase, and its resilience will decrease due to Climate Change.[25] What does this mean for most African countries, particularly those already seeing the positive results of economic measures? This means that many would experience drawbacks in development that would result in hardships and frustration for the citizens. Already, some regions are experiencing record rainfall that has displaced hundreds of thousands of people from their homes and farmlands. In 2022, a report called African Shifts by the African

[24] UNDP and Climate Change in Africa (2023).

[25] AfDB (2020).

Climate Mobility Initiative shows that up to 2.5 million people are at risk of forced relocation from Africa's coastal area because of sea level rise.

In its bid to increase intra-African, the African Union Heads of State and Government, on December 5, 2020, agreed to launch the African Continental Free Trade Area (AfCFTA) from January 1, 2021. Since 2021, the implementation and other protocol negotiations have been ongoing simultaneously. While Africa contributes a very minimal share of the global greenhouse gas (GHG) emissions, increased trade ambitions stimulated by the desire of the state to take advantage of the benefits of the trade agreement are causing an increase in emissions.[26] As the countries continue to expand trade frontiers, there are indications that they will be wary of making commitments that could affect national trade policies and industrial strategies. To ensure that Africa's intra-trade expansion does not increase GHG emissions, the African Union seeks sustainable development by growing the economy without increasing the carbon footprint.

10 African Union Strategy for Sustainable Economic Growth

As Africa continues to drive for a better life for its citizens by creating investment-friendly policies and attracting innovative approaches to stimulating accelerated and competitive growth in the national economies, the African Union has launched a vision that will guide the continent in building a citizen-centred future. Therefore, Agenda 2063 was launched for sustainable economic growth. The document tagged "Africa We Want" is a comprehensive framework that provides details of the leaders' aspirations with a master plan on how Africa can collectively build itself to become a secure, inclusive, developed, and prosperous continent for its people. The plan spans 50 years, from 2013 to 2063. As part of the Agenda, strategic continental frameworks were conceived and designed to guide the African Union Commission and the national governments in developing critical sectors such as agriculture, energy, health, and transport infrastructure. However, the economy takes a center position amongst other areas that the Agenda 2063 aspires to transform for the good of the citizens. The strategy is that stakeholders will employ a coordinated approach in initiating policies and cross-sectoral actions, with implementation ultimately geared at improving the economy to the extent that annual growth will be at least 7% of GDP and real per-capita income would be a third more than in 2013. Those are part of the key transformational outcomes of the Agenda 2063.

Several policies and agreements have been initiated and signed as part of the effort to achieve the agenda for 2063. The agreement establishing the African Continental Free Trade Area is a cardinal document already being implemented by member states. But while much focus is on growing the African economy, experts warned that aggressive pursuit of economic development without having policies and structures

[26] De Melo and Solleder (2023, 3392–3418).

that will ensure environmental standards are not relegated would make Africa a haven for investments that could make us become net emitters of greenhouse gases. Former President Uhuru Kenyatta, the coordinator of the Committee of the African Heads of State and Government on Climate Change, expressed the same concern. In the Forward to the African Union Climate Change and Resilient Development Strategy and Action Plan (2022–2023), "The realization of Africa's Agenda 2063 cannot be possible without proactive collective continental efforts aimed at addressing the impacts and encumbrances of climate change, which hampers our integration and development."

Given Africa's economies' reliance on climate-sensitive sectors and the potential to grow, the principle of intergenerational equity can be consciously breached to satisfy the aspirations of the current generation. The African Union Climate Change and Resilient Development Strategy and Action Plan provides a roadmap for effectively responding to climate change and building climate-resilient economies to ensure sustainable development.

Some scholars may argue that the above strategy lacks the political force required to have the impact that could substantially contribute to balancing development with environmental necessities. The African Union has negotiated trade agreements without emphasizing ecological or climate change factors. However, the AfCFTA agreement's protocol on investment in Article 7 and Article 8 mandates state parties to ensure investments under the treaty comply with the state sustainable development laws. Traditionally, many believe trade agreements are not expected to be favourably aligned with climate-friendly policies because the deal aims to increase activities contributing to greenhouse gas emissions. That belief was reinforced by the World Trade Organization (WTO) when it reported that the production and transport of traded goods and services account for an average of 20–30% of global GHG emissions, with the international transport sector generating 12% of global emissions.[27] The AfCFTA protocol on investment, apart from mandating the States to make sustainable development part of the standard for receiving and approving investment, Article 8 of the protocol also permits State Parties to introduce incentives that could attract, retain and expand investments that have the potential of or focused on sustainable development of State Parties. The subsidies can be in the form of financial and fiscal incentives, such as investment insurance, grants or loans at concessionary rates[28]; subsidized infrastructure or services, and market preferences[29]; development-oriented incentives to encourage preferential market schemes and specific investments in Africa, especially in the sectors related to attaining sustainable development[30]; incentives for technology, technical assistance, technology transfer, and

[27] Brenton and Chemutai (2021).

[28] Article 8 (1) a.

[29] Ibid (1) b.

[30] Ibid (1) c.

research and development[31]; e. investment guarantees[32]; incentives for low-carbon investments[33]; Or incentives to encourage investors' responsible business conduct.[34]

11 Comparative Summary of the ASEAN and AU Approaches

The vulnerability of Southeast Asia and African regions is a significant factor determining the economic environment's receptivity to investment and trade in these regions. To the extent of the observations of experts, stakeholders, and decision-makers, the regions have designed strategies and plans to make the States build systems amenable to the climate realities and drive the social and economic aspirations of the supervening organizations for their people. However, both organizations are constrained in implementing the policies they created, and their members cannot commit themselves to global efforts to address the challenge of climate change due to economic and developmental factors. Despite being signatories to international climate treaties, such as the Paris Agreement, members of the ASEAN find it difficult to optimally implement climate policies in line with the demands of the commitment they have made under the various treaties and agreements they have ratified in their respective states. Their dependence on non-renewable energy sources to power their economy and their deforestation practices substantially contribute to greenhouse gas emissions. Yet the ASEAN Way principle precludes the regional body from introducing measures or pressing the states to introduce and implement policies that will help reduce GHG emissions to an acceptable and safe level over a target period. The principle complicates the situation by emphasizing non-intervention and consensus, frustrating regional cooperation and adherence to internationally agreed environmental policies.

Though African countries also suffer from the adverse effects of climate change, they have demonstrated a more precise approach to addressing its impact on the continent's development. The African Union and its members have designed their strategies to adopt mitigation, adaptation, resilience, and cooperation in their measures against the effects of climate change. However, financial and technological limitations are the main barriers to implementing climate measures to build a climate-resilient continent. Despite these challenges, African countries have achieved notable strides in pooling initiatives to design and launch the African Union's Climate Change and Resilient Development Strategy and Action Plan. The document places much emphasis on sustainable development and climate-resilient growth.

Asian and African countries share similar challenges in political dynamics, economic needs, and high development and technology deficits. However, while

31 Ibid (1) d.

32 Ibid (1) e.

33 Ibid (1) f.

34 Ibid (1) g.

the ASEAN Way principle often impedes effective policy implementation, Africans are more willing to contribute to global efforts to address environmental issues by implementing climate change policies at the national level and coordinating the integration of regional climate change policies into national strategies. Also, Africa has created institutions that enhance the delivery of objectives that align to ensure the environment is protected from further exploitation and that clean energy resources are explored for sustainable development. For example, environmental courts exist in some African countries. At the regional level, the African Court on Human and Peoples' Rights has recognized the right to a clean and healthy environment as enshrined. It has delivered several judgements on enforcing the right.

References

African Union Agenda 2063. https://au.int/en/agenda2063

African Union Green Recovery Action Plan. https://au.int/sites/default/files/documents/40790-doc-AU_Green_Recovery_Action_Plan_ENGLISH1.pdf

African Union, Climate change and resilient development strategy and action plan (2022–2032). https://repository.uneca.org/handle/10855/47738

Asian Development Bank, The economics of climate change in southeast Asia: a regional review (ADB, Metro Manila, Philippines, 2009) at 4–5; Ding DK, Beh SE (2022) Climate change and sustainability in ASEAN countries. Sustainability 14:999, at 1

Brenton P, Chemutai V (2021) The trade and climate change nexus. The Urgency and Opportunities

Chapter 9, IPCC Report for Africa. https://www.ipcc.ch/report/ar6/wg2/chapter/chapter-9/

Climate Change Impacts on Africa's Economic Growth | African Development Bank Group (afdb.org)

De Melo J, Solleder JM (2023) The landscape of CO_2 emissions across Africa: a comparative perspective. World Econ 46(11):3392–3418

For Developing Countries. World bank group. See also Making trade work for climate change mitigation (unctad.org). https://unctad.org/system/files/official-document/ditctab2022d7_en.pdf

Hoste JC, Anderson A (2022) African dynamics at the climate change negotiations. Egmont Institute. Also, see online chrome-extension://efaidnbmnnnibpcajpcglclefindmkaj/https://www.uneca.org/sites/default/files/ACPC/State-of-the-Climate-in-Africa/WMO_State%20of%20the%20Climate%20in%20Africa%202020_Report_en.pdf

https://devinit.org/cd3d27#fa2302fd

https://unfccc.int/news/innovative-technology-key-to-climate-action

https://wmo.int/publication-series/state of-climate-africa-2022

ICJ, Case concerning sovereignty over Pulau Ligitan and Pulau Sipadan (Indonesia/Malaysia), Judgement of 17 December 2022

Institute for Economics & Peace (2020) Ecological threat register 2020: understanding ecological threat. https://www.visionofhumanity.org/wp-content/uploads/2020/10/ETR_2020_web1.pdf

Intergovernmental Panel on Climate Change (2021) Climate change 2021: the physical science basis. Contribution of working group I to the sixth assessment report of the intergovernmental panel on climate change. Available online at https://www.ipcc.ch/report/ar6/wg1/

Kyoto Protocol to the United Nations Framework Convention on Climate Change, Dec. 10, 1997, 2303 U.N.T.S. 162

Meattle C et al (2022) Climate policy initiative, landscape of climate finance in Africa. Available online at https://www.climatepolicyinitiative.org/publication/landscape-of-climate-finance-in-africa/

Paris Agreement to the United Nations Framework Convention on Climate Change, Dec. 12, 2015, T.I.A.S. No. 16-1104

Reported by Leadership Newspaper, 6th August 2024, available online https://leadership.ng/nigerias-830kms-coastline-vulnerable-to-climate-change-impacts-nesg/

State of the Global Climate (2023) Report by the world metrological organization. https://wmo.int/publication-series/state-of-global-climate-2023

Tamasiga P et al (2023) Is Africa left behind in the global climate finance architecture: redefining climate vulnerability and revamping the climate finance landscape—a comprehensive review. Sustainability 15(17):13036

UN General Assembly, United Nations framework convention on climate change: resolution/adopted by the general assembly, 20 January 1994, A/RES/48/189

UNDP and Climate Change in Africa. It is available online at https://www.undp.org/africa/publications/fast-facts-undp-and-climate-change-africa

Woetzel J et al (2020) Climate risk and response in Asia. McKinsey Global Institute, at 3

Implementing Climate Standards in the African Continental Free Trade Area (AfCFTA): Progressive and Precautionary Lessons from the European Union

Deborah O. Ayodele-Olajire, Olaitan O. Olajire, and Adeniyi S. Gbadegesin

Abstract Pursuing economic development is arguably the bane of the global climate crisis. Moral considerations have risen in the form of 'ecological debt' expectations of the global North and an avoidance of the 'treadmill of accumulation' by the global South. Despite the differences in contributions to and benefits from past emissions, there remains a global responsibility to stem global temperature increases, adapt to current or anticipated climate change effects and build resilience. In Africa, the African Continental Free Trade Area (AfCFTA) agreement was introduced by the African Union (AU) to promote economic development across the continent. In the preamble to AfCFTA's Protocol on Trade in Services, member states of the AU agreed to recognize "the right of State Parties to regulate in pursuit of national policy objectives, and to introduce new regulations, on the supply of services, within their territories…without compromising consumer protection, environmental protection and overall sustainable development". However, the agreement does not provide an overarching framework for member states to key into to prevent cross purposes, nor is there a mechanism to ensure that member states protect the environment. The AU appears to be vested in preserving member autonomy with a common resource—the environment, specifically, the climate. This deliberate omission of an integrated system to apply climate standards within AfCFTA remains a gap in the current agreement, which we argue that the AU still has a good opportunity to integrate. In this chapter, we identify precautionary and progressive lessons that can be learned from the European Union's Single Market and Green Deal and suggest ways to implement climate standards within the AfCFTA.

D. O. Ayodele-Olajire (✉)
Department of Geography, University of Ibadan, Ibadan, Nigeria
e-mail: debbyodel@yahoo.com; do.ayodele-olajire@mail.ui.edu.ng

O. O. Olajire
Olawale Olajire & Co., Lagos, Nigeria
e-mail: forolaitan@gmail.com; olaitan@simplyolajire.org

A. S. Gbadegesin
Department of Geography, University of Ibadan, Ibadan, Nigeria
e-mail: as.gbadegesin@mail.ui.edu.ng

D. Oghenerobor Agelebe (ed.), *Climate Ethics in Africa*, Global Perspectives on Law and the Environment 1, https://doi.org/10.1007/978-3-032-08757-7_11

1 Introduction

The AfCFTA, which entered into force in May 2019, is the world's largest free trade area, given the number of participating countries since the formation of the World Trade Organisation.[1] To place Africa on the road to operationalizing its continental free trade area, the agreement seeks to bring together 1.3 billion people with a combined Gross Domestic Product (GDP) valued at 2.5 trillion US dollars. The main objectives of the agreement are to create a single market, deepen the continent's economic integration, achieve market liberalization, allow the movement of capital and people, and foster investment. The agreement's implementation aims to achieve sustainable and inclusive socio-economic development.

Although the overarching aim of the agreement is to maintain sound macroeconomic policies and a business environment that is friendly to domestic and foreign investors, it is imperative that policymakers also carefully monitor AfCFTA's impacts on the environment. Trade discussions are gradually beginning to take account of the relationship between achieving sustainable and inclusive socio-economic development and the environment—a trend now widely acknowledged in the international community. Thus, considering the potential impacts of the AfCFTA on industrialization and structural transformation in Africa, it is essential to balance the economic and environmental aspects of the AfCFTA. More specifically, environmental issues—particularly the impacts of climate change, which have become more pronounced in the continent—and efforts to promote economic growth and development through trade and industrialization may increase CO_2 emissions if there are no commensurate efforts to change the continent's energy mix towards renewables.

While there has been a growing body of literature on the economic impacts of the AfCFTA, the implementation of climate standards within the AfCFTA remains a glaring knowledge gap. In adding to the body of literature, this study posits that the agreement does not provide an overarching framework for member states to key into to prevent cross purposes, nor is there a mechanism to ensure that member states protect the environment. We further argue that the deliberate omission of an integrated system to apply climate standards within AfCFTA remains a gap in the current agreement, which the AU still has a good opportunity to rectify. To put these ideas into perspective, in this study, we identify precautionary and progressive lessons that can be learned from the European Union's Single Market and Green Deal and suggest ways to implement climate standards within the AfCFTA.

[1] African Union Commission 'AfCFTA' (2018) at 31.

2 Standard Organisations and Climate Standards

Many organizations are involved in developing standards globally. These include organizations focused on different geographical or jurisdictional scales across global, continental or regional, and national and local scales. Membership of these organizations also ranges across national government agencies, corporations, industries, small businesses, specialized cooperatives (for example, farming), consumer groups, and individuals, among others. These standard organizations include the International Telecommunication Union (ITU), International Electrotechnical Commission (IEC), Institute of Electrical and Electronics Engineers (IEEE), American National Standards Institute (ANSI), International Commission on Illumination (CIE), International Institute of Welding (IIW), International Union of Leather Technologists and Chemists Societies (IULTCS), among others.

In this chapter, we define climate standards as standards that promote the preservation of the environment and help combat climate change. Standard-like initiatives that have developed partly from the business community (corporations and non-corporations, respectively) but outside the scope of this chapter include Environmental Social and Governance (ESG), Taskforce on Climate-related Disclosures (TCFD) and Fairtrade. However, the chapter will discuss standard organizations with varying geographical focus and membership developing or potentially developing climate standards, namely the International Organization for Standardization (ISO), the European Committee for Standardization (CEN) and the Pan African Quality Infrastructure.

2.1 International Organization for Standardization (ISO)

The International Organization for Standardization (ISO) is a non-governmental organization of national standard bodies from 167 of the 195 countries that are members and non-member states of the United Nations. Established in 1947 and headquartered in Geneva (Switzerland), the organization has developed standards on over 250 specific subjects through its technical committees (comprising subject area experts).

Individuals or companies who desire to participate in standardization work can only do so through the national standard body in their country—the only ISO member from each member country. The organization's membership levels include member bodies, correspondent and subscriber members (see Fig. 1). Member bodies can vote in technical and policy meetings and sell ISO standards nationally. Correspondent members can only be observers in ISO technical and policy meetings but can sell standards nationally within their territory. Subscriber members follow ISO's activities but cannot vote in its meetings or sell standards nationally.

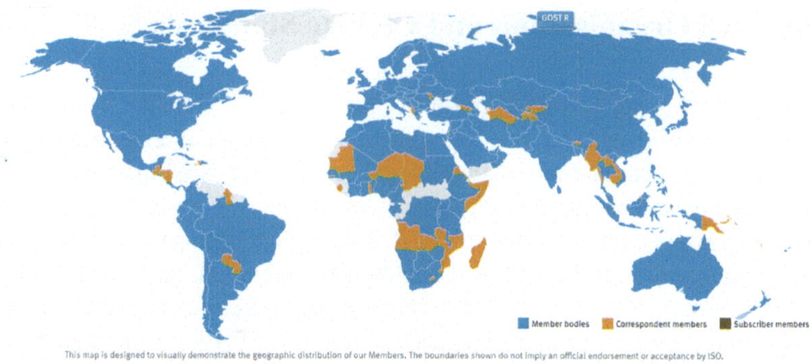

Fig. 1 Country memberships of the ISO[2]

Close to 24,000 standards developed by ISO relate to various supply chains and thus influence trading activities.[3] Also, ISO claims to have 1189 standards that are helping to meet the Climate Action goal of SDG 13.[4] Of particular interest is the ISO 14 000 family of standards, an example of ISO standards that generally relate to the environment and climate adaptation or resilience in certain respects.[5] For example, ISO 14 001 provides guidance for the use of environmental systems. ISO 14 064 provides guidance on the quantification, monitoring and validation or verification of greenhouse gas emissions; ISO 14 067 guides on quantifying and reporting the carbon footprint of products; ISO 14 080 provides a framework for developing improved methodologies for combatting climate change.[6]

With ISO's global membership, the global North and South are involved in developing international standards. The organization aims to involve national standard bodies, industry experts, civil society and consumers, and, more recently, vulnerable populations in developing standards.[7] Meanwhile, ISO boasts substantial membership from the global South. According to the organization, more than three-quarters of its members are from developing countries.[8] ISO has a Committee on Developing Country Matters (DEVCO) to involve developing nations better and ensure their needs are reflected in the developed standards. Also, the needs of consumers are reflected in the development of ISO Standards through Consumers International (an NGO), the consumer representatives of national members, and ISO's Committee on Consumer Policy (COPOLCO). Given the diverse stakeholders involved in developing ISO standards, the organization could be a platform for developing climate standards that can be integrated into AfCFTA trade and development activities. In

[2] International Organization for Standardization 'Members'.

[3] British Standards Institution (2024).

[4] International Organization for Standardization 'Goal 13: Climate Action'.

[5] International Organization for Standardization 'ISO 14000'.

[6] International Organization for Standardization (n 4).

[7] British Standards Institution (n 3).

[8] International Organization for Standardization 'Who develops standards'.

other words, the platform of policy development committees within the ISO, a great opportunity to institute, integrate and implement climate standards to guide AfCFTA, exists.

2.1.1 ISO: The London Declaration

In September 2021, during the ISO week in London—a meeting of the organization's 163 national body members—ISO committed to combating climate change through standards to achieve its climate agenda by 2050. In other words, ISO's new and revised standards will consider the climate.[9] The Declaration reads:

> ISO hereby commits to working with its members, stakeholders, and partners to ensure that ISO International Standards and publications accelerate the achievement of the Paris Agreement, the United Nations Sustainable Development Goals, and the United Nations Call for Action on Adaptation and Resilience.[10]

The London Declaration supposedly embodies the goals of the Paris Agreement, the United Nations Sustainable Development Goals and the United Nations Call for Action on Adaptation and Resilience.[11] It is poised to ensure that future and revised standards consider the climate.[12]

2.2 *European Committee for Standardization (CEN)*

The CEN is one of only three European Standards Organizations (ESOs) recognized by the European Union and the European Free Trade Association (EFTA).[13] The other standard organizations are the European Committee for Electrotechnical Standardization (CENELEC) and the European Telecommunications Standards Institute (ETSI). CEN members include the national standard organisations of 34 European countries. Its standardization activities cover sectors and cross-cutting areas such as the environment.

Through the activities of CEN's Strategic Advisory Body on the Environment (SABE) and in collaboration with CENELEC, the organization published the CEN-CENELEC Guide 32 for addressing climate change adaptation in standards—a guide for helping standard organizations develop and revise standards that require climate adaptation measures.[14] By 2020, the ISO scaled up this guide to publish ISO Guide

---

[9] British Standards Institution (n 3).

[10] International Organization for Standardization 'London Declaration: ISO commits to climate agenda'.

[11] International Organization for Standardization 'Climate action kit'.

[12] International Organization for Standardization (n 10).

[13] European Committee for Standardization (2024).

[14] International Organization for Standardization 'Climate action kit Case Study European Community'.

84: Guidelines for addressing climate change in standards. The CEN-CENELEC Guide 32 already represents a climate standard emanating from the EU. At the same time, the ISO Guide 84 is a similar international climate standard that can be adopted by national (and continental) members of ISO.

2.3 Pan African Quality Infrastructure

The Pan African Quality Infrastructure (PAQI) comprises the African Organisation for Standardisation (ARSO), the African Accreditation Cooperation (AFRAC), the Intra-Africa Metrology System (AFRIMETS) and the Africa Electrotechnical Standardisation Commission (AFSEC). ARSO, AFRAC, AFRIMETS and AFSEC are independent organizations that signed a memorandum of understanding in 2013 to form PAQI. In addition, Africa currently has over 40 national standard organizations also members of ISO. With these institutions, Africa does not lack standardization institutions. However, we would argue that the African Union Commission has not sufficiently deployed (where possible) these institutions to make Africa competitive and sustainable in its trade activities. For example, African products are still being rejected at global North borders for not meeting quality standards. At the same time, the AfCFTA agreement seems to intentionally shy away from coordinated enforcement of standards (climate or environmental and otherwise) in intra-African trade.[15]

3 Climate Ethics, Trade and Development

During the early development of the climate ethics field, the main discussion areas were scientific uncertainty, responsibility for past emissions, mitigation targets, adaptation and geoengineering.[16] Afterwards, concerns about the responsibility for the cost of climate change, market solutions, and geoengineering or non-market solutions dominated research between 1992 and 2020, while ethics relating to people and place are emerging research areas.[17]

Emissions continue to occupy a central place in development and trade and influence the arguments of climate ethics between the global North and South. The global North has achieved the level of development it has today, mainly due to the Industrial Revolution and emissions dating back to 1750.[18] Development was the focus of those times, not an overriding goal to protect the environment. Arguably, having developed on the back of carbon emissions, the global North is demanding the global South

---

[15] FO Kareema et al. (2018).

[16] SM Gardiner (2010, 54–66).

[17] DJ Bridge (2022, 2651–2665).

[18] Gardiner (n 16).

not to develop 'the way they had.' Unfortunately, no alternative route is sufficiently tested, and the suggestion of renewable energy sources to replace fossil fuels is not affordable for developing countries. Even now, coalitions in the global North are resistant to changing the old fossil-based pathway to economic growth that is being emphasized for the global South. Despite not being constrained by development concerns like Africa, developed continents are, to a large extent, yet to achieve fully climate-proof trading. Research by Arabesque—a sustainable finance firm, suggests that Europe will miss its 2030 climate targets by at least 21 years.[19]

Meanwhile, constraining the pathway to development for African countries, specifically—after years of colonization and post-independence years characterized by foreign aid and neoliberalism—may only perpetuate the vicious cycle of the continent's economic dependence. Particularly, the global North, which is partly paying for the global South's climate adaptation, may become an extension of the unfavourable foreign (economic) control often exerted through foreign aid. Thus, climate ethics reveal some disadvantages for economic development and agency in the global South.

Another aspect of climate ethics that is important to global North and South relations is responsibility for the cost of climate change. The global North needs to be willing to fund the South with the riches they had earlier derived from so flagrantly violating the environment. European countries have amassed wealth from environmental degradation. Yet, along with other developed countries, they are not completely forthcoming in granting some of those funds to Africa.[20] Meanwhile, even when there is a desire to toe the line of green development, the fact remains that Africa and others cannot afford the alternative sources that are being promoted now. In any case, Africa should not follow the development template of currently developed nations to preserve the environment for our joint posterity (African and European). However, the many unfulfilled climate funding promises from the global North invariably hinder the global target and race to net zero emissions.

Having argued against 'pay-outs' from the global North to the South, climate financing for developing countries is essential in promoting the capacity of developing nations even to attempt a 'greener' pathway to development. However, the relational power dynamics need to reflect partnership rather than the serfdom experienced in the past.[21] For example, the EU is Africa's biggest trading partner. Yet, trade terms have often been unjust, with many calling for Europe to play on the field and relate with Africa as partners.[22]

Though critical in preserving lives, the climate change discourse is essential for protecting development and the global economy. Given the level of development already attained, many countries of the global North are now in pursuit of 'global dominance' rather than growth. Yet, development remains a significant goal of Africa as a continent and as an individual African nation. It is reasonable to expect 'means

[19] H Ziady '21 April 2021'.

[20] C Lopes (2022)

[21] O Rwabwogo '3 February 2022; Luke et al. (2021)'.

[22] Lopes (n 20).

to develop' to feature first in considerations within the global South. In contrast, the global North is concerned not with developing but with 'protecting their development.' Unpacking this dichotomy in the make-up of Africa (global South) and Europe (global North) is essential to showing that comparisons between both continents are not necessarily 'apples to apples' and serve as an important backdrop for the rest of this chapter's discussion. 'Common but differentiated responsibilities' is a good way of summarising the ethics involved in the issue of climate change between the global North and the South.[23]

4 The European Union (EU) Single Market

The EU Single Market agreement, which came into effect in 1993, was established to transform the entire EU into a European Economic Area. This agreement has allowed EU countries to trade freely among themselves, without all the erstwhile bottlenecks that each individual EU country had put in place for their particular benefit—but which were considered, by foreigners, as impediments to easy trade. Therefore, establishing the Single Market brought about integrating all EU member states' markets (economies) into a 'single market.'[24] Countries' economies attract adequate—and long-lasting—interests from everywhere. The EU has capitalized on the single market's innate opportunity to maximize its sustainable economic agenda by promoting it alongside its crucial and unavoidable economic agenda.

Using legislation, directives, frameworks and targets, the EU has been able to entrench sustainability consciousness in its member states by directly influencing the ways businesses carry out their activities. For instance, the EU has guidelines for all companies within the Single Market for the packaging of goods in a manner that ensures the protection of the environment and promotes a sustainable economy. For example, the Packaging and Packaging Waste Directive (PPWD) is a directive that applies to member nations. This packaging directive by the European Parliament and Commission was set in 1994 with amendments up to 2018, while further revision plans began in 2020.[25] This directive's effect on trade is that national legislators must ensure their laws comply with the Single Market principles and Article 18 of the Packaging and Packaging Waste Directive (PPWD).[26] Complying with such legislations, directives, frameworks, and targets has ensured that countries inevitably take deliberate and bold steps to address environmental issues. In effect, through the EU Single Market, the Union can influence the direction, entrenchment, and growth of a sustainable economic culture within the EU, displacing individual countries.

The EU has not yet achieved a perfect implementation of Directives. As captured in the following quote by COWI & Eunomia (2019)

[23] E Lisinge-Fotabong et al. (2016).

[24] European Commission (2020) and Ploae (2010).

[25] Council of the European Union (2018) Directive (EU) 2018/852.

[26] Europen (2021) *Single Market.*

> The effectiveness of EU environmental law depends on its implementation at member state, regional, and local levels. Complaints concerning non-compliance with EU environmental law and many infringement cases indicate that implementation can be improved.

However, while its implementation is not yet perfect, successes have been recorded. For example, 2021 research by Arabesque suggests that among Sweden, Germany, Switzerland, Finland, Japan, London, the US, Hong Kong, and Australia, only the first five countries had the highest number of listed companies on track to meeting the 2050 1.5 degrees Celsius target.[27]

The Single Market has thus become the EU's most excellent tool in mitigating environmental degradation. In essence, ecological sustainability issues have been brought to the fore and given a place of pride in trending news because it has become intertwined with the EU's regulations, policies and/or directions to drive the economies of EU member states via the Single Market. Since countries take the issue of their economies—which has witnessed tremendous boost since the creation of the Single Market—very seriously, there exists a willingness by member states to abide by such regulations, policies and/or directions from the EU. Of course, the fear of adequate sanctions, in the case of violation, has also acted as a good deterrence to members that may have, otherwise, been lax about the problematic issues of the environment.

4.1 *The European Green Deal*

The European Green Deal (EGD), announced in December 2019, represents the direction the EU has adopted to meet climate targets and protect the environment effectively.[28] According to the European Commission, the EGD aims to make Europe climate-neutral by 2050 by reviewing the continent's climate and energy policies. The Commission has also proposed a European Climate Law, which targets at least 55% of greenhouse gas emissions by 2030 compared to 1990 (European Commission, n.d.b). The EGD's lofty goals include moving to a circular economy, reversing biodiversity loss, cutting pollution, and combatting climate change. It covers all sectors of the economy.[29]

The Green Deal is expected to be integrated into the Single Market via policies, directives, and laws to guide and regulate how businesses conduct their activities, ultimately affecting how the environment is treated.[30] Even though there have been criticisms of how similar EU policies of the past have been implemented or not implemented, there is considerable suspicion that the EU may be unable to achieve a holistic success of the Green Deal policy across its member states because of its lack

---

[27] Ziady (n 19).

[28] European Commission (2019).

[29] WECOOP 'EU Environment'.

[30] M Siddi 16 July 2020.

of coordination and enforcement of its policies,[31] There is no denying that the EU has, at least, a firm policy that attempts to guide member states towards a sustainable economy within a targeted period. The potential to achieve some result is preferable to achieving nothing.

4.2 Climate Standards Within the EU Single Market

As discussed earlier, CEN, CENELEC and ETSI are the continental standards organizations of the EU. These standard organizations are already involved in developing climate standards, as seen in the development of the CEN-CENELEC Guide 32 for addressing climate change adaptation in standards. The presence of these active continental organizations provides a structure that EU member states can key into and provides uniformity in standard adoption within the trade and other activities. Meanwhile, after many years of ineffectiveness, the EU is starting to focus on businesses to ensure that environmental standards such as the Packaging and Packaging Waste Directive (PPWD) are implemented across member states of the EU Single Market. Through sanctions on businesses, the EU can enforce standard adoptions that are environmentally beneficial and climate responsive.

5 The African Continental Free Trade Area

The AU introduced the AfCFTA agreement to promote economic development across the continent and create a single market for goods and services.[32] AfCFTA is expected to connect up to 1.3 billion people across 54 African countries. According to the AU, its objectives include creating a single market for goods and services, providing opportunities for intra-African trade, achieving economies of scale in production, and gaining continental market access.[33]

The AfCFTA Secretariat, based in Ghana, oversees the implementation of AfCFTA. Its broad responsibilities include overseeing trade in goods and services, dispute settlement, customs, and trade facilitation.[34] The ninth meeting of the AfCFTA Council of Ministers in Accra announced the first set of trading countries (in a pilot phase) to be monitored by the Secretariat. These countries are Cameroun, Egypt, Ghana, Kenya, Mauritius, Rwanda, and Tanzania. In comments attributed to the Secretariat, this pilot phase of AfCFTA implementation will assess the agreement's environmental, trade, and legal policy basis for intra-African trade.[35]

[31] M Reese et al. (2021).

[32] African Union Commission (2018).

[33] African Union 'Agenda 2063: The Africa we want'.

[34] African Union 'Continental Free Trade Area'.

[35] F Adekoya 'AfCFTA secretariat commences pilot trading with seven countries' 27 July 2022.

Within the agreement establishing AfCFTA and in the preamble to AfCFTA's Protocol on Trade in Services, member states of the AU agreed to recognize "the right of state parties to regulate in pursuit of national policy objectives…without compromising consumer protection, environmental protection, and overall sustainable development".[36] However, the agreement does not provide an overarching framework for member states to key into to prevent cross purposes, neither are there structures or policies to ensure that member states protect the environment. Currently, with the non-inclusion of environmental policies, frameworks and/or regulations in the AfCFTA, the AfCFTA can be said to have taken a dangerous departure from what could be termed the more recent 'progressive' approach of dealing with environmental sustainability matters—the deliberate inclusion of environmental-protecting policies within trade agreements to help boost environmental sustainability. It is disappointing that an opportunity of this magnitude was missed with the AfCFTA. Indeed, as well stated by the United Nations Conference on Trade and Development (UNCTAD),

more explicit inclusion of environmental provisions could play an essential role for Africa, as these provisions 'can be a way to ensure policy coherence, promote sustainable development and make sure that countries do not lower their environmental standards or derogate from them to gain trade and investment advantages.'[37]

This omission of a climate-responsive system for protecting the environment remains a gap in the current agreement. However, we argue elsewhere in this chapter that the AU still has a good opportunity to include these essential yet glaringly absent considerations. It remains to be seen how countries involved in the pilot testing of AfCFTA will implement climate change concerns of reduced carbon emissions, adaptation and resilience building alongside intra-African trade activities.

5.1 *Agenda 2063*

AfCFTA is a 'brainchild' of the African Union's Agenda 2063. Meanwhile, Agenda 2063 results from African leaders' commitment to refocus Africa's agenda from the fight for political independence and against apartheid at the celebration of the 50th anniversary of the Organisation of African Unity/African Union in 2013. In celebration of this anniversary, leaders signed a Solemn Declaration rededicating Africa towards achieving the Pan-African vision of "an integrated, prosperous and peaceful Africa, driven by its citizens, representing a dynamic force in the international arena."[38] Agenda 2063 represents the 50-year plan to achieve this vision.

Agenda 2063 was developed through a consultative process involving civil society, African think tanks, research institutions, government planners, inter-faith leaders,

[36] African Union Commission (2018) *Agreement establishing the Africa Continental Free Trade Area* at 31.

[37] United Nations Conference on Trade and Development (2021).

[38] African Union 'Agenda 2063: The Africa we want'.

and the private sector, a forum for former African Heads of State and Government, among others, with a final document published in 2015.[39] The document outlines seven aspirations, divided into 20 goals and 41 priority areas. As outlined in the published document, the aspirations of the African Agenda 2063 are[40]:

- A prosperous Africa based on inclusive growth and sustainable development
- An integrated continent, politically united, based on the ideals of Pan-Africanism and the vision of Africa's Renaissance
- An Africa of good governance, democracy, respect for human rights, justice and the rule of law
- A peaceful and secure Africa
- Africa with a strong cultural identity, common heritage, values and ethics
- An Africa whose development is people-driven, relying on the potential offered by African people, especially its women and youth, and caring for children
- An Africa as a strong, united and influential global player and partner

The Agenda 2063 document also outlines 12 flagship programmes, one of which is AfCFTA. AfCFTA falls within the second aspiration of the Agenda.

5.2 *The African Union's Green Recovery Action Plan*

The African Union's Green Recovery Action Plan (GRAP) is similar to the European Green Deal (discussed in the next section) as an economic strategy for Africa. However, GRAP is also a post-COVID-19 recovery plan, as it was developed after the COVID-19 pandemic. The priorities of GRAP include climate finance, renewable energy, biodiversity and nature-based solutions, climate-resilient agriculture, and green and resilient cities.[41] The Action Plan appears to limit the extent of its influence on intra-continental trade to the agricultural sector following this singular mention of AfCFTA in its document: "Under Climate Resilient Agriculture, the Action Plan will aim to… 3. Invest in green industrialization of the agricultural sector, the opportunities available for intra-regional trade under the AfCFTA".[42] Thus, the AU GRAP is insufficient and does not necessarily implement a climate change adaptation agenda in AfCFTA.

[39] African Union Commission (2015) *Agenda 2063.*

[40] African Union Commission (n 39) at 105, 106.

[41] African Union Commission (2021) *African Union Green Recovery Action Plan 2021–2027.*

[42] African Union Commission (n 41) at 20.

5.3 Climate Standards Within AfCFTA

The Pan African Quality Infrastructure is Africa's foremost continental standards institution comprising four individual and specialized organizations: ARSO, AFRAC, AFRIMETS and AFSEC. Within the AfCFTA agreement, Articles 21 (Technical Barriers to Trade) and 22 (Sanitary and Phytosanitary Measures) of the Protocol on Trade in Goods provide the agreement's text on Product Standards and Regulations, which limit implementation to Annex 6 and 7, respectively. Article 16 in Annex 4 of the agreement reflects the sovereignty of member states concerning the standards used. In 2021, the Pan African Quality Infrastructure Joint Committee (PAQI JC) published the African Quality Policy (AQP)—the AU's standards policy. Overall, the publication of the AQP is a step that could positively impact AfCFTA depending on the political will of the AU to coordinate, support and sanction (as necessary) member states in maintaining quality and meeting safe and climate-responsive standards.

6 Learning from the European Union's Single Market and Green Deal: Precautionary and Progressive Lessons

The concept of the AfCFTA closely resembles the European Union's Single Market. In this section, we discuss some lessons (both precautionary and progressive) that Africa should either be wary of or consider from the European implementation of the Single Market and the 'climate change combatting' Green Deal.

6.1 Precautionary Lessons

As the African Union develops its implementation of AfCFTA, ethics require that the Union beware of using its bloc economic strength to undermine the sustainable economy of other regions. The European Union's Single Market framework, among other institutions, allows it to conduct business with other areas as a bloc. However, in promoting the economic benefits of member states, European foreign trade policy currently appears to undermine the sustainable economy across Africa through AfCFTA. This is due to the Economic Partnership Agreements (EPAs) the Union makes with individual countries.[43] Specifically, the European demand for secure access to the markets of partner African countries puts these countries in a dilemma regarding intra-African versus European trade activities. Another area is

[43] D Luke et al. (2021).

the high tariff on processed goods, which mostly limits African exports to raw agricultural products with low economic returns.[44] Meanwhile, agricultural standards from the EGD's farm-to-fork policy may further limit African farmers' access to the European market due to their financial costs and accessibility. These last two barriers, in particular, are significant because it seems that the EU is paying lip service to their support of AfCFTA, Africa's economic development and climate resilience. While applauding Europe's Single Market, the AfCFTA Secretariat needs to approach an inter-continental relationship critically due to its effect on intra-Africa trades. Overall, similar to an uncritical adoption of the global North's industry development pathway, Africa cannot afford to be uncritical when adopting trade and environmental policies from the global North.

Through some aspects of the EGD, such as its transition to low-carbon energy, the European Union is exporting environmental degradation, particularly carbon emissions and undermining environmental sustainability abroad. The sourcing of critical raw materials (such as platinum, cobalt, and bauxite) from African countries like South Africa, the Democratic Republic of Congo, and Guinea, among others, could lead to higher carbon emissions due to increased raw materials being sourced.[45] In implementing climate standards within AfCFTA, the Secretariat must seek holistic solutions that are sustainable in the long term and do not transfer an environmental burden. In this regard, optimal—rather than best—solutions can be sought in pursuing climate adaptation, resilience, and reducing carbon emissions.

Due to the power asymmetry that still permeates the relationship between the European Union and Africa,[46] EGD strategies could become counterproductive. For example, NaturAfrica is the EGD's biodiversity strategy. But as long as the European Union continues to impose its conservation plans on African partners, it may unconsciously perpetuate human rights abuses and environmental degradation. Within AfCFTA, developing and least-developing countries should relate as partners, as cooperation based on such a relationship is needed to move the continent forward in fulfilling its development and climate resilience goals.

The challenge of continental and national policies undermining an environmental sustainability strategy such as the EGD. For instance, several national and continental moves within the European Union challenge the EGD's success. Examples include Spain's increase in wind farm levies and pesticide use and the threat of divergent packaging policies from EU member countries.[47] These examples show that there is no perfect system. Even well-thought-out plans must constantly be reviewed to ensure they still fulfill their purpose.

[44] Rwabwogo (n 21).

[45] Z Usman et al.

[46] Lopes (n 20).

[47] Pesticide Action Network Europe 'EU Commision draft Sustainable Use of Pesticides Regulation undermines the objectives of the European Green Deal' 4 February 2022.

6.2 Progressive Lessons

The EGD has a target, and work has begun to pursue it. As a result of the EU's overall direction, EU Member States are developing compatible legislation and policies within their jurisdiction to achieve sustainability goals.

> Such legislation and policies include environmental taxation, green public procurement, integrated product policy, eco-labelling, and eco-auditing. Green public procurement requires companies to meet specific environmental performance standards (for example, ISO 14 000) to be eligible to work for and with the government.[48]

Through its EGD, the EU integrates climate adaptation and resilience building into its continental trading activities, especially by including mandates for companies and businesses. By directing its legal instruments at firms rather than countries specifically, the EU has found a way to shape the thoughts and actions of countries in ways that support sustainability and, at the same time, reduce non-compliance.[49] Countries have had to put up legislation and policies that have allowed them to comply with EU laws and policies to ensure the ease of doing business in their countries. That is the only way that EU countries can guarantee the enjoyment of the prosperity of the Single Market. Therefore, as long as the economic issue remains germane to EU countries—and it would be—so too would the issue of sustainability automatically. Since it has cleverly intertwined its economy and sustainability matters, the EU has ensured that the fight against environmental degradation has finally become winnable and quick. This strategy could prove very useful in Africa in aiding the implementation of climate standards within the AfCFTA.

The availability of the CEN alongside other specialty standards organizations creates a platform for developing continental standards. CEN includes the national standard organizations of 34 European countries and is a member of the ISO. Furthermore, CEN and CENELEC have been vocal in contributing to the EGD.[50] The PAQI is an institution similar to the CEN by the AU. The AU would do well to deploy PAQI to develop climate and trade standards for the AfCFTA.

7 Implementing Climate Standards in the AfCFTA

The AfCFTA, which the AU introduced to promote economic development across the continent, is—in form—very similar to the European Union's Single Market. According to the AU, the objectives of the Continental Free Trade Area (CFTA) include to:

---

[48] L Čekanavičius et al. (2014) 78.

[49] C Hey (2005).

[50] European Committee for Standardization, European Committee for Electronical Standardization 'Standards in support of the Green Deal' 08 July 2020.

> Create a single continental market for goods and services, expand intra-African trade, and enhance competitiveness at the industry and enterprise level by exploiting opportunities for scale production, continental market access, and better resource reallocation.[51]

Quite surprisingly, AfCFTA does not have a robust environmental protection strategy and/or action, which some older regional trade treaties in Africa have.[52] Environmental sustainability measures are not weaved into the AfCFTA or given serious consideration beyond what appears to be a lame acknowledgment of its existence, which—one could argue assuredly—has been nothing more than an attempt to be politically correct in these modern times.

In the preamble to AfCFTA's Protocol on Trade in Service (para. 8), the AU states agreed to recognize "the right of State Parties to regulate in pursuit of national policy objectives…without compromising consumer protection, environmental protection, and overall sustainable development."[53] Interestingly, the only other mention of the environment is contained in the "objectives" of the AfCFTA protocol in Art. 3(2)(b), where it was provided that one of "[t]he specific objectives of this Protocol [is] to…promote sustainable development by the Sustainable Development Goals (SDGs)". However, the agreement does not provide an overarching framework for State Parties to key into to prevent cross purposes, nor are there known structures or policies to ensure that State Parties protect the environment. In other words, the AfCFTA contains mere words without concrete action points that would help drive the accomplishment of the stated words. As equally observed by van der Ven and Signé,[54] "These preambular references do not set out legally binding obligations and cannot, therefore, be interpreted as reflecting a strong commitment to the environment on the part of AfCFTA State Parties."

Notably, the General Exceptions are contained in Art. 28(b) and (g)—to the AfCFTA Protocol allow, under some restricted circumstances, State Parties to adopt or enforce measures that are "necessary to protect human, animal or plant life or health" and/or "relat[e] to the conservation of exhaustible natural resources if such measures are made effective in conjunction with restrictions on domestic production or consumption." However, UNCTAD observed that the lack of specific provisions may prove daunting and/or detrimental in certain instances since the exceptions "pertain only to justifying trade-restrictive measures; they cannot form the sole basis for incorporating BioTrade into the implementation of the AfCFTA Agreement."[55]

In this case, continental-level policy—is essential to any interventions to protect the climate in trade and development activities. A good example can be seen in the

---

[51] African Union 'Continental Free Trade Area'.

[52] The World Bank '27 July 2020'.

[53] African Union Commission 'Agreement establishing the African Continental Free Trade Area'.

[54] van der Ven and Signe (2021) at 7.

[55] UNCTAD 'Policy and Regulatory Recommendations' 2021 at 13.

development and implementation of Rwanda's RS 407-1:2019, Emission limits—Specification. The standard supports Rwanda's air quality and the country's policies. It is recording some success in phasing out old technology vehicles and introducing those with engine technology of at least EURO 4 generation.[56] The foregoing is to reiterate the critical relationship between policy and standards. Despite not being the site of sufficient climate change research,[57] African leaders participate in global climate change discussions and have created several institutions to participate in climate change deliberations and action.[58] However, the continent lacked a final climate action plan for many years.[59] The African Union Climate Change and Resilient Development Strategy and Action Plan (2022–2032) was published in 2022 to represent the continent's common position on climate change.[60] This Strategy and Action Plan directs the continent's climate adaptation and resilience. The next right step should be integrating the strategy with AfCFTA to achieve sustainable economic development that is climate responsive.

Without a doubt, the recently developed African Union Climate Change Strategy and Resilient Development Strategy and Action Plan will become easier to enforce, create and observe if it can be interwoven with the AfCFTA. The provisions of Art.8(3) of AfCFTA, which states that "[a]ny additional instruments, within the scope of this Agreement, deemed necessary, shall be concluded in furtherance of the objectives of the AfCFTA and shall, upon adoption, form an integral part of this Agreement" has provided the perfect opportunity for such strategy and action plan to be integrated into AfCFTA for a much more effective result. The continent stands to gain a whole lot more with such integrations than when they are left stand-alone. The Action Plan itself lends credence to this argument when it admits the colossal role the AfCFTA can play in enabling sustainability in the continent by focusing on the choice of goods while referring to the AfCFTA agreement and the Africa Quality Policy for climate standards (African Union Commission 2022). However, that strategy is insufficient. Instead, it could have been better to point toward continental institutions such as the Pan African Quality Infrastructure (PAQI) for implementing, monitoring and ensuring climate-responsive trade practices.

Specifically, the African Union needs a continental enforcement and Quality support framework for intra-African trade (AfCFTA) and inter-continental trade (between African nations and other nations) such as PAQI. The current reliance on nation-states to direct and implement measures to protect the environment is insufficient and ineffective. On the other hand, a continental framework of quality and standards enforcement may, at a continental scale, help the generally low Regulation

---

[56] International Organization for Standardization 'Climate action kit Case Study Rwanda' https://www.iso.org/files/live/sites/isoorg/files/developing_standards/resources/climate_action_kit/documents/case_studies/Climate_Case_Study_RSB.pdf.

[57] Ayodele-Olajire and Olusola (2022, 7).

[58] Lisinge-Fotabong (n 23).

[59] Usman (n 45).

[60] African Union (2022) *African Union Climate Change and Resilient Development Strategy and Action Plan (2022–2032).*

Quality scores published by the World Bank for African Nations[61] Despite the presence of national standard organizations. Meanwhile, African nations' experience of having their products rejected at, for example, European borders due to not meeting the EU's quality standards reflects the relative inability of national standard organizations to ensure and enforce quality standards in trade.[62] AfCFTA and the need for climate standards allow PAQI to fill a vast and existing gap in Africa's climate-neutral and sustainable economic development. Collaborations between CEN-CENELEC and the ISO (reported earlier) in developing ISO Guide 84 should be promoted between the PAQI and ISO in developing climate standards for trade and development across the continent, within AfCFTA and worldwide. Also, standard-setting and enforcement at the continental level will help prevent fragmentation and support sustainable trade.[63]

It is not like the AU, outside the AfCFTA, has no plan to encourage environmental sustainability amongst its Member States. It does. For instance, there is the recent African Union Climate Change and Resilient Development Strategy and Action Plan 2022–2032. However, as it has become obvious worldwide, expecting countries to become thoroughly interested in enforcing policies to ensure environmental sustainability is unrealistic. Countries are not easily persuaded to tackle ecological degradation seriously; they would prefer to make a splash here and there—to save face—whilst giving themselves an infinite amount of time to tackle the scourge sometime in the unknown future finally successfully. More is, therefore, required to get countries to do what is necessary to protect the environment. Nothing beats introducing environmental standards into trade agreements, excellently exemplified by the EU's Single Market. At this time, trade agreements between countries, regions, and blocs can no longer be seen as successful merely when trade successes are achieved—much more is now expected.[64]

Thus, member state autonomy in directing environmental protection within AfCFTA is another challenge in ensuring intra-African trade and economic development that is climate responsive. It appears that the AU favours this ineffective approach concerning climate and environmental protection, as seen in the following quote from the African Union Climate Change and Resilient Development Strategy and Action Plan (2022–2032):

> It is recognized that each Member State will direct its climate response in a self-determined manner based on its unique national circumstances and capacities. While recognizing differences in national contexts and circumstances, this Strategy builds upon the shared challenges and opportunities for the continent, encouraging African-led and African-owned innovation, social and cultural values, and leveraging nature-based solutions….[65]

---

[61] D Kaufmann & A Kraay 'Worldwide Governance Indicators'.

[62] Kareema (n 15).

[63] Poole (n 48).

[64] Van der Ven & Signe (n 55).

[65] African Union (2022) at 2.

Even though such a clause appears to promote member states' agency, it limits the effectiveness of such plans in ensuring climate and environment protection, especially within intra-African trade. Rather than a broad brush approach, if Africa was serious about achieving results, it may be better for the AU to identify those 'circumstances' and enforce them, especially in sanctioning and penalizing member countries (or companies) as appropriate. Using Regional Economic Communities (RECs) for implementation (as indicated at the continental level) is insufficient, as different RECs may impose unequal sanctions across regions. However, for effectiveness, RECs can report defaulting countries to the continental AUC, while this central body then metes out the required sanctions or penalties. There may be no better way to achieve this than to ensure that the AUC takes charge of such oversight functions—by providing policies, guidelines, laws, etc.—through the instrumentality of the AfCFTA.

Deliberate actions are required to tackle climate change challenges. Climate standards cannot be merely wished for or talked into existence. For a continent that has become synonymous with paying lip service to the issues of sustainable environment, if not an out-right nonchalance and/or wilful disregard for the matter, Africa appears to have lost a golden chance to lay down concrete guidelines to assist African countries, and—in particular—businesses in Africa, develop the requisite appetite that would help Africa effectively combat climate change problems, which are seriously ravaging the continent. This appears to be a lost opportunity because the AfCFTA represents Africa's boldest economic policy yet. The AfCFTA can revolutionize the way business and commerce operate in the continent, so without a doubt, it would have been a great vehicle to help Africa make a significant leap in the global push for a sustainable world. Had climate standards been woven into the fabric of AfCFTA, it would have ensured that the expected boost in economic development that the treaty would surely bring to Africa would have also ensured that climate and sustainable development principles would equally flourish simultaneously.

AfCFTA is still in the early stages of its implementation as only recently (25 July 2022), the 9th meeting of the AfCFTA Council of Ministers announced the commencement of pilot trading with seven countries.[66] However, continental-wide stated environmental and climate-responsive standards have yet to be with the necessary technical support to ensure environmentally sustainable trading activities. According to the results of the World Bank's Worldwide Governance Indicators project on Regulation Quality, only 4 African countries showed index values above zero to indicate measures of good quality governance.[67] The foregoing suggests that the capacity to utilize standards may be problematic for many African countries. However, all hope is not lost as there are a few country examples to show that African standards experts can develop and utilize climate-responsive standards given the political will, incentives and support. For instance, since the initial revision in 2018, the Kenya Bureau of Standards has been successful in implementing

[66] Adekoya (n 35).

[67] Kaufmann & Kraay (n 62).

Minimum Energy Performance Standards (MEPS) for air conditioners and refrigerators that have led to the registration of air conditioner models that contain the hydrofluorocarbon R-410A, which does not deplete the ozone.

Thus, there is sufficient room for implementing climate standards within AfCFTA with the necessary institutional support and enforcement. Africa can still include frameworks and policies for a sustainable economy in AfCFTA. AfCFTA only came into effect on 1st January 2021. It is still in its infancy, and many State Parties have yet to implement it fully. Africa can still cleanly incorporate frameworks and policies for a sustainable economy into AfCFTA. Countries are still negotiating different aspects of the AfCFTA, which means some earlier wrongs can be corrected.[68] Indeed, there is no better time to correct the lost opportunity than now.

References

Adekoya F (2022) AfCFTA secretariat commences pilot trading with seven countries. https://guardian.ng/business-services/afcfta-secretariat-commences-pilot-trading-with-seven-countries/

AfCFTA Secretariat (2022) The secretariat. https://au-afcfta.org/secretariat/

African Union Commission (2015a) Agenda 2063: the Africa we want. https://au.int/sites/default/files/documents/36204-doc-agenda2063_popular_version_en.pdf

African Union Commission (2015b) Agenda 2063: the Africa we want. Framework document. https://au.int/sites/default/files/documents/33126-doc-framework_document_book.pdf

African Union Commission (2018) Agreement establishing the African continental free trade area. https://au.int/sites/default/files/treaties/36437-treaty-consolidated_text_on_cfta_-_en.pdf

African Union Commission (2021) African union green recovery action plan 2021–2027. https://au.int/sites/default/files/documents/40790-doc-AU_Green_Recovery_Action_Plan_ENGLISH1.pdf

African Union Commission (2022) African union climate change and resilient development strategy and action plan (2022–2032)

Ayodele-Olajire D, Olusola A (2022) A review of climate change trends and scenarios (2011–2021). In: Mohammad Z, Abdul Wahid M, Chatterjee U (eds) Current directions in water scarcity research. Elsevier, London, pp 7–545

Bridge DJ (2022) The ethics of climate change: a systematic literature review. Account Finance 62(2):2651–2665. https://doi.org/10.1111/acfi.12877

BSI (2024) Understanding the London declaration: what you need to know. https://www.bsigroup.com/en-GB/about-bsi/uk-national-standards-body/london-declaration/

Čekanavičius L, Bazytė R, Dičmonaitė A (2014) Green business: challenges and practices. Ekonomika 93(1):74–88

CEN-CENELEC (2020) Standards in support of the green deal: CEN and CENELEC#s new position paper. https://www.cencenelec.eu/news-and-events/news/2020/briefnews/2020-07-08-standards-green-deal-position-paper/; European Committee for Standardization, European Committee for Electronical Standardization 'Standards in support of the European Green Deal Commitments' https://www.cencenelec.eu/media/CEN-CENELEC/Areas%20of%20Work/CENELEC%20sectors/Accumulators,%20Primary%20cells%20and%20Primary%20Batteries/Documents/standardsinsupporteuropeangreendealcommitments.pdf

CEN-CENELEC (2024) About CEN. https://www.cencenelec.eu/about-cen/

[68] Van der Ven & Signe (n 55).

Council of the European Union (2018) Directive (EU) 2018/852 of the European parliament and of the council of 30 May 2018 amending directive 94/62/EC on packaging and packaging waste. https://eur-lex.europa.eu/legal-content/EN/TXT/?uri=celex%3A32018L0852
COWI, Eunomia (2019) Study: the costs of not implementing EU environmental law
European Commission (2019) Communication from the commission to the European parliament, the European council, the council, the European economic and social committee and the committee of the regions
European Commission (2020) A single market that delivers for businesses and consumers
European Commission (2021) Climate action and the green deal. https://ec.europa.eu/info/strategy/priorities-2019-2024/european-green-deal/climate-action-and-green-deal_en
Europen (2021) Single market. https://www.europen-packaging.eu/policy-area/single-market/
Gardiner SM (2010) Ethics and climate change: an introduction. Wiley Interdiscip Rev Clim Change 1(1):54–66. https://doi.org/10.1002/wcc.16
Hey C (2005) EU environmental policies: a short history of the policy strategies. In: Scheuer S (ed) EU environmental policy handbook, European Environmental Bureau. Brussels, Belgium
ISO (2021a) Climate action kit case study European community. https://www.iso.org/files/live/sites/isoorg/files/developing_standards/resources/climate_action_kit/documents/case_studies/Climate_Case_Study_CEN.pdf.
ISO (2021b) Climate action kit case study Rwanda. https://www.iso.org/files/live/sites/isoorg/files/developing_standards/resources/climate_action_kit/documents/case_studies/Climate_Case_Study_RSB.pdf
ISO (nda) Members. https://www.iso.org/members.html
ISO (ndb) Goal 13: climate action. https://www.iso.org/sdg/SDG13.html
ISO (ndc) ISO 14000 family: environmental management. https://www.iso.org/standards/popular/iso-14000-family
ISO (ndd) Who develops standards. https://www.iso.org/who-develops-standards.html
ISO (nde) London declaration: ISO commits to climate agenda. https://www.iso.org/news/ref2726.html
ISO (ndf) Climate action kit. https://www.iso.org/ClimateAction.html
Kareema FO, Martinez-Zarzosob I, Brümmerc B (2018) Entry barriers in the European Union and rejection of Africa's exports: the roles of trade related logistic factors. https://www.researchgate.net/profile/Fatima-Kareem-2/publication/323779905_Entry_Barriers_in_the_European_Union_and_Rejection_of_Africa's_Exports_The_Roles_of_Trade_Related_Logistic_Factors/links/5da4f0a2a6fdcc8fc3527ed2/Entry-Barriers-in-the-European-Union-and-Rejection-of-Africas-Exports-The-Roles-of-Trade-Related-Logistic-Factors.pdf
Kaufmann D, Kraay A (2022) Worldwide Governance Indicators. https://info.worldbank.org/governance/wgi/#doc
Lisinge-Fotabong E, Diakhite M, Ababio K et al (2016) Climate diplomacy in Africa
Lopes C (2020) Building an EU-Africa partnership of equals. https://jordantimes.com/opinion/carlos-lopes/building-eu-africa-partnership-equals
Lopes C (2022) The rich world's empty climate promises. https://jordantimes.com/opinion/carlos-lopes/rich-worlds-empty-climate-promises
Luke D, Desta M, Mevel S (2021) The European Union is undermining prospects for a free trade agreement with Africa. African Trade Policy Programme, London School of Economics
PAN (2022) EU commission draft sustainable use of pesticides regulation undermines the objectives of the European Green Deal. https://www.pan-europe.info/press-releases/2022/02/eu-commission-draft-sustainable-use-pesticides-regulation-undermines
PAQI (2013) Integrated quality infrastructure memorandum of understanding concerning standardisation, accreditation and metrology. Available from: https://www.intra-afrac.com/Documents/A016-01.pdf
Ploae C (2010) The revitalisation of the EU's single market. Roman Econ J 13(36):79–93

Poole J (2022) Unite behind circularity: single market fragmentation undermining EU green deal, warns industry. https://m.packaginginsights.com/news/unite-behind-circularity-single-market-fragmentation-undermining-eu-green-deal-warns-industry.html
Reese M, Jendroska J, Squintani L (2021) The European green deal—new horizons in environmental policy and regulation. ICLG environmental and climate laws and regulations
Rwabwogo O (2022) Lopsided EU trade agreements are harming Africa. https://african.business/2022/02/agribusiness-manufacturing/lopsided-eu-trade-agreements-are-harming-africa/
Siddi M (2020) What are the prospects for the European Green Deal? https://blogs.lse.ac.uk/europpblog/2020/07/16/what-are-the-prospects-for-the-european-green-deal/
UNCTAD (2021) Implications of the African continental free trade area for trade and biodiversity: policy and regulatory recommendations. https://unctad.org/system/files/official-document/ditctedinf2021d3_en.pdf
Usman Z, Abimbola O, Ituen I (2021) What does the European green deal mean for Africa? https://carnegieendowment.org/files/202110-Usman_etal_EGD_Africa_final.pdf
Van der Ven C, Signe L (2021) Greening the AfCFTA: it is not too late. https://www.brookings.edu/research/greening-the-afcfta-it-is-not-too-late/
WECOOP (2012) EU environment policy brief. https://wecoop.eu/regional-knowledge-centre/eu-policies-regulations/
World Bank (2024) The African continental free trade area: economic and distributional effects. https://www.worldbank.org/en/topic/trade/publication/the-african-continental-free-trade-area
Ziady H (2021) Companies are crucial to solving the climate crisis. 75% are falling short https://edition.cnn.com/2021/04/21/business/corporate-emissions-climate-crisis/index.html

Intellectual Property Rights and Access to Green Technology in Economic Community of West African States: Issues, Challenges and the Way Forward

Solomon Ezike and Dennis Oghenerobor Agelebe

Abstract Using a legal history methodology, this paper critically examined the international intellectual property rights regime as embodied in the Agreement on Trade-Related Aspects of Intellectual Property Rights, 1994 (TRIPS Agreement) to evaluate its effectiveness in aiding the transfer of climate change mitigation and adaptation technologies from developed to the Economic Community of West African States ("**ECOWAS**") and proposes approaches to promoting regional capacities and access to green technologies. Drawing on the historical antecedents underpinning the TRIPS Agreement, this paper argues that the TRIPS Agreement is not well-suited to adequately address the issue of access to green technologies faced by developing countries. As such, developing countries cannot benefit from the TRIPS Agreement for climate change mitigation and adaptation. Further, the paper posits that the legislative history around the TRIPS flexibilities for essential medicines may point to a path for climate change technologies. The paper's primary contribution will be to take advantage of this historical perspective, which has been largely neglected in the existing literature on the subject. Building on this and situating intellectual property rights in the context of what climate change means today, the paper recommended necessary legal and policy interventions to improve access to essential technological innovations for climate change mitigation and adaptation in the ECOWAS. These recommendations, which will be fair and balanced, comprise a combination of options focused on possible exploitation, circumvention, and potential modification of the TRIPS Agreement.

S. Ezike (✉)
Faculty of Law, University of Ottawa, Ottawa, Canada
e-mail: sezik016@uottawa.ca

D. O. Agelebe
Environmental Law Centre, Faculty of Law, University of Cologne, Cologne, Germany
e-mail: Dageleb1@uni-koeln.de

D. Oghenerobor Agelebe (ed.), *Climate Ethics in Africa*, Global Perspectives on Law and the Environment 1, https://doi.org/10.1007/978-3-032-08757-7_12

1 Introduction

Climate change is a growing concern and will remain so for many decades. This concern is believed to be more urgent today than at any other point in history. To limit global average temperature increases to 1.5 °C, global greenhouse gas ("**GHG**") emissions must be halved by 2030 and netted by 2050. Addressing climate change will neither come cheap nor easy. The challenges climate change presents no doubt require innovation models to address the various challenges attendant to climate change. If climate change is to be addressed effectively in the long run, nations of all descriptions must pursue mitigation and adaptation strategies. Therefore, an effective response to climate change must involve deploying and using technological innovations that address climate change. These include adaptation and mitigation technologies that facilitate clean energy and reduce GHG emissions. With technological invention comes the inseparable role and issues of intellectual property rights ("**IPRs**"). The impact and role of IPRs in mitigating climate change remains a subject of extensive and intense debate and deep controversy, with no accurate or universal position taken on the issue. And herein lies the conundrum between climate change and intellectual property, especially regarding climate change mitigation and adaptation. IPRs have never been more economically and politically important and controversial.

Developing countries are at the crossroads of these debates (which the Economic Community of West African States ("**ECOWAS**") falls into). Developing countries need access to environmentally sound technologies to meet their climate change mitigation and adaptation obligations. Technology transfer to developing countries is an essential component of global efforts to deal successfully with the adverse effects of climate change. Most of these nations rely primarily on external sources of knowledge and technology to support domestic production processes. Most climate change mitigation and adaptation technologies developed in response to climate change originate from in developed countries.

Thus, as the world struggles to grapple with the scourge of climate change, there is a collective understanding that rapid development and diffusion of green technologies, especially to developing countries, is an essential component of the global response to climate change. The African continent is typically vulnerable to the impact of climate change, as global warming is expected to increase droughts, desertification, and flooding across the continent.[1] Most of these countries rely primarily on external sources of knowledge and technology to support domestic production processes.[2] Most climate change-related technologies remain concentrated in industrialized countries.[3] However, despite several international declarations and commitments to promote the transfer of green technologies to developing countries, the pace of such transfers has been too slow for these nations to have the required capacity

[1] White & Case LLP, "Climate change litigation in Africa: Current status and future developments" (2021).

[2] Maskus and Okediji (2010).

[3] World Bank (2008) and Sarnoff (2011, 300).

for building the necessary climate change mitigation and adaptation capabilities and measures.[4] One of the critical points of discussion has been, and continues to be, the licensing of IPRs and technology transfer, especially to developing countries needing "accelerated access to critical mitigation and adaptation technologies."[5] IPRs must be a veritable tool for transferring critical technologies to help developing countries mitigate and adapt to climate change. It is to be noted that technology transfer is not merely the import or purchase of machines and other hardware at commercial rates. A central aspect of technology development and transfer is the building of local capacity so that people and institutions in developing countries can design and make technologies which can be diffused into the domestic economy. The global efforts on green technology transfer seem not to have provided a holistic and practical solution to these needs and concerns.

The Agreement on Trade-Related Aspects of Intellectual Property Rights (1994) ("**TRIPS Agreement**"),[6] which introduced IPRs into the international trading system and remains the most comprehensive global agreement on the topic, is significant to the ongoing discussions on the transfer of green technologies to developing countries. The TRIPS Agreement establishes the most comprehensive minimum standards of IPRs protection regarding covered areas and their applicability to all World Trade Organization ("**WTO**") members. The TRIPS Agreement is focused on the promotion of innovation through the provision of commercial incentives. This, amongst others, has given rise to debates on the utility of the TRIPS Agreement, especially whether it achieves a balance between intellectual property protection and the transfer of technology and whether the TRIPS Agreement is an ideal or sufficient basis for building climate change solutions.

Using a legal history methodology, this paper examines the international IPRs regime embodied in the TRIPS Agreement to evaluate its effectiveness in aiding the transfer of climate change mitigation and adaptation technologies from developed to developing countries. The paper focuses on the ECOWAS and proposes approaches to promoting regional capacities and access to green technologies. It must be mentioned that other factors pose barriers to transferring green technology to ECOWAS. These factors may include—trade secrets, capacity, financing, wealth distribution, corruption, etc. This paper addresses only the issues relating to IPRs, particularly patents, by addressing the following questions: (i) To what extent do IPRs promote or hinder the development, commercialization and distribution of green technologies in developing countries, especially the ECOWAS? (ii) Does the TRIPS Agreement balance protecting IPRs and transferring technology as an ideal, suitable or sufficient option to build regional climate change solutions? (iii) Are the national intellectual property legislations of ECOWAS members couched in a manner that will allow member states and the region to take full benefit of the TRIPS flexibilities for boosting access to climate change mitigation and adaptation? (iv) How can ECOWAS countries develop

[4] Littleton (2009, 233–244).

[5] AIPPI Standing Committee on Intellectual Property and Green Technology (2014).

[6] Agreement Establishing the World Trade Organization, Annex 1C: Agreement on Trade-Related Aspects of Intellectual Property Rights, 15 April 1994, 1869 U.N.T.S. 299, WTO,

regional manufacturing capacity and facilitate easy access to green technologies in the region?

Drawing on the historical antecedents underpinning the TRIPS Agreement and its application over the years, the paper argues that the TRIPS Agreement is not well-suited to adequately address the issue of access to green technologies faced by developing countries. As such, developing countries cannot benefit from the TRIPS Agreement for climate change mitigation and adaptation. Further, the paper posits that the legislative history around the TRIPS flexibilities for essential medicines may point to a path for climate change technologies. The paper's primary contribution will be to take advantage of this historical perspective, which has been largely neglected in the existing literature on the subject. Building on this and situating IPRs in the context of what climate change means today, the paper will recommend necessary legal and policy interventions to achieve improved access to essential technological innovations to mitigate and adapt to climate change in the ECOWAS. These recommendations, which will be fair and balanced, will comprise a combination of options focused on possible exploitation, circumvention, and potential modification of the TRIPS Agreement.

2 The TRIPS Agreement: Historical Developments, Overview and Evolution

2.1 The Uruguay Round and the Commencement of Negotiations Leading to TRIPS

Two major international agreements on intellectual property existed before the harmonization efforts that culminated in the TRIPS Agreement. The weaknesses and shortcomings these agreements resulted in many developed countries pushing for a more holistic agreement. This push was also predicated on minimizing losses incurred by Western IPRs holders occasioned by unauthorized use and gains from these IPRs by developing countries.[7] Thus, despite strong opposition from developing countries,[8] the United States ("**US**") successfully spearheaded the call to include the subject of intellectual property protection in the Uruguay Round negotiations.[9] Simultaneously, the US successfully pressured some of its trading partners to improve their enforcement of IPRs—the initiation of sanctions backed up this pressure.[10] As a result, countries like Brazil, Singapore, South Korea, Taiwan and Thailand either adopted new or revised their IPRs laws. Additionally, the US pressured various developing countries to strengthen their IPRs laws by threatening to withdraw their benefits

---

[7] Gervais (2012) at 12.

[8] See Gervais, *supra* note 7 at 33; David M. Fox (2019) 1.

[9] See Gervais, "*The TRIPS Agreement*," supra note 7 at 12.

[10] Jackson et al. (West, 2021) at 1017.

under the Generalized System of Preferences. These aspects of the historical background—the opposition from developing countries and the US threat and sanctions approach—are critical because they influenced some of the techniques incorporated into the TRIPS Agreement and how they have been applied over the years. For instance, the TRIPS Agreement reflects the approach of sanctions. With the threat of trade sanctions behind it, the TRIPS regime poses a formidable barrier to the diffusion of critical technologies necessary for climate change mitigation and adaptation. In addition to the opposition by developing countries, there were palpable tensions among the developed countries. These differences arose partly from the need to map the exact contours and terms of patents, copyrights, trademarks and other forms of intellectual property protection.[11]

The European Communities, Japan, the United States, Australia, Switzerland, and 14 developing nations contributed texts that served as the foundation for the TRIPS Agreement.[12] The fact that few developing countries actively engaged in the TRIPS debates and that some of those who participated may not have had the necessary degree of the understanding of intellectual property is crucial to note. They were hindered by a lack of resources and sufficient information to participate in the negotiations properly and were not fully aware of the far-reaching repercussions of the issues discussed. Additionally, developing countries believed that increased intellectual property protection would not augur well for their development interests. Some thought TRIPS concessions would be made up for by lower tariffs on goods like tropical fruit and textiles, but this belief has since been substantially discredited.[13] This view may have led several countries to pay little attention to the language of the TRIPS Agreement and its commitments as it was drafted.

Some important patent-specific issues that dominated the negotiations of the TRIPS Agreement included: (i) whether the US should be permitted to maintain its "first to invent" rule, which awards a patent to the first person to have invented a product, as opposed to the practice elsewhere in the world, where the first party to file a patent application for an invention is typically awarded the patent; (ii) what subject matter could be patented, and the extent to which specific products could be excluded from patent coverage, such as pharmaceutical products or the results of biotechnology research; and (iii) the question of compulsory licensing and under what conditions should the use of the patent be allowed without the voluntary consent of the holder.[14]

---

[11] Jackson et al., *supra* note 10 at 1017.

[12] Gervais (2012).

[13] Gervais, "Climate Change," *supra* note 12 at 3.

[14] Jackson et al., *supra* note 10 at 1019.

2.2 Brief Overview of the TRIPS Agreement

The TRIPS Agreement negotiations attempted to address the above issues and many other prevalent concerns. While it did not solve all the problems and challenges raised, it trumped previous efforts at harmonizing the IPRs regime. The TRIPS Agreement recognizes that a growing tension in international economic relations has been the widely divergent standards for protecting and enforcing IPRs and the absence of a multilateral framework of rules, principles, and disciplines addressing international trade in counterfeit goods. Rules and disciplines were required to deal with these conflicts. To that end, the TRIPS Agreement covers the applicability of fundamental principles of the General Agreement on Tariffs and Trade ("**GATT**") and pertinent international IPRs agreements, as well as the provision of adequate IPRs, multilateral dispute settlement, and effective enforcement measures for those rights and transitional arrangements.

There is also a requirement to comply with the substantive provisions of the Paris Convention for the Protection of Industrial Property (1883) (as amended on September 28, 1979). Additionally, the TRIPS Agreement mandates that inventions in practically all technological disciplines, whether products or processes, be eligible for 20-year patent protection. It excluded from patentability inventions that their commercial exploitation is prohibited for moral or public policy considerations. Specific conditions are outlined for compulsory licencing or governmental use of patents without the owner's consent. Sometimes, a court may force putative infringers to demonstrate that they have not exploited the patented process. Rights granted about process patents must encompass the items obtained directly through the procedure. The TRIPS Agreement provides detailed conditions for compulsory licensing or governmental use of patents without the patentee's authorization. Rights conferred regarding patents for processes must extend to the products directly obtained by the process; under certain conditions, a court may order alleged infringers to prove that they have yet to use the patented process. Part III of the TRIPS Agreement outlines the obligations of member states to provide procedures and remedies under their domestic law to ensure that foreign rights holders and their nationals can effectively enforce IPRs. Such procedures should permit effective action against infringement of IPRs but should be fair and equitable, not unnecessarily complicated or costly, and not entail unreasonable time limits or unjustifiable delays. Provisions should also be made for judicial review of final administrative decisions.[15]

The civil and administrative procedures and remedies should include provisions on evidence and proof, injunction, damages and other remedies, which would embody the right of judicial authorities to order the disposal or destruction of any infringing goods. Judicial authorities must also be able to order prompt and effective provisional measures, mainly where any delay is likely to cause irreparable damage to the right holder or where evidence is expected to be destroyed. Further provisions relate to measures to be taken at the border for the suspension by customs authorities of release, into domestic circulation, of counterfeit and pirated goods. Finally, parties should

[15] See generally Jackson et al., supra note 10.

provide for criminal procedures and penalties, at least in cases of wilful trademark counterfeiting or copyright piracy on a commercial scale. Remedies should include imprisonment and fines sufficient to act as a deterrent. Developed countries had a one-year transition period to conform to their legislation and practices. On the other hand, developing countries and countries in the process of transformation from a centrally planned into a market economy had a five-year transition period, and least-developed countries had eleven years. Developing countries that did not provide product patent protection in technology would have up to ten years to introduce such protection. However, in the case of pharmaceutical and agricultural chemical products, such countries must accept the filing of patent applications from the beginning of the transitional period. Though the patent need not be granted until the end of this period, the invention's novelty is preserved as of the application date. Where authorization for marketing the relevant pharmaceutical or agricultural chemical is obtained during the transitional period, the developing country concerned must offer an exclusive marketing right for the product for five years or until a product patent is granted, whichever is shorter.

Articles 7 and 8 of the TRIPS Agreement provide, respectively, that the 'protection and enforcement of intellectual property rights should contribute to technological innovation and the transfer and dissemination of technology, to the mutual advantage of producers and users of technological knowledge and in a manner conducive to social and economic welfare, and to a balance of rights and obligations'; and that WTO Members may adopt regulatory measures 'necessary to protect public health.'

2.3 Putting the TRIPS Agreement to the Test: The Road to Doha

The impact of the TRIPS Agreement on the costs of pharmaceuticals in developing nations has been one of the most contentious concerns in connection with the TRIPS Agreement. The issue has received a lot of attention, especially regarding HIV/AIDS medication. Prices of some HIV/AIDS medications either increased or were anticipated to increase dramatically as the obligations of developing countries under the TRIPS Agreement phased in, and they were required to give effect to all pharmaceutical patents, including process patents, raising concerns that poor patients in developing countries affected by the HIV/AIDS pandemic would not be able to afford them.[16] Two significant possibilities were explored to address this contentious problem. The most appealing of these options was the possibility of "compulsory licenses." Compulsory patent licenses are used for various purposes under the domestic laws of WTO Members, such as sanctioning patent holders who have abused their patents for anti-competitive purposes.[17] Thus, the limitation of obligatory licencing to the domestic market of the relevant WTO Member was a

[16] Jackson et al., *supra* note 10.

[17] Jackson et al., *supra* note 10.

significant topic of discussion during the Doha Round on access to medications in developing countries.

Compulsory licenses under the TRIPS Agreement are governed by Article 31, which contains three provisions of particular relevance here: (i) the prospective issuer of the compulsory license or its licensee must have made efforts (and failed) over a reasonable period to obtain authorization to use the patent from the right holder on reasonable commercial terms, except that this requirement may be waived in the case of a "national emergency or other circumstances of extreme urgency;" (ii) the compulsory license must be "predominately for the supply of the domestic market" of the country that grants it, and (iii) the right holder is to receive "adequate remuneration…taking into account the economic value of the authorization." These provisions open up the possibility that a developing country faced with a severe health crisis can declare a "national emergency" and issue compulsory licenses for the production of patented pharmaceuticals with little or no prior negotiation with the patent holder. The country or licensee would still be required to pay "sufficient remuneration." Still, they could contend that a commensurate level of remuneration was only minimal if the drugs were being used by underprivileged persons who could not afford to pay for them. However, the requirement that production under a compulsory license be predominantly for the domestic market of the country that issues the license is potentially problematic. Developing countries without the technical manufacturing capacity might seem to be foreclosed from the compulsory license option by this principle. The compulsory licensing process in the pharmaceutical setting was a particular focus of the Doha ministerial meeting in 2001. The meeting resulted in a Declaration on the TRIPS Agreement and Public Health, which provides in pertinent parts:

> Each member has the right to grant compulsory licences and the freedom to determine the grounds upon which such licences are granted. Each member has the right to decide what constitutes a national emergency or other circumstances of extreme urgency, it being understood that public health crises, including those relating to HIV/AIDS, tuberculosis, malaria and other epidemics, can represent a national emergency or other circumstances of extreme urgency.

Two essential documents emerged from the Doha Ministerial Meeting: the Doha Ministerial Declaration[18] ("**Ministerial Declaration**") and the Declaration on the TRIPS Agreement and Public Health[19] ("**Doha Declaration**"). Paragraph 4 of the Doha Declaration states that member states "agree that the TRIPS Agreement does not and should not prevent Members from taking measures to protect public health. The Doha ministerial meeting underscored the notion that compulsory licenses could be used to address pressing public health matters. However, it sidestepped the issue of "adequate remuneration" and the problem of developing countries that lack domestic manufacturing capability.[20] The latter issue was revisited in 2003, resulting in a waiver of the obligation to require the licensee to produce primarily

[18] World Trade Organization, Ministerial Declaration of 14 November 2001 (2002).

[19] World Trade Organization, Declaration on the TRIPS Agreement and Public Health (2002).

[20] Jackson et al., *supra* note 10.

for the domestic market, applicable only in cases involving pharmaceutical products. A member wishing to avail himself of the waiver must notify the WTO and demonstrate that it lacks domestic manufacturing capacity. The waiver puts in place several safeguards designed to ensure that production under the compulsory license is limited to what is needed to supply the market of the member invoking the waiver and that the output is not diverted to other markets.[21]

"Parallel imports" afford another possible option for lowering pharmaceutical prices in developing countries. Article 28.1 of the TRIPS Agreement gives the patent holder the exclusive right to control "making, using, offering for sale, selling, or importing" the patented product (or products made directly by a patented process). Standing alone, Article 28 seemingly allows the holder of a valid patent to prevent others from importing the patented product from another nation. If the patent holder could do that effectively, it would be able to engage in price discrimination across national markets. Article 6 of the TRIPS Agreement provides that the agreement does not resolve "the issue of the exhaustion of intellectual property rights." This cryptic statement, cross-referenced in Article 28, means that the TRIPS Agreement does not dictate to national governments the circumstances under which the right holder may be deemed to have exhausted its rights through, for example, the sale of the patented product. If the product's initial sale "exhausts patent rights, then the patent holder would not have the right to prevent another party from reselling the product, including for exportation. Such a situation opens up the possibility of price arbitrage across national markets. The patent holder's exports to a particular market might compete with its products sold initially in another national market at a lower price—a situation of "parallel imports." Where the patent holder has no right to prevent parallel imports into a developing country, consumers of pharmaceutical products in that country may import them from the foreign country where they are cheapest.

In summary, the Doha Ministerial Declaration emphasized the significance of centring the work programme outlined in that Declaration on the needs and interests of developing nations. The WTO specifically emphasized in an accompanying declaration the importance of interpreting and implementing the TRIPS Agreement in a way that supports the right of WTO Members to protect public health and, in particular, to advance access to medications for everyone.

3 The Unending Debate: Does Intellectual Property Rights Impact Access to Green Technology

Having provided the history of the TRIPS Agreement and its application over the years, it is critical to address the role of IPRs in access to and transfer of green technology. As has already been highlighted, the TRIPS Agreement contains comprehensive patent provisions. Though it is not settled that IPRs promote or hinder the

[21] See the General Council Decision on Implementation of Paragraph 6 of the Doha Declaration on the TRIPS Agreement and Public Health, 2003.

development, commercialization, and distribution of green technologies in developing countries, some argue that a ban on IPRs related to green technologies could potentially create a disincentive to companies investing in research and development for green technology innovation.[22] Accordingly, developing countries have repeatedly questioned the relationship between the global intellectual property system and technology transfer and mostly believe that IPRs hinder access to green technologies. This has led to calls in the past by some developing countries that countries should not be allowed to patent green technologies so that their transfer and use can be easily facilitated.[23] There are irreconcilable differences between those who believe IPRs are an inherent barrier to the transfer of green technologies and those who argue they are an essential incentive to innovation.[24] Increasingly, IPRs appear to be a pressing challenge for the emerging climate technology regime, which might compromise its mission if not adequately addressed.[25] Therefore, the global transfer of clean development technologies, particularly from industrialized to developing countries, is complicated by the IPRs regime. These debates and disagreements still rage and will likely undermine the efforts of developing countries in addressing climate change if a way forward is not fashioned. Unfortunately, no agreement has been reached on this issue, and even the path to a meaningful policy dialogue seems elusive.[26]

Patents are typically granted to non-obvious inventions with actual or potential commercial applications. A patent confers on the inventor many rights, including the right to prevent others from (i) producing or selling the invention for the patent's life and (ii) impermissibly developing products that incorporate the invention. This time-limited monopoly allows the patent holder to gain an opportunity to recoup returns from the production and sale of the product. Thus, a patent is an incentive for innovation. Because inventors bear the cost of research, they should be allowed the leeway to recover their cost and appropriate some profits from their invention. Patents make this possible. Without the benefit of patents, inventors may not have the incentive to invent. Thus, IPRs strive to foster innovation by protecting assets and rewarding innovators. Others argue for increased intellectual property protection because it will encourage knowledge transfer between developed and developing nations and higher levels of research and development.[27] However, the foregoing comes with the cost of delayed access, reduced competition, and higher prices.[28] Developing countries bear the brunt of these problems, as many green technologies remain unavailable at reasonable prices. As such, these technologies cannot be readily and quickly deployed in developing countries where they are most needed.

---

[22] Hutchison (2012, 517–537).

[23] Webb (2016)

[24] Abdel-Latif (2015, 103–126).

[25] Abdel-Latif, *supra* note 24 at 104.

[26] Abdel-Latif, *supra* note 24 at 104.

[27] See Kanwar and Evenson (2001, 831).

[28] Littleton, *supra* note 4 at 234.

On the flip side, a growing body of literature shows that IPRs have hindered rather than encouraged innovation and that innovation has historically thrived under competition.[29] In his significant empirical research analysis, Branstetter Most falls short of demonstrating a substantial increase in domestic innovation that could be appropriately attributed to the impacts of more significant IPRs. He asserted further that extensive patent rights might even slow innovation because they allow initial patent holders to block crucial "follow-on" research that would have improved product quality and reduced costs. With such a view, IPRs only exist to entrench monopoly and stifle competition, leading to 'rent extraction' from poor developing countries.[30] For instance, in the case of a patent, the patent holder will typically charge a monopoly price throughout the patent term if the patented product has no or few good alternatives, which introduces the distortion of monopoly pricing into the economy. Furthermore, the patent system may encourage excessive investment in research due to the potential size of the benefits from a patent, thereby leading to a "patent race."[31]

Yet some other scholars advocate for context-based IPRs regulation. According to this school of thought, intellectual property policies should be modified to fit specific contexts and circumstances. They contend, for instance, that the reward of bringing a product to market first, especially if the expenses of reverse engineering the invention are significant, could provide certain inventors with sufficient motivation to innovate. The patent system is the most effective solution to the problem if reverse engineering costs are relatively low.[32] Indeed, the practice of tailoring intellectual property laws and policies to specific priorities and objectives is expressed by including public health-related flexibilities in the TRIPS Agreement.

Without diving deeper into these debates, the intellectual property regime poses a barrier to transferring green technologies[33] to ECOWAS on both sides of the divide—the supply and demand sides. For instance, without asset protection, businesses may refuse to transfer, while weak IPRs reduces demand for formal technology imports.[34] Other factors that vary by industry also come into play, such as market dynamism, technical sophistication, the value of research and development, and the ease of imitation.[35] Developed and transition economies benefit from increased technology transfer due to their strong IPRs protection.[36] Conversely, developing countries do not enjoy such increased technology transfer due to their perceived poor IPRs governance.[37] However, as China and Brazil show, substantial foreign investment does not

[29] See, for instance, Michele Boldrin and David K. Levine, (Cambridge UP, 2008); Albert G.Z. Hu and Adam B. Jaffe, (Oxford UP, 2014) 95.

[30] McCalman (2001, 161–86).

[31] Jackson et al., *supra* note 10 at 1010.

[32] See Roger Blair and Tom Cottier, (Cambridge UP, 2005), at 15.

[33] Krishna Ravi Srinivas, (July 29, 2009); Yu (2009), 797–1046.

[34] Littleton, *supra* note 4 at 234.

[35] Littleton, *supra* note 4 at 234.

[36] Littleton, *supra* note 4 at 234.

[37] Pamela Smith (2001) 411–439; Juan Blyde and Christina Acea (2003) 135–152.

necessarily depend on strong IPRs protection.[38] This, by extension, confirms historical empirical data that stronger IPRs do not always result in more technology transfer and absorption.[39] This undermines the argument that developing countries should provide more substantial protection. However, strong IPRs protection in underdeveloped nations could have "threshold effects" that encourage technology transfer only in specific circumstances.[40] This may dovetail with further research on the ideal IPRs regime for achieving the maximum balance between IPRs holder protection and the transfer/diffusion of technology to developing countries.

The TRIPS Agreement has also been criticized for its Western-centric outlook and for placing undue burdens on developing countries. As noted earlier in this chapter, the threat of sanction underlying the TRIPS Agreement makes it a source of concern, and its utility in providing an ideal balance between IPRs protection and building climate solutions is questionable. Apart from a broken promise of technology transfer, the TRIPS Agreement fundamentally changed the global market structure, making it more oligopolistic.[41] The TRIPS Agreement has restricted the options of developing countries to use compulsory licencing and competition policy.[42] It is, therefore, preposterous to think that the TRIPS Agreement will increase the transfer of green technologies in these circumstances. Moreover, far from the hope and realization of actual technology transfer, developing countries are burdened with higher prices of patented technology and products from the Global North.[43]

4 Beaming the Light on ECOWAS IPRs Regime and Climate Change

4.1 Brief Overview of ECOWAS

The ECOWAS, which covers an area of 5.2 million square kilometres, was established in May 1975 to promote cooperation and integration, raise the living standards of its peoples, maintain and enhance economic stability, foster relations among Member States, and contribute to the progress and development of the African continent.[44] ECOWAS comprises the following Member countries: Benin, Burkina Faso, Cabo Verde, Côte d'Ivoire, The Gambia, Ghana, Guinea, Guinea Bissau, Liberia, Mali, Niger, Nigeria, Sierra Leone, Sénégal and Togo. ECOWAS has a combined GDP

---

38 Jerome Reichman and Keith Maskus (2004) 279–320.

39 Srinivas, "Climate Change, Technology Transfer and Intellectual Property Rights," *supra* note 33 at 17.

40 Littleton, note 7 *supra* 234.

41 Arvind Panagariya (1999).

42 Srinivas,, supra note 33 at 23.

43 Panagariya, *supra* note 50 at 6.

44 See ECOWAS, "About ECOWAS".

of USD 734.8 billion, with economic activities revolving around industry, transport, telecommunications, energy, agriculture, natural resources, commerce, monetary and financial issues, and social and cultural matters.[45] Among its numerous objectives, ECOWAS aims to (i) harmonize and co-ordinate policies for the protection of the environment; (ii) harmonize national investment codes leading to the adoption of a single Community investment code; and (iii) liberalize trade by the abolition, among Member States, of customs duties, levied on imports and exports, and the abolition, among Member States, of non-tariff barriers to establish a free trade area at the Community level.[46]

4.2 ECOWAS and Climate Change

As a region highly exposed to climate change,[47] ECOWAS has been at the forefront of taking steps to tackle climate change. ECOWAS focuses on renewable energy, institutional capacity building, and early warning and resilience mechanism development to sustainably address vulnerability, primarily environmental and climate change challenges.[48] ECOWAS has been collaborating and forging partnerships to advance its climate goals.[49] In furtherance of its Regional Climate Strategy ("**RCS**"),[50] the ECOWAS has recently earmarked USD 294 billion over the next 10 years for climate change mitigation and adaptation targets for 2030, with a review planned for 2050.[51] Through the RCS, ECOWAS intends to take action in three areas: lowering its carbon footprint to serve as an example of what institutions can accomplish, incorporating mitigation and adaptation objectives into regional policies and sectoral targets, and enhancing political dialogue among member states to raise their ambitions for emission reduction.[52] The RCS further acknowledges that green technologies are vital to achieving its objectives. Unfortunately, these technologies have not been deployed on a massive scale due to the low level of development and poor regulatory landscapes of several countries in the ECOWAS zone.[53]

---

45 ECOWAS, "About ECOWAS," *supra* note 44.

46 ECOWAS, "About ECOWAS," *supra* note 44.

47 ECOWAS, "ECOWAS Vision 2050," 30.

48 ECOWAS, "ECOWAS Vision 2050," *supra* note 47 at 30 and 47.

49 See "ECOWAS, WASCAL Move to Mitigate Effects of Climate Change in West Africa".

50 Kebba Jeffang, "West Africa's climate strategy breaks new ground but is vague" (2022).

51 Madhumita Paul, "Climate change: West African countries mark $294 billion to meet challenge" (2022).

52 Jeffang, "West Africa's climate strategy," *supra* note 50

53 ECOWAS, "Press Kit: ECOWAS Regional Climate Strategy".

5 ECOWAS Intellectual Property Regulatory Regime: Nigeria as a Case Study

As has been acknowledged by ECOWAS, the inadequate regulatory regime prevailing in its Member countries has been a barrier to accessing the required green technologies needed to combat climate change in the region. Chief among these are the intellectual property laws of the ECOWAS Members. There are two prominent regional intellectual property organizations in Africa: the African Regional Intellectual Property Organization ("**ARIPO**") and the African Intellectual Property Organization ("**OAPI**"). All ECOWAS Member States, except Nigeria and Cape Verde, belong to one or the other. All French-speaking ECOWAS Member States belong to OAPI, while Ghana, Liberia, Sierra Leone and The Gambia belong to ARIPO. The two systems are significantly different, with OAPI being its member states' only intellectual property organization. In contrast, ARIPO Member States have maintained national systems for obtaining IPRs.

A critical examination of the ECOWAS Member States' national legislation reveals that provisions affect the TRIPS Agreement flexibilities and safeguards. However, these provisions address the prominent wiggle room about patent exhaustion and using patented products for experiments and studies. Also worthy of note is that several countries had provisions for some form of compulsory licensing to prevent the abuse of IPRs. Few ECOWAS countries, on the other hand, had provisions dealing with new uses or known or previously patented subject matter.[54] ECOWAS Members do not use the flexibilities and are lax in incorporating the TRIPS Agreement flexibilities fully in their national laws to enable them to take advantage of the patent system, reflected in the lack of access to essential medicines regarding pharmaceuticals covered by patents.[55] Thus, the vital flexibilities have not been fully incorporated in the provisions of some national and regional intellectual property legislations of the ECOWAS Members.[56] Many ECOWAS countries have yet to amend their laws to fully utilize these flexibilities. Findings from a 2007 United Nations Development Programme ("**UNDP**") study showed that only six countries in Sub-Saharan Africa had adopted the flexible regime of international exhaustion of IPRs in their legislation.[57]

Nigeria, for instance, has not updated its intellectual property laws since the TRIPS Agreement, including its patent regime, came into effect. The patent grant, protection and enforcement in Nigeria are primarily governed by the Patents and Designs Act 1970 ("**PDA**"). The PDA is complemented by the civil rules of the Federal High Court of Nigeria (the "**Court**"), which provides for procedures on sundry patent-related proceedings. Patent litigation in Nigeria centres primarily on patentability, infringement and invalidation of patents, and references/challenges of exercising the

[54] West African Health Organization (WAHO), "Development of a Harmonised TRIPS Policy" (2012).

[55] West African Health Organization, *supra* note 54 at 6.

[56] West African Health Organization (WAHO) (2012).

[57] West African Health Organization, *supra* note 54 at 8.

powers of Nigeria's Registrar of Patents.[58] The Court usually decides these cases by making orders of declaration, awarding damages, or granting injunctive relief.

Under the PDA, a patent is granted concerning an invention for a product or process that is (i) new or constitutes an improvement on a patented activity and results from inventive activity and (ii) capable of industrial application. These requirements are conjunctive. An invention must satisfy the requirements of novelty and industrial applicability. Under the PDA, the following are unpatentable: (a) plants and animal varieties or biological processes for the production of plants and animals; (b) inventions whose publication or exploitation would be contrary to public order or morality; and (c) principles and discoveries of a scientific nature.[59] A patent confers on the patentee the right to preclude any other person where: (1) the patent has been granted in respect of a product from making, importing, selling or using the product, or stocking it for sale and use; and (2) the patent has been granted in respect of a process, from applying the process of doing, in respect of a product obtained directly by the process, any of the acts in (1).[60]

Although the PDA stipulates the conditions that must be satisfied before an invention can be the subject of a patent grant, the Nigerian Patent Registry (the "**Patent Registry**") does not, in practice, insist on strictly satisfying the conditions. The Patent Registry merely examines inventions for compliance with the following formal requirements: (a) the application for the patent must be made in the prescribed form; (b) there must be a description of the invention with any drawings and plans; (c) the applicant must make a claim or claims about the invention; (d) there must be a declaration by the actual inventor (where required); (e) a power of attorney (where required); and (f) the payment of the prescribed fee.[61] Where these requirements are met, the Patent Registry will grant a patent to the inventor without subjecting the invention to a substantive examination for compliance with the patentability requirements of Section 1 of the PDA. The absence of a substantive examination regime has encouraged the grant of patents for inventions that are typically not patentable, resulting in a high volume of patent litigation.[62] A patent grant subsists for 20 years from the application's filing date. Thereafter, it goes into the public domain. However, a patent may lapse before the expiry of 20 years if the patentee fails to renew it by paying the prescribed annual fees. In practice, the Patent Registry has not been known to allow a patent to lapse for non-payment of annual fees. Instead, the Patent Registry charges a penalty for late payment of the annual fees.[63]

Nigeria is also a party to the Patent Cooperation Treaty of 1970 ("**PCT**"). Thus, a resident or citizen of any other state party to the PCT may file an international application for the grant of a patent in Nigeria. However, the Patent Registry may approve or reject such an application as it deems fit. Residents and citizens of other

[58] Solomon Ezike et al. (2020) 175.

[59] Ezike et al. "Patent Law and Litigation in Nigeria," *supra* note 58 at 176.

[60] Ezike et al. "Patent Law and Litigation in Nigeria," *supra* note 58 at 181.

[61] Ezike et al. *supra* note 58 at 176.

[62] Ezike et al. "*supra* note 58 at 176.

[63] Ezike et al. *supra* note 58 at 176.

state parties to the PCT may also make priority applications for patents granted earlier by another state party to the PCT. If approved by the Patent Registry, a priority application would give the applicant a right to the patent commencing on the date the earlier application was granted. If the Patent Registry approves international and priority applications, it affords the same protection as those filed directly under the PDA.[64]

Border control agencies also play a role in enforcing patent rights. The Nigerian Customs Service ("**NCS**") is empowered to search for, seize, detain and destroy counterfeit goods brought into Nigeria. A patent owner may take advantage of this provision by writing a petition to the Comptroller General of the NCS. Patent owners collaborate with the NCS to enforce their rights against infringers. The role of the NCS in such collaborations includes following up on any information from the owner of a patent on the possible entry of counterfeit goods into Nigeria, seizing such counterfeit goods and alerting rights owners of any suspected counterfeit brands that might be of interest to such right owners, discovered in the course of a routine check on goods entering or leaving Nigeria.[65]

The doctrines of equivalents and file wrapper estoppel do not apply in Nigeria. Similarly, Nigerian patent law is silent on the exhaustion of rights. However, in practice, a defendant can assert that the patentee cannot claim the exclusivity conferred by the patent to oppose further commercialization of genuine products protected by such rights once they have been brought to the market by the patentee or with its consent.[66] The PDA also provides for administrative[67] and court-sanctioned[68] Compulsory licensing on specified grounds.

Notwithstanding that Nigeria has not updated its patent law post-TRIPS Agreement, its patent law appears to satisfy the TRIPS Agreement to a reasonable degree. Conforming to TRIPS Agreement standards and incorporating the TRIPS Agreement flexibilities fully would require only a minimal upgrade. The country, however, requires increased, consolidated, and effective IPRs enforcement.[69] However,

[64] Ezike et al. "*supra* note 58 at 176.

[65] Ezike et al. "*supra* note 58 at 177.

[66] Ezike et al. "Patent Law and Litigation in Nigeria," *supra* note 58 at 181.

[67] The Minister of Industry, Trade and Investment may, if satisfied that it is in the public interest to do so, authorise any person to purchase, make, exercise or vend any patented article or invention for the service of a government agency in Nigeria. See Ezike et al., "Patent Law and Litigation in Nigeria," *supra* note 58 at 184.

[68] The Court has the power to hear applications for and to grant compulsory licences in respect of patents. An application for compulsory licence may be brought on several grounds. These grounds include that: (a) the patented invention, being capable of being worked in Nigeria, has not been so worked; (b) the existing degree of working the patented invention in Nigeria does not meet, on reasonable terms, the demand for the product; (c)the working of the patented invention in Nigeria is being hindered by the importation of the patented article; or by reason of the refusal of the patentee to grant licences on reasonable terms, the establishment or development of industrial or commercial activities in Nigeria is unfairly and substantially prejudiced. See Ezike et al. "Patent Law and Litigation in Nigeria," *supra* note 58 at 183 and 184.

[69] Chineze Sophia Ibekwe and Celestine Nnagozie Obunne, "The WTO-Trips Agreement and Intellectual Property Rights Enforcement in Nigeria," (2020) 11 *NAUJILJ* 84.

legislation-wise, it is safe to conclude that Nigeria is a TRIPS Agreement-compliant nation that needs a few legal provisions to be fine-tuned or updated to achieve greater consonance with the requirements of the TRIPS Agreement. Unfortunately, although Nigeria has shown substantial compliance in providing for most of the TRIPS Agreement provisions in its national patent laws, green technologies are still scant in Nigeria.[70]

6 Finding a Pathway for Access to Green Technology in ECOWAS: Lessons from Doha

As it should have been apparent from this chapter so far, the issue of green technology transfer from developed to developing countries is often characterized by uncertainties, conflicts and divergent views. Many factors also complicate the subject, with IPRs being the chief one. For instance, Nigeria's substantial, robust, TRIPS Agreement-compliant patent law has not translated to access to green technologies. A recent study has attributed Nigeria's lack of national goals or actions to grow its green economy to the lack of capacity to create the necessary technologies and the non-transfer of technical aid and technologies from developed nations as envisioned by the United Nations Framework Convention on Climate Change ("**UNFCCC**").[71]

Developing countries have proposed that the UNFCCC adopt an international compulsory licencing system to facilitate access to green technologies. As expected, this proposal sparked immediate and vehement opposition from the US.[72] The TRIPS Agreement contains several flexibilities, principles and provisions that could provide a pathway to access green technologies in Africa. Thus, working within TRIPS and leveraging these flexibilities is more straightforward than attempting to modify or abandon the TRIPS Agreement at this stage. The TRIPS Agreement expressly and impliedly supports environmental, public health, and development objectives and grants Members some latitude in deciding when those objectives should take precedence over the standard TRIPS limitations. The foregoing assertion finds justification in Articles 7 and 8 of the TRIPS Agreement.[73] Article 7 of the TRIPS Agreement delineates the objectives of the TRIPS Agreement. It provides that "*the protection and enforcement of intellectual property rights should contribute to the promotion of technological innovation and the transfer and dissemination of technology, to*

[70] See Stephen Enyinnaya Eluwa et al., "Renewable Energy and Green Technology Adoption: A Viable Option for Efficient Energy Supply in Nigeria," (2022) 10:1 *Environ Anal Eco stud.* 1104 <https://crimsonpublishers.com/eaes/pdf/EAES.000728.pdf>; Samuel Ajala, "On Nigeria's Clean Energy Plans and Improved Energy Access by 2030," (2022) <https://www.dataphyte.com/latest-reports/climate/on-nigerias-clean-energy-plans-and-improved-energy-access-by-2030/>.

[71] Agaptus Nwozor et al. "Transition to Green Energy and Sustainable Development in Nigeria: A Prospective and Evaluative Analysis," (2021) *IOP Conf. Ser.: Earth Environ. Sci.* 665.

[72] Charles McManis and Jorge L Contreras, (Edward Elgar, 2014) 113.

[73] Yu (2009, 797–1046).

the mutual advantage of producers and users of technological knowledge and in a manner conducive to social and economic welfare, and to a balance of rights and obligations."

Article 8 of the TRIPS Agreement provides the interpretative or normative principle of the TRIPS Agreement. It reflects the Agreement's Preamble by recognizing "*the special needs of the least-developed country Members in respect of maximum flexibility in the domestic implementation of laws and regulations to enable them to create a sound and viable technological base.*" Yu posits that Articles 7 "*confirms the broad and unfettered discretion that Members have to pursue public policy objectives.*"[74] Indeed, this broad and unfettered discretion (or, more aptly, flexibilities) has already been used to support developing countries' access to affordable essential medicines and can provide some lessons for climate change.[75] Since the TRIPS Agreement does not support the transfer of green technologies, the Doha Declaration needs to be widened, or a waiver similar to those for public health issues should be included to allow for the transfer of green technologies for climate change mitigation and adaptation in ECOWAS.[76] This could necessitate a new "Declaration on TRIPS Agreement and Climate Change" that clarifies existing flexibilities and offers new incentives for green technologies.

Because there will likely be no "one-size-fits-all" remedy to this impasse regarding access to green technologies, the ECOWAS should pursue many options or a combination of them. These should include flexibility in enforcement, substantive protection, and implementation. Beyond the quest to increase the transfer of green technologies to the region, efforts should also be made to build regional and domestic manufacturing capacity. Technology transfer includes not only the transmission of tools and machines but also the transfer of knowledge and skills and the development of the capacity to use and modify the technology.

The remainder of this chapter will focus on one aspect of the TRIPS flexibilities—compulsory licensing under Article 31 of the TRIPS Agreement to aid the transfer of green technologies to ECOWAS, which is mandatory licensing. The Doha Declaration declares that each Member has the right to grant compulsory licences and the freedom to determine the grounds upon which such licences are granted. Considering the growing threat and urgency of climate change, this option offers a very effective path to accessing green technologies. Compulsory licensing in green technologies may not be as easy as essential medicines. As such, the difficulties that may arise are also highlighted and addressed. Compulsory licencing allows the government to grant authorization to a third party, against or despite the will of the patent owner, to carry out tasks that would otherwise be legally required to obtain a licence or other authorization from the patentee and to do so at prices that are typically lower than prevailing market rates.[77]

---

[74] Yu, "*supra* note 73 at 1010.

[75] Frederick M. Abbott (July 13, 2009).

[76] Srinivas, *supra* note 33 at 28;

[77] McManis and Contreras, *supra* note 72 at 110.

Historically, international recognition of compulsory patent licensing as a permissible policy mechanism to curb potential patent abuse can be traced back to Article 5 A. (2) of the Paris Convention for the Protection of Industrial Property, which authorizes member countries of the Paris Union to provide for the grant of compulsory licenses "*to prevent the abuses which might result from the exercise of the exclusive rights conferred by the patent, for example, failure to work.*"[78] Building on this, Article 31 of the TRIPS Agreement recognizes that the practice of compulsory licensing is itself subject to potential abuse and as such, should be subject to more stringent legal constraints than the Paris Convention imposes. Article 31 of the TRIPS Agreement provides 12 specific procedural safeguards designed to curb any potential abuse inherent in compulsory licensing, as well as four key circumstances in which particular of the procedural safeguards may be waived, modified or supplemented in the interest of either more effectively curbing or more precisely defining potential patent abuse.[79]

The types of situations under which a compulsory licence may be granted are not restricted by the TRIPS Agreement (except for semiconductor technology and trademarks and the new special regime for pharmaceuticals discussed below). Historical facts show that negotiators prefer to establish safeguards in a detailed procedure that must be followed before a compulsory licence can be granted. The first procedural safeguard prescribed by Article 31 of the TRIPS Agreement is that each compulsory licence application must be treated individually. This provision would appear violated by compulsory licencing that automatically makes a specified class of inventions eligible for a licence. Thus, the following requirements must be satisfied to grant a compulsory licence, with a few significant exceptions: (i) prior negotiation with the right holder has taken place, (ii) the licence is of limited duration, (iii) the licence is limited scope, (iv) as regards semi-conductor technology, a compulsory license may only be granted for public non-commercial use or to remedy an anti-competitive practice, (v) the license must be non-exclusive and non-assignable, (vi) the licenses must be issued predominantly to supply the domestic market, (vii) adequate remuneration of the right holder is required, and (viii) judicial or similar review must be available.[80]

The limitation of compulsory licencing to mainly the domestic market of the relevant WTO Member was a significant topic of discussion during the Doha Round on access to medications in developing countries. To address this, a new Article 31bis and accompanying Appendix were added to TRIPS in 2005, and under certain circumstances, countries are permitted to supply (i.e., export) pharmaceutical products, such as those used to treat HIV/AIDS, to nations that lack the necessary manufacturing capacity and would not profit from issuing compulsory licences in their territory. The sole goal of TRIPS's most recent revision was to lessen the degree of protection by loosening the constraints on patent compulsory licencing.[81] Additionally, while the

[78] McManis and Contreras, *supra* note 72 at 119.

[79] McManis and Contreras, *supra* note 72 at 120.

[80] See Article 31 of the TRIPS Agreement.

[81] Gervais, "Climate Change," *supra* note 7 at 8.

TRIPS Agreement refers to national emergencies or other circumstances of extreme urgency in connection with mandatory licensing [Article 31(b)], this is only to indicate that, in these circumstances, the usual condition that efforts must first be made to seek a voluntary licence is inapplicable. In this context, the term does not refer to the substantive grounds for issuing the compulsory license. This lends further credence to the effectiveness of mandatory licences for green technology access in ECOWAS.

Indeed, climate change bears characteristics sufficient to place it within the precincts of national and health emergencies and the public good. Addressing climate change has been framed as a public good. For instance, some scholars have argued that "*reducing GHG emissions is a global public good that is difficult to achieve because there are extensive free-riding incentives, cross-border effects that are hard to value, and political failures to price the use of carbon appropriately.*"[82] This has spurred countries like India and Ecuador to argue that patented green technologies should be made available to the rest of the world through compulsory licensing to achieve the public good of environmental sustainability and climate change mitigation through environmentally sound practices and technologies.[83] In support of Ecuador, India argued stating that:

> It is high time that the role of intellectual property is addressed constructively and balancedly to address greenhouse gas emissions and climate change adaptation and mitigation efforts. Since any effort in this direction is independent of the diffusion of environmentally sound technologies in developing countries, barriers to accessing these technologies must be suitably addressed.[84]

Preserving the environment is beneficial to the entire globe, and the technologies that facilitate it are valuable to all. There have also been attempts to frame climate change as a human health issue[85] and as a national emergency.

As has been suggested in this paper, there is a need for a new "Declaration on TRIPS and Climate Change" that could clarify existing flexibilities and offer new incentives for green technologies. Additionally, the flexibilities provided by TRIPS can only be utilized if incorporated into a country's domestic legislation. This is the first and most crucial step in any effort to employ the flexibility to access green technologies in ECOWAS. Therefore, it is paramount that ECOWAS Members update or pass laws that enable them to take full advantage of the TRIPS flexibilities and safeguards. Nigeria's patent law has robust provisions for compulsory licensing. There is an urgent need for other countries in the ECOWAS region to measure up and make their intellectual property legislation fully TRIPS-compliant.

This chapter has highlighted the historic stalemate over intellectual property and climate change in international climate talks. Since the industrialized nations, the major contributors to climate change, possess the technical know-how and the technologies necessary to combat climate change, it behooves them to play their part as enshrined in numerous relevant international agreements by fostering the transfer

---

[82] Lewis (2015, 131).

[83] Lewis, supra note 82 at 131.

[84] WTO, General Council, Minutes of Meeting (held on 25–26 February 2014),

[85] Verena Rossa-Roccor et al. (2021, 5).

and diffusion of green technologies to developing countries. This includes setting into motion processes that would ameliorate the obstacles posed by IPRs. Developing countries have argued that, given the necessity of climate change mitigation, intellectual property and, by extension, the TRIPS Agreement pose barriers to easy access to green technologies. The TRIPS Agreement allows developing country governments to overcome specific obstacles created by IPRs, such as through compulsory and government-use licensing. The starting point is an international consensus that the TRIPS flexibilities that have been applied to essential medicines are extendable to climate change mitigation and adaption technologies to enable the ECOWAS nations to exploit those flexibilities through their national intellectual property laws to access requisite green technologies for climate change mitigation and adaptation.

References

Abdel-Latif A (2015) Intellectual property rights and the transfer of climate change technologies: issues, challenges, and way forward. Clim Policy 15:103–126

Abdel-Latif A et al (2015) Climate mitigation and intellectual property in tension <http://thebulletin.org/climate-mitigation-and-intellectual-property-tension7997>

Agreement Establishing the World Trade Organization, Annex 1C: Agreement on Trade-Related Aspects of Intellectual Property Rights, 15 April 1994, 1869 U.N.T.S. 299, WTO <http://www.wto.org/english/docs_e/legal_e/27-trips_01_e.htm>

AIPPI Standing Committee on Intellectual Property and Green Technology (2014) Climate change and environmental technologies—the role of intellectual property, esp. Patents <https://apeb.eu/wp-content/uploads/2015/02/Report198-Climate-Change-and-Environmental-Technologies-The-Role-of-IP-esp.-Patents.pdf>

Brown A (ed) (2013) Environmental technologies, intellectual property and climate change: accessing, obtaining, and protecting. Cheltenham (UK) and Northampton (MA): Edward Elgar

Copyright Act (2004) Cap. C28, Laws of the Federation of Nigeria

Correa CM (2005) Can the TRIPS agreement foster technology transfer to developing countries? In: Maskus KE (ed) International public goods and transfer of technology under a globalized intellectual property regime. Cambridge UP, Cambridge, pp 227–305

Ezike S et al (2020) Patent law and litigation in Nigeria. In: Cook T (ed) The patent litigation law review. Law Business Research Ltd., London, p 175 Chap. 13

Gordon RW (2012) Critical legal histories revisited: a response. Law Soc Inquiry 37(1):200–201

Hu AGZ, Jaffe AB (2014) Lessons from the economics literature on the likely consequences of international harmonization of IPR protection. In: Cimoli M (ed) Intellectual property rights: legal and economic challenges for development. Oxford UP, Oxford, pp 89–110

Hutchison C (2012) Does TRIPS facilitate or impede climate change technology transfer into developing countries? Univ Ottawa Law Technol J 3:517–537

Ibbetson D (2004) What is legal history a history of? In: Lewis A, Lobban M (eds) Law and history. Oxford UP, Oxford, pp 33–40

Ibbetson D (2005) Historical research in law. In: Tushnet M, Cane P (eds) The Oxford handbook of legal studies. Oxford UP, Oxford, pp 863–879

Khor M (2014) Climate change, technology and intellectual property rights: context and recent negotiations. Clim Policy Brief South Centre 15:11–12

Lewis L (2015) Applicability of TRIPS flexibilities to the developing world for climate change mitigation as a public good in green energy projects. Asper Rev Int'l Bus Trade Law 15:129–169

Littleton M (2009) The TRIPS agreement and transfer of climate-change-related technologies to developing countries. Nat Resour Forum 33:233–244

Maskus KE, Okediji RL (2010) Intellectual property rights and international technology transfer to address climate change: risks, opportunities and policy options. Issue Paper no. 32 International Centre for Trade and Sustainable Programme on IPRs and Sustainable Development, Geneva, Switzerland

Maskus KE, Okediji RL (2014) Legal and economic perspectives on international technology transfer in environmentally sound technologies. In: Cimoli M (ed) Intellectual property rights: legal and economic challenges for development. Oxford UP, Oxford, pp 392–414

Nicol D, Owoeye O (2013) Using TRIPS flexibilities to facilitate access to medicines. Bull World Health Organ 91(17):533–539

Odilov P (2017) Are developed and developing countries in a legal stalemate over the WTO TRIPs agreement? A study on whether the TRIPs agreement impedes the transfer of environmentally sound technologies to developing countries. Yonsei Law Rev 27(1):293–335

Owoeye O (2020) Intellectual property and equitable access to COVID-19 vaccines and therapeutics. Eur Intell Prop Rev 42(9):584–588

Patents and Designs Act (2004) Cap. P2, laws of the federation of Nigeria

Pugatch MP (2010) Mitigating climate change through the promotion of technology transfer and the use of environmentally sound technologies (ESTs): the role of intellectual property rights. Eur J Risk Reg 1(4):408–414

Reichman J et al (2014) Intellectual property and alternatives: strategies for green innovation. In: Cimoli M (ed) Intellectual property rights: legal and economic challenges for development. Oxford UP, Oxford, pp 356–384

Rose J (2010) Studying the past: the nature and development of legal history as an academic discipline. J Legal Hist 31:101–119

Saha PK (2011) Technology transfer in TRIPS agreement: implications for developing countries. Dehradun Law Rev 3(1):5–26

Serrano O, Burri M (2016) Making use of TRIPS flexibilities: implementation and diffusion of compulsory licensing regimes in Brazil and India. World Trade Institute, Working Paper No. 1 March 2016

Trademarks Act (2004) Cap T13, laws of the federation of Nigeria

United Nations Climate Change News (2020) Climate change is an increasing threat to Africa <https://unfccc.int/news/climate-change-is-an-increasing-threat-to-africa>

Webb R (2016) Protecting intellectual property while mitigating climate change: Can we do both? <https://news.climate.columbia.edu/2016/10/19/protecting-intellectual-property-while-mitigating-climate-change-can-we-do-both/>

West African Health Organization (2012) Guidelines for Implementation of trips flexibilities in national legislation to improve access to medicines in the west African region <https://www.healthresearchweb.org/files/ECOWAS_TRIPsGuidelines_English.pdf>

White & Case LLP (2021) Climate change litigation in Africa: current status and future developments <https://www.jdsupra.com/legalnews/climate-change-litigation-in-africa-6294585/>

The Relevance of the African Court on Human and Peoples' Rights in Addressing Climate Change in Africa: A Critical Assessment

Bhavna Mahadew

Abstract In Daniel Billy and Others v. Australia 2022 (the Torres Strait Islanders Petition), the UN Human Rights Committee determined in September 2022 that Australia had violated the International Covenant on Civil and Political Rights by failing to adequately protect Torres Strait Islanders from the adverse effects of climate change. This article focuses on the possibility of such litigation before the African Court on Human and Peoples' Rights because it has lately surfaced before regional human rights judicial organizations such as the European Court of Human Rights. Article 24 of the African Charter on Human and Peoples' Rights (African Charter) guarantees a generally agreeable environment favourable to the development of all peoples. Thus, in contrast to other regional courts for human rights, a State's compliance with its duties to respect and preserve the human right to a healthy environment, which can include climate change challenges, may be directly reviewed by the African Court. To date, there have not been many climate change-related trials. This could be because of domestic roadblocks like shoddy legal systems, backlogged courts, or inadequate funding. Under normal circumstances, this would prevent potential litigants from using all available domestic remedies before taking advantage of the African human rights system. Furthermore, in the African context, climate change has most likely been less important than more general environmental disputes that focus more on land, property rights, or natural resources (as evidenced by the most recent African Court case, Ogiek, which is African Commission on Human and Peoples' Rights v. Republic of Kenya 2017. Africa is expected to join the global climate lawsuit movement, facing procedural challenges in litigation before regional and international courts. This chapter informs non-state actors about admissibility issues and whether to seek advisory opinions.

B. Mahadew (✉)
Department of Business Management and Law, University of Technology Mauritius, Port Louis, Mauritius
e-mail: bhavna.mahadew@utm.ac.mu

D. Oghenerobor Agelebe (ed.), *Climate Ethics in Africa*, Global Perspectives on Law and the Environment 1, https://doi.org/10.1007/978-3-032-08757-7_13

1 Introduction

The World Meteorological Organization (WMO) has warned that responding to climate extremes costs African nations between 2 and 5% of their GDP. Over the next 10 years, the cost of adaptation in sub-Saharan Africa is expected to be between $30 and $50 billion yearly, or 2–3% of the region's GDP. Up to 118 million people living in extreme poverty in Africa may experience drought, flooding, and intense heat by 2030, which will make attempts to reduce poverty even more complex and severely impede growth.[1] To save lives and livelihoods, research highlights the critical need for additional funding for the Early Warnings For All initiative and the National Meteorological and Hydrological Services.[2] They also emphasize the essential requirement of funding meteorological services and early warning systems to aid in climate change adaptation and resilience building across Africa.[3] On September 2, 2024, at the 12th Climate Change for Development in Africa (CCDA) Conference in Abidjan, Cote d'Ivoire, the study will be released in cooperation with partners.[4]

In addition to funding, sometimes litigation is also a priority to bring governments to terms with their legal obligations towards the combat against climate change. This is equally relevant to the African continent. Climate change is causing severe impacts on the planet and its people, leading to varying degrees of government response and ineffective action. Industries responsible for causing climate change, such as carbon extractors and emitters, have generally taken less action. Civil society groups and individuals have turned to litigation to remedy and redress government and industry's inaction and ineffective action.[5] The growing number of cases raising climate change issues can be categorized in various ways, such as seeking to enforce governments' legislative and policy commitments and establishing liability for failures to adapt to climate change.[6]

It is not new to rely on rights to achieve social or environmental goals.[7] Numerous claimants have taken legal action to remedy interferences with various human rights, including the right to life, a fair trial, and respect for family and private life, as a result of the European Court of Human Rights interpretation of several substantive human rights as including environmental considerations on an international scale. National courts have taken the lead in defining the right to life as encompassing a "qualitative life… free from environmental hazards" or a right to a good environment, especially in the South Asian region.[8] Litigation in the South Asian region has shifted from

[1] World Meteorological Organisation (2024).

[2] Watts, N., Adger, W.N., Ayeb-Karlsson, S., Bai, Y., Byass, P., Campbell-Lendrum, D., Colbourn, T., Cox, P., Davies, M., Depledge, M. and Depoux, A. (2017) 1151–1164.

[3] Cools et al. (2016, 117–122).

[4] Cools et al. (2016, 117–122).

[5] Wilensky (2015, 131).

[6] Peel and Osofsky (2018, 37).

[7] Pavoni (2015, 69).

[8] Razzaque (2002).

relying on generic constitutional rights to specific rights related to the environment, such as the right to a quality environment in recent years.[9]

Since it recently appeared before regional human rights judicial bodies like the European Court of Human Rights, this chapter focuses on the potential for such litigation before the African Court on Human and Peoples' Rights. Article 24 of the African Charter on Human and Peoples' Rights (African Charter) guarantees a generally acceptable environment favourable to all peoples' development. Therefore, unlike other regional human rights courts, the African Court has the authority to directly evaluate a State's adherence to its obligations to uphold and protect the human right to a healthy environment, which may encompass issues related to climate change. There haven't been many trials about climate change thus far.

This can result from domestic obstacles such as poor legal systems, backlogged courts and inadequate funds. Under normal conditions, this would stop prospective litigants from utilizing the African human rights system before exhausting all domestic remedies. Furthermore, as demonstrated by the most recent African Court on Human and Peoples' Rights (hereafter referred to as the African Court) case of the *African Commission on Human and Peoples' Rights v Republic of Kenya*.[10] In 2017, also known as the Ogiek case, climate change was probably not as significant in Africa as general environmental disputes centring more on land, property rights, or natural resources. Due to procedural issues in litigation before regional and international courts, Africa is anticipated to join the global climate lawsuit movement. Against this background, this chapter presents a human rights-based litigation approach to climate change. It then provides an overview of the African Union and climate change to understand the degree of the enshrinement of aspects of the combat against climate change within the framework of the continental African organization. This is followed by underlining the connection between the African Court and climate change in section three. It assesses the potential of the African Court as a platform for litigating climate change-related matters and its probable success.

2 Human Rights-Based Litigation of Climate Change Challenges

2.1 Litigation Under the European Convention on Human Rights

Climate change lawsuits may find a basis for human rights as provided under international conventions and instruments. The impact of climate change on human rights has not yet been specifically and directly addressed in litigation under the European Convention for the Protection of Human Rights and Fundamental Freedoms; instead,

[9] May and Daly (2014).

[10] *African Commission on Human and Peoples' Rights v Kenya* (merits) (2017) 2 AfCLR 9.

the focus has been more broadly on the environmental effects of projects and activities on human rights. Under the European Convention for the Protection of Human Rights and Fundamental Freedoms, the European Court for Human Rights (ECtHR) has affirmed rights such as the right to life, a fair trial, and the right to privacy and family. The European Court of Human Rights is considering four claims involving air pollution's violation of human rights. If and when a significant climate change issue is brought before the Court, these cases provide insight into the path the Court will take.[11]

The applicant in *Fadeyeva v. Russia*[12] claimed that the State violated Article 8 of the Convention because it failed to protect her home and private life from the plant's severe environmental nuisance, as a result of which her health and well-being were endangered. The steel plant was the most significant iron smelter in Russia. The ECtHR ruled that although there isn't a right to nature preservation under the Convention per se, Article 8 might be applicable if the negative consequences of environmental contamination have intensified to a particular degree. Since the local courts recognized the applicant's right to resettle and the average pollution levels far surpassed the allowable amounts of harmful substances, this criterion had been attained. The ECtHR granted damages and costs after finding that Russia had violated Article 8. The Court also directed the Russian government to "take appropriate measures to remedy" her circumstances. The Court noted that it had no authority to "dictate precise measures which States should adopt to comply" with their duties under the human rights doctrine. However, in 2007, the ECtHR Department for Execution of Judgments concluded that, despite Russia's assertions, Russia had not produced proof that any suitable actions had been taken (Para. 134, 138, 149 and 150).

The applicants in *Okyay v. Turkey*[13] Aimed to shut down three thermal power units in Turkey's Aegean area. Low-quality lignite coal was used in the plants. Large-scale air pollution was caused by sulphur and nitrogen emissions from the sites, and plant-related activities negatively impacted biodiversity in the vicinity. The applicants said the factories lacked the licenses to operate legally, so they filed a lawsuit in local courts to stop the operations. They cited the Environment Act's clauses requiring authorities to avoid pollution or ensure its effects are minimized and Article 56 of the Turkish Constitution's guarantee of a healthy, balanced environment. The factories were ordered to cease operations after the local courts upheld their appeal, concluding they lacked the necessary licenses. The Turkish authority did not enforce the rulings of the local courts. The applicants protested to the ECtHR that the authorities' failure to uphold the rulings of the local courts to stop the power plant's operations violated their right to a fair hearing under Article 6 of the Convention. The ECtHR granted the applicants compensation after concluding that Turkey had breached Article 6 (Para 74, 75 and 79).

---

[11] Preston (2018, 131).

[12] *Fadeyeva v. Russia* [2005] ECHR 376 (2007) 45 EHRR 10.

[13] *Okyay v. Turkey* Application No. 36220/97, ECHR 2005-VII.

In *Giacomelli v. Italy*,[14] a waste treatment plant that handled hazardous waste, was the subject of a complaint regarding noise and emissions. The applicant, who lived 30 m from the plant, requested compensation and the closure of the establishment. She claimed that the plant's continuous noise and toxic emissions violated Article 8 of the Convention (the right to respect one's private and family life) by seriously disrupting her surroundings and posing a long-term risk to her health and house (para. 68). An operating license to treat non-hazardous waste was given to the plant's operator in 1982, and 1989, additional authorization was given to process toxic and hazardous waste. The corporation was not obliged to do an environmental impact assessment until 1996, and suitable environmental research preceded neither of these decisions. The Ministry of the Environment discovered that the plant's operations violated environmental standards twice. Furthermore, a Regional Administrative Court ruled that the plant's operations were unlawful and were to be promptly halted. Nevertheless, the administrative authorities never ordered the facility's closure (para. 89–92). According to the ECtHR, Article 8 may be applied to environmental issues regardless of whether the State is directly responsible for the contamination or if its failure to control private-sector activities results in State responsibility adequately.

According to the court, the community's interests and each person's right to privacy and respect for their house must be pretty balanced (para. 82). According to the ECtHR, this individual right extends beyond tangible or physical violations like breaking into someone's home without permission. It also includes intangible violations like noise, emissions, odours, or other types of interference (para. 76). After taking into account the administrative authorities' actions, the European Court of Human Rights (ECtHR) determined that the State failed to effectively balance the applicant's right to respect for her home and private property with the community's interest in having a plant for the treatment of toxic industrial waste. Article 8 was found to be violated (para. 97).

The European Commission of Social Rights ruled in *MFHR v. Greece*,[15] another case involving air pollution that the Greek government had failed "to strike a reasonable balance between the interest of persons living in the lignite mining areas and the general interest" in controlling and regulating air pollution from lignite mining operations, thereby violating Article 11 of the European Social Charter, which guarantees the right to health protection.

2.2 Litigation Under the American Convention on Human Rights

The effect of climate change on human rights has been directly brought up in litigation under the American Convention on Human Rights. The Inuit, indigenous to the Arctic, petitioned the US in 2005, claiming that the US had violated their human

[14] *Giacomelli v. Italy* Application No. 59909/00, ECHR 2006-XII.

[15] *MFHR v. Greece European Commission of Social Rights, No 30/05, 6 December 2005.*

rights by failing to restrict its greenhouse gas emissions, which would have lessened the effects of climate change. The petitioners cited the rights to property, culture, and preserving life, health, and bodily integrity. In 2006, the petition was denied by the Inter-American Commission for Human Rights without explanation.[16] Nonetheless, the Commission granted the petitioners a hearing on the issue in 2007 at their request, but no additional ruling was issued.[17]

According to Article 1(1) (Obligation to Respect Rights), Article 4(1) (Right to Life), and Article 5(1) (Right to Humane Treatment) of the American Convention on Human Rights, states must protect the environment, including by preventing environmental harm from happening outside their borders. The Colombian government asked the Inter-American Court of Human Rights to provide an opinion on this matter in 2016.[18] The proposed Nicaraguan canal and other significant infrastructure projects in the Caribbean were of particular concern due to their potential transboundary environmental impact. The Court's opinion from November 15, 2017, was released in February 2018.[19]

Since environmental deterioration and climate change impact the enjoyment of human rights, the Court determined an "irrefutable relationship" between ecological conservation and human rights (para. 21–22). According to the Court, State Parties are required by Article 1 of the Convention to protect the human rights of those under their "effective authority or control" inside and outside their borders (para 3). This duty involves ensuring that activities inside a state's borders don't lead to environmental damage or human rights abuses beyond the state's borders. The Court ruled that states must prevent significant environmental damage, act by the precautionary principle, regulate and supervise activities causing considerable damage, conduct environmental impact studies, establish contingency plans, cooperate in good faith, notify other affected states of potential transboundary damage risks, guarantee access to information related to environmental effects, ensure public participation, and ensure access to justice about environmental protection obligations stated in the Opinion. States must also provide good faith cooperation and consultation with other affected states (para 95–96).

2.3 Litigation Under Domestic Human Rights Legislation

National human rights commissioners can also look into the effect of climate change on human rights. As part of the National Inquiry on the Impact of Climate Change on the Human Rights of Filipino People, a public interest petition was filed with the Philippine Commission on Human Rights in May 2015, asking it to look into

[16] *Supra* note 11.

[17] Osofsky (2007, 675).

[18] Republic of Colombia, Request for Advisory Opinion of the Inter-American Court of Human Rights (14 March 2016).

[19] Inter-American Court of Human Rights Advisory Opinion OC-23/17 of 15 November 2017.

the allegations of various fundamental human rights violations by 50 large, publicly traded, multinational corporations that produce fossil fuels and contribute to climate change.[20] These 50 companies are accused of being responsible for 21.71% of all carbon dioxide emissions from 1751 to 2010.

The petitioners are human rights organizations, typhoon victims, and other concerned individuals. The Commission declared the start of the aforementioned investigation in December 2015, and by July 2016, it had sent the petition to these 47 'carbon majors', requesting a response within 45 days.[21] In March 2018, hearings in the Philippines got underway. Gascon, the Chairperson of the Commission, remarked that '*among those who are suffering the most from the effects of climate change is the Philippines. Nowhere has it been more dramatically demonstrated than in November 2013, when our country was visited by Typhoon Haiyan or Yolanda*'.[22] "The Philippines is among those who are suffering the most from the effects of climate change," said Commission Chairperson Jose Luis Martin Chito Gascon in his opening remarks. It has never been more vividly displayed than in November 2013, when Typhoon Haiyan, also known as Yolanda, made landfall in our nation.

Since people who have contributed the least to climate change and are least equipped to respond are disproportionately affected, the demand for climate justice lies at the heart of climate change litigation. Both adaptation and mitigation are part of the response to climate change, and striking the correct balance between the two is a matter of justice. Climate change and the government's mitigation and adaptation measures negatively impact fundamental rights, environmental rights, and access to shared natural resources.[23] Litigation related to climate change is increasingly using rights to reveal and address these injustices and imbalances. To tackle the social and environmental emergencies brought on by climate change, litigation has broadened the scope of conventional legal claims and causes of action. These include broadening the right to life, not being deprived of life, and developing collective natural resources held in public trust. Globalization also influences this trend in rights-based climate change mitigation, as arguments and claims are being modified and used in other jurisdictions.[24]

[20] Savaresi (2021, 231).

[21] Vidal (2016).

[22] Philippines Commission on Human Rights (2018) PHL at the forefront of seeking climate justice with CHR's landmark inquiry on the effects of climate change to human rights (28 March 2018) http://chr.gov.ph/phl-at-the-forefront-of-seeking-climate-justice/.

[23] Preston (2018, 131–164).

[24] Preston (2018, 131–164).

3 The African Union, the African Commission and Climate Change

Climate-resilient communities and economies are a crucial part of the continental vision for an integrated, wealthy, and peaceful Africa that is led by its people and represents a powerful force in the world, as stated in the Africa Union's Agenda 2063. By outlining the guiding principles, top goals, and action areas for improved climate cooperation and long-term, climate-resilient development, the African Union's Climate Change and Resilient Development Strategy and Action Plan (2022–2032) aids in implementing this vision. Planning for the continent's low-emission future is aided by the Strategy, which offers a framework for coordinated and harmonized responses to the effects of climate change.[25]

The Strategy outlines the key guidelines and objectives for enhancing African resilience in adaptation and maximizing the continent's mitigation capacity. As a key component of attaining sustainable development, as defined by Agenda 2063 and the UN's Agenda 2030, it aims to guarantee that institutions, plans, and choices for climate risk management and climate-resilient development are integrated and implemented.[26] The Strategy emphasizes how critical it is to assist the communities and groups most at risk. While acknowledging the difficulties women and young people experience in adapting to the effects of climate change, it also recognizes and works to support the vital role that these groups play as change agents at the local, national, sub-regional, and continental levels. This Strategy aims to improve collaborative development, engagement, implementation, and monitoring to advance an inclusive climate approach. The Strategy offers explicitly a centralized platform that may be used to establish collaborations with the commercial sector, civil society organizations, and non-governmental organizations.[27]

According to the African Charter on Human and Peoples' Rights (African Charter), 'all peoples shall have the right to a general satisfactory environment favourable to their development,' a unique provision. As a result, in contrast to other regional human rights courts, the African Court may be asked to personally examine whether a State is fulfilling its duties to uphold and defend the human right to a healthy environment, which may include concerns about climate change.[28] For instance, the Ogoni claimed in the *Social and Economic Rights Action Centre (SERAC) and the Centre for Economic and Social Rights (CESR) v. Nigeria.*[29] The Nigerian government had violated Article 24 by allowing oil companies to operate in Ogoniland and failing to prevent pollution; although the Court's ruling and the parties' pleas made no mention of the environment, there is a link between the two since oil firms are responsible for greenhouse gas emissions.

[25] Chevallier and Chesterman (2022).

[26] Karani et al. (2023).

[27] Emupenne and Small (2024, 223–251).

[28] Buys and Lewis (2022, 949–977).

[29] The Social and Economic Rights Action Centre and the Centre for Economic and Social Rights v Nigeria African Commission on Human and Peoples' Rights, Communication 155/96.

Suede and Fall (2023) have argued that it is possible that internal barriers such as legal systems, inadequate legislative frameworks, or a lack of funding are to blame for the dearth of climate trials to date. Usually, such situations would prevent potential litigants from using up all domestic remedies before attempting to use the African human rights system. Furthermore, the 2017 Ogiek case[30] before the African Court is a recent example of how climate change has likely been a secondary consideration in the African context of environmental disputes that are more widespread and focus more on land, property rights, or natural resources.[31]

The African Commission's duties, which fall under the categories of protection and promotion, are outlined in Article 45 of the African Charter. Activities like state reporting, specific mechanisms, and contacts with national human rights organizations and NGOs are all part of promotional functions. Based on claims of rights abuses under the African Charter, the protective obligation is carried out through individual and interstate communications. However, there is insufficient evidence linking climate change and human rights. State reporting procedures typically do not address climate change, which enables the African Commission to carry out the African Charter and investigate obstacles to its goals.[32] The African Commission does not take a stance on how to solve the problem or provide suggestions on how the African Charter relates to climate change. In contrast, petitions about climate change have been filed within the Inter-American human rights system, indicating the Commission's alleged responsibility for enforcing accountability for the negative consequences of climate change.[33]

Although the African Commission has passed several resolutions on human rights and climate change, their substance is unimpressive. Resolution 342 of 2016 highlights the necessity of regional collaboration for climate action that respects human rights. Resolutions 153 of 2009, 271 of 2014, 342 of 2016, and 342 of 2016 concentrate on the UNFCCC and Kyoto Protocol. However, there is currently no comprehensive analysis of the effects of climate change, which contrasts negatively with the substance and procedure at the UN level.

Despite its promotional purpose, the African Commission has not started or commissioned any particular climate change programs, conversations, or debates. A barrier to broader participation in activities about the impact of climate change on human rights in Africa is created by its lack of engagement with its monitoring role, its limited connection of climate change to particular human rights, and the limited involvement of stakeholders in climate change-related activities. The OHCHR, UNFCCC, UNEP, and UNDP serve as the UN-level anchors for the Commission's efforts to increase knowledge, involvement, and access to information on climate change, its impacts, and appropriate responses.[34]

[30] The African Commission on Human and Peoples' Rights versus Republic of Kenya Application 006/2012.

[31] Ako (2014, 160–176).

[32] Jegede (2018).

[33] Jegede (2018).

[34] Jegede (2018).

The African Committee of Experts has organized no similar events on the Rights and Welfare of the Child. The absence of clarity makes it difficult for more people to get involved in initiatives about how climate change affects African human rights. Although there are particular procedures, such as working groups and special rapporteurs, to support the Commission's promotional goals, there is no hard proof that these processes have influenced official guidance on the connection between human rights and climate change.[35]

The African Commission has paid minimal attention to climate change, as seen by its resolutions and state reports. The precise steps involved in creating the application and link are likewise unclear. In addition, unique mechanisms such as the Special Rapporteur on Women and the Working Group on the Rights of Older Persons and People with Disabilities in Africa are two unique bodies that have not yet addressed climate change. The connection between human rights and climate change must be strengthened. Against this backdrop, it is apposite to discuss the potential of the African Court in combating climate change through litigation in Africa.

4 The African Court and Climate Change Litigation

This section of the chapter will focus on the admissibility criteria and challenges as well as the mechanism of the African Court's advisory opinion. These two aspects can be viewed as being determined in any African judicial attempt to combat climate change. Suedi and Fall have argued that before going over particular admissibility requirements, it is crucial to remember that potential plaintiffs who have chosen a government to sue should be informed that certain climate impacts can be attributed to specific States.[36] Both the cases of *State of the Netherlands v. Stichting Urgenda*[37] and *Chiara Sacchi* et al. *v. Argentina* et al.[38] They have demonstrated that, even though climate change is a global collective issue, each state bears personal responsibility for its actions or inactions. This is true even though the Paris Agreement reaffirms the idea of common but differentiated responsibilities. Other judicial organizations will probably confirm this conclusion. Furthermore, alleged violations should have happened after the African Charter was enforced so the state could be sued. However, they can still be admitted in contentious proceedings if they occurred before that date but had ongoing effects ("continuous violations"). This is especially true in a climate context.[39]

---

[35] Jegede (2018).

[36] Suedi and Fall (2024, 146–159).

[37] *The State of the Netherlands v. Stichting Urgenda* (20 December 2019) No. 19/00135, Supreme Court of the Netherlands (Netherlands).

[38] Sacchi, C. et al. v. Argentina et al. (8 October 2021) UN Committee on the Rights of the Child, CRC/C/88/D/104/2019.

[39] African Commission on Human and Peoples' Rights (ACmHRP). 2020. Rules of Procedure.

In the context of climate change, some of the admissibility standards are not very contentious. For example, Article 56(1) of the African Charter requires that the names of the "authors," or applicants, be disclosed; Article 56(3) prohibits the use of derogatory or offensive language in communications; Article 56(4) requires the presentation of evidence other than news reports; and Article 56(7) stipulates that the case cannot have already been resolved elsewhere. Nonetheless, victimhood and the exhaustion of home remedies are two criteria for inclusion that have emerged from contemporary international law practice and are particularly contentious in the context of climate change.[40]

4.1 *Exhaustion of Local Remedies*

The African Commission and Court are only authorized to handle complaints that have been submitted to them after all domestic remedies have been exhausted under Article 56(5) of the African Charter and Article 6(2) of the Protocol to the African Charter on Human and Peoples' Rights on the Establishment of an African Court on Human and Peoples' Rights (Ouagadougou Protocol). The purpose of the African Charter's exhaustion of domestic remedies criterion is to guarantee that the State in question must have had a chance to resolve the issue within its local system before actions are filed before an international authority.[41] Before appealing to the African Commission or Court, applicants may have a better chance of success in the admissibility phase if they sincerely try to exhaust local remedies. In the *Sacchi* case, Greta Thunberg and fifteen other children claimed that pursuing separate domestic lawsuits would be 'futile,' 'unlikely to secure any effective relief,' and 'unduly burdensome,' especially in light of a climate emergency. They also claimed that they had not tried to exhaust domestic remedies.

The UNCRC rejected the petition, however, precisely because domestic remedies had not been exhausted.[42] In addition to being attacked for its lack of realism and as a tactic to 'fend off a backlash from states.',[43] Nolan contended that this choice 'reflected a strong grasp of principle, procedure, and pragmatism.'[44] Wewerinke-Singh argued that It appears that the Committee will only consider complaints after it is too late to stop the most severe abuses of these children's rights if they are expected to try out mainly unproven, extremely complicated, and costly transnational litigation tactics to meet the exhaustion of domestic remedies requirement.[45] Applicants in the ongoing Duarte Agostinho case before the ECtHR have made similar arguments

[40] Suedi and Fall (2024, 146–159).

[41] Udombana (2003, 1–37).

[42] Sacchi, C. et al. v. Argentina et al. (8 October 2021) UN Committee on the Rights of the Child, CRC/C/88/D/104/2019. para 10.12.

[43] Çali (2021).

[44] Nolan (2021).

[45] Wewerinke-Singh (2021).

that, given the urgency of climate change, it would not be feasible to pursue domestic proceedings against each state.[46]

Suede and Fall have argued that the claimants' argument in *Sacchi* and *Duarte Agostinho* is unlikely to be upheld by the African Commission and Court.[47] The Court described the exhaustion of domestic remedies as 'not a matter of choice' but rather a legal requirement in international law' (Peter Joseph Chacha v. The United Republic of Tanzania 2014). It is also the 'first requirement considered' (Law Offices of Ghazi Suleiman v. Sudan 2003), examined *proprio motu* by both bodies even in the absence of any objection thereto (Urban Mkandawire v. Republic of Malawi 2012), and not a matter of choice.

Concerns about the time it takes for cases to be decided are valid in the climate crisis. Applicants may justify exempting themselves from exhausting domestic remedies for other reasons developed in the Commission's and Court's jurisprudence. Domestic remedies under Article 56(5) of the Charter must be available and adequate.[48] If domestic remedies do not meet these criteria, victims may not have to exhaust them before complaining to an international body. Applicants must provide concrete evidence and demonstrate their apprehensions are well founded. The African Commission and Court adopt an *in-concreto* approach, considering the case's circumstances and procedure. They can also be guided by the common law doctrine of a reasonable man's test.[49]

The burden of proof for applicants is to present evidence of their exhaustion of domestic remedies, which is relatively high. If applicants are not exempt from exhausting domestic remedies, the acceptable time frame between exhaustion and the seizing of the Commission or Court is a final matter. The African Court determines the period on a case-by-case basis. Still, the lack of predictability, clarity, and precision may negatively affect the legal security of applicants before the Court. You are advised to file a case as soon as domestic remedies have been exhausted. If domestic remedies are exhausted, applicants may be granted admission; however, they may request an exemption due to the climate emergency. In the African context, the admissibility requirement can be evaded because local remedies might not be adequate or accessible.

4.2 Victimhood

Establishing victimhood—people harmed by actions or inactions that contravene international human rights law—is a barrier to climate litigation in the African human rights system. The state must have already violated the victims' rights, or the victims'

[46] Stanculescu (2021).

[47] Suedi and Fall (2024, 146–159).

[48] Musila (2006, 442–464).

[49] Zimbabwe Lawyers for Human Rights and the Institute for Human Rights and Development in Africa v. Zimbabwe (22 May 2008) 293/04, African Commission on Human and Peoples' Rights.

risk of being impacted must be more than just a hypothetical potential. Because some harms may be collective and communications may be submitted by an individual or group acting in the public interest, climate litigation before the African Court and Commission may be more straightforward. Since climate change is regarded as a worldwide issue that affects everyone equally, this strategy is especially beneficial in the context of climate change.[50]

However, it must be demonstrated that the applicant's injury was caused by the State's climate action—or lack thereof. This necessitates interaction with sophisticated scientific information and climate research, which can be challenging given the climate's uncertainties. By using the Intergovernmental Panel on Climate Change's (IPCC) scientific expert findings, courts have been able to get beyond this obstacle. It is recommended that applicants use this climate science in their submissions to the Commission or Court. In conclusion, the admissibility of collective victimhood and public interest litigation facilitates climate litigation, providing substantial scientific proof that links the applicant's harm to the State's climate action or inactivity.[51]

The question of which body should be seized is whether the Commission or the Court. The Commission is more accessible, while the Court is limited to NGOs and individuals who consented to its decisions, as stated in a declaration under article 34(6) of the African Court's Protocol. Between 2019 and 2020, several states withdrew their declarations, leaving only eight out of 30 parties with consent. However, individuals or NGOs may consider seizing the Court due to its binding decisions.

4.3 Advisory Opinion of the African Court

Courts such as the International Court of Justice and the Inter-American Court of Human Rights (IACtHR) have been asked to provide their opinions on governments' environmental commitments through advisory opinions. Only four of the fifteen petitions for advisory opinions the African Court of Human Rights has received have included answers to the legal issues presented.[52] Restrictive procedural restrictions for NGOs, the duration of procedures, or the non-binding nature of opinions could cause the dearth of advisory opinions, notably in the climate domain. There are comparatively more uncomplicated ways to submit such a request within the African human rights framework. According to Article 4(1) of the Ouagadougou Protocol, the Court may express an opinion on any legal issue about the Charter or any other pertinent human rights instruments, so long as the problem is unrelated to one the Commission is considering. NGOs must, however, meet specific criteria, including having a recognized reputation in the field in which they operate, sharing the African Union's goals and values, and being registered in an AU Member State for at least 3 years.

[50] Suedi and Fall (2024, 146–159).

[51] Suedi and Fall (2024, 146–159).

[52] Van der Mei (2005, 27–46).

Insofar as the African Charter and "any other relevant human rights instrument" are included in the material scope of the Court's advisory jurisdiction, it would be preferable for an advisory opinion to be submitted by a recognized NGO with observer status. The Court would have the chance to expand on UN treaty bodies' recent jurisprudence regarding the International Covenant on Civil and Political Rights (Torres Strait Islanders Petition) in climate change cases, as well as the IACtHR's 2017 advisory opinion on States' extraterritorial duties to limit environmental harm. There are only two restrictions on the Court's authority in this area: the requested opinion can only address a 'legal question,' and its subject matter cannot be related to a matter.'

5 Conclusion

This chapter examined the African human rights system's climate litigation options, highlighting obstacles and opportunities in advisory and contentious cases. Although climate litigation is growing quickly, it is possible to overcome procedural obstacles using scientific facts, well-supported arguments, and civil society organizations' applications for African Union Observer Status. While some legal paths might not lead to legally binding rulings, judicial institutions' legitimacy comes from their reputation and superior expertise. Multiple rulings and opinions from different international legal bodies could pressure states to alter their conduct. Notwithstanding the difficulties, every option discussed in this practice note is worthwhile to pursue as public interest litigation since it frequently results in a "success without victory" and encourages systemic change. Given the procedural limitations, victimization, and probable difficulties in gaining an advisory opinion on climate change, climate litigation on the African continent warrants more investigation and discussion.

References

Literature

African Commission on Human and Peoples' Rights (ACmHRP) (2020) Rules of procedure

Ako R (2014) Environmental justice in Nigeria's oil industry: recognizing and embracing contemporary legal developments. In: Percival RV et al (eds) Global environmental law at a crossroads. Edward Elgar, Cheltenham, pp 160–176

Buys E, Lewis B (2022) Environmental protection through European and African human rights frameworks. Int J Hum Rights 26(6):949–977

Çali B (2021) A handy illusion? Interpretation of the 'Unlikely to Bring Effective Relief'Limb of Article 7 (e) OPIC by the CRC in Saachi et al. Blog of the European Journal of International Law

Chevallier R, Chesterman S (2022) Enhancing the operationalization of the African Union's climate change and resilient development strategy and action plan (2022–2032)

Cools J, Innocenti D, O'Brien S (2016) Lessons from flood early warning systems. Environ Sci Policy 58:117–122
Emupenne O, Small M (2024) Global challenges, regional interventions: exploring the climate change adaptation strategies of the African and European union. In: Africa-EU relations and the African continental free trade area: redefining the dynamics of power and economic partnership in a complex global order. Springer, Cham, pp 223–251
Jegede AO (2018) Climate change in the work of the African commission on human and peoples' rights. SSRN
Karani P, Failler P, Gilau AM (2023) Framework for mainstreaming climate change into African blue economy strategies to enhance adaptation, mitigation, and resilience in sustainable development. Am J Clim Change 12(3):1
May J, Daly E (2015) Global environmental constitutionalism. Cambridge University Press, Cambridge
Musila GM (2006) The right to an effective remedy under the African charter on human and peoples' rights. Afr Hum Rights Law J 6(2):442–464
Nolan A (2021) Children's rights and climate change at the UN committee on the rights of the child: pragmatism and principle in Sacchi v Argentina. EJIL: Talk, 20
Osofsky H (2007) Inuit petition as a bridge? Beyond dialectics of climate change and indigenous peoples' rights. Am Indian Law Rev 31:675
Pavoni R (2015) Environmental jurisprudence of the European and inter-American courts of human rights: comparative insights. In: Boer B (ed) Environmental law dimensions of human rights. Oxford University Press, Oxford, p 69
Peel J, Osofsky HM (2018) A rights turn in climate change litigation? Trans Environ Law 7(1):37–67
Philippines Commission on Human Rights (2018) PHL at the forefront of seeking climate justice with CHR's landmark inquiry on the effects of climate change to human rights, 28 March 2018 <http://chr.gov.ph/phl-at-the-forefront-of-seeking-climate-justice/>
Preston BJ (2018) The evolving role of environmental rights in climate change litigation. Chin J Environ Law 2(2):131–164
Razzaque J (2002) Human rights and the environment: the national experience in south Asia and Africa. Background paper no. 4, joint UNEP-OHCHR expert seminar on human rights and the environment, Geneva, 14–16 January 2002
Savaresi A (2021) Human rights and the impacts of climate change: revisiting the assumptions. Oñati Socio Legal Ser 11(1):231–253
Ştanculescu D (2021) The requirement to exhaust domestic remedies and the future of climate change litigation before the ECtHR. Public International Law Policy Group, London
Suedi Y, Fall M (2024) Climate change litigation before the African human rights system: prospects and pitfalls. J Human Rights Pract 16(1):146–159
Udombana NJ (2003) So far, so fair: the local remedies rule in the jurisprudence of the African commission on human and peoples' rights. Am J Int Law 97(1):1–37
Vidal J (2016) World's largest carbon producers face landmark human rights case' guardian (28 July 2016) <https://www.theguardian.com/environment/2016/jul/27/worlds-largest-carbon-producers-face-landmark-human-rights-case>
Watts N, Adger WN, Ayeb-Karlsson S, Bai Y, Byass P, Campbell-Lendrum D, Colbourn T, Cox P, Davies M, Depledge M, Depoux A (2017) The lancet countdown: tracking progress on health and climate change. Lancet 389(10074):1151–1164
Wewerinke-Singh M (2021) Communication 104/2019 Chiara Sacchi et al. v. Argentina et al. Leiden Children's Rights Observatory
Wilensky M (2015) Climate change in the courts: an assessment of non-U.S. climate litigation. Duke Environ Law Policy Forum 131.1
World Meteorological Organisation, Africa faces disproportionate burden from climate change and adaptation costs. Available at https://wmo.int/news/media-centre/africa-faces-disproportionate-burden-from-climate-change-and-adaptation-costs#:~:text=The%20African%20continent%20has%20been,decade%20between%201961%20and%201990

Van der Mei AP (2005) The advisory jurisdiction of the African court on human and peoples' rights. Afr Human Rights Law J 5(1):27–46

Case Law

Chiara Sacchi et al. v. Argentina et al. (8 October 2021) UN Committee on the Rights of the Child, CRC/C/88/D/104/2019

Fadeyeva v. Russia [2005] ECHR 376 (2007) 45 EHRR 10

Giacomelli v. Italy Application No 59909/00, ECHR 2006-XII

Inter-American Court of Human Rights (Advisory Opinion OC-23/17 of 15 November 2017)

MFHR v. Greece European Commission of Social Rights, No 30/05, 6 December 2005

Okyay v. Turkey Application No 36220/97, ECHR 2005-VII

The State of the Netherlands v. Stichting Urgenda (20 December 2019) No. 19/00135, Supreme Court of the Netherlands (Netherlands)

Zimbabwe Lawyers for Human Rights and the Institute for Human Rights and Development in Africa v. Zimbabwe (22 May 2008) 293/04, African Commission on Human and Peoples' Rights